Data Science, Interactive Visualizations, and Generative AI Tools for the Analysis of Qualitative, Mixed-Methods, and Multimodal Evidence

Data Science, Interactive Visualizations, and Generative AI Tools for the Analysis of Qualitative, Mixed-Methods, and Multimodal Evidence

Manuel S. González Canché University of Pennsylvania, PA, United States

MK
MORGAN KAUFMANN PUBLISHERS
AN IMPRINT OF ELSEVIER
ELSEVIER

Morgan Kaufmann is an imprint of Elsevier
50 Hampshire Street, 5th Floor, Cambridge, MA 02139, United States

ISBN: 978-0-443-21961-0

For information on all Morgan Kaufmann publications visit our website at https://www.elsevier.com/books-and-journals

Publisher: Mara Conner
Acquisitions Editor: Chris Katsaropoulos
Editorial Project Manager: Soumya Yadav
Production Project Manager: Gomathi Sugumar
Cover Designer: Mark Rogers

Typeset by Integra Software Services

Contents

PART I

Democratizing Data Science for Textual, Relational, and Multimodal Inquiry

1

Democratizing data science, relational, and multimodal scholarly inquiry

Introduction

This book is organized around a simple but far-reaching proposition: researchers working with textual, relational, visual, spatial, affective, and overall multimodal evidence should have broad and meaningful access to rigorous data science and artificial intelligence tools. Moreover, this access must not be conditioned on (a) having to master computer programming, (b) depending on proprietary black-box ecosystems, or (c) relinquishing control of sensitive materials to external servers. This proposition gives coherence to the methodological frameworks, software tools, and epistemic commitments developed across the book. The project is therefore best understood not as a loose collection of analytic procedures, but as a connected ecosystem of seven primary analytic frameworks together with a local interpretive orchestration environment through which outputs generated by those frameworks can be compared, memoed, synthesized, and translated into evidence-grounded scholarly writing.

At the center of this ecosystem lies the democratization of data science (DDS). Although open-source environments such as R and Python make advanced statistical, computational, and machine-learning procedures available free of charge, meaningful access to those procedures is still restricted by substantial barriers in computer-language and coding literacy. In practice, this means that the analytic power of cost-free, open-source data science remains unevenly distributed across the research community. The software environments presented throughout this book address that gap by making advanced procedures available through no-code interfaces that can be used by researchers whose expertise lies in qualitative inquiry, mixed methods, and interpretive scholarship rather than in programming. This is a democratizing intervention because it expands who can participate in computationally informed research and how they can participate in it. In a sense, then DDS builds on the idea that qualitative and mixed methods researchers should not have to learn to code to benefit from rigorous open-source, cost-free software that uses AI, ML, and interactive visualization tools—just as people do not need to know C++ or TypeScript to benefit from Microsoft Word.

At the same time, the book is not only about access. It is also about epistemology. Across all chapters, the guiding principle is that the computational, statistical, and visual outputs generated by these tools should not be treated as substitutes for interpretation, nor is it our argument that "truth" can only be discovered through data science or mathematics. Instead, these outputs should be treated as structured maps that guide researchers back into the original evidence

Data Science, Interactive Visualizations, and Generative AI Tools for the Analysis of Qualitative, Mixed-Methods, and Multimodal Evidence. https://doi.org/10.1016/B978-0-443-21961-0.00008-5

and, when used properly as illustrated throughout the book, may help build stronger understandings through the integration of quantitative and qualitative forms of knowing. This logic is grounded in what we call a **Truly Equal Status Design (TESD)**, by which we refer to a genuinely equal-status mixed methods and multimodal design in which qualitative and quantitative elements are not subordinated to one another, but integrated so that each becomes necessary for the intelligibility of the other. The point is not to replace narrative richness with quantification, but to use computational structure to deepen, clarify, and make inspectable the interpretive process.

The architecture of the book makes this claim concrete. The methodological ecosystem presented here includes software frameworks for network modeling, temporal interaction mapping, latent code identification, machine-driven classification of open-ended responses, sentiment and emotion analysis, geo-contextualized storytelling, and local generative/reasoning AI-supported synthesis. Taken together, these frameworks handle a wide range of evidence forms: interviews, focus groups, essays, ethnographic observations, open-ended survey responses, social media posts, literature corpora, co-authorship networks, coded qualitative databases, temporal event sequences, geospatial narratives, charts, scanned pages, screenshots, notes, diagrams, illustrations, and other visual research artifacts. The result is not a loose accumulation of methods but a connected analytic ecosystem.

Within that ecosystem, Intelligent Systems for Academic Research Integration (ISARI), our local AI brainstorming partner, occupies a crucial place. It provides the infrastructure through which many of the outputs generated by the other frameworks can be brought into a new phase of scholarly work: brainstorming, memoing, synthesis, comparison, and evidence-grounded writing. If the earlier chapters of this book (chapters two to eight to be precise) help researchers identify structure, generate visual outputs, classify themes, model relationships, and contextualize evidence, the ISARI chapter helps researchers **think with those outputs**. Although the use of ISARI is optional, for each chapter will provide ample examples about how to interpret and analyze each output without relying on AI, ISARI's role matters. Intelligent systems are here to stay, but currently their use requires uploading data to external servers, which may not be ethical or allowed when dealing with sensitive data. Moreover, there still remains a lack of guidance or understanding about how to incorporate AI tools into our scholarly endeavors. From these perspectives, our discussion of ISARI is centered around the idea that scholarly work does not end when a topic model, a network graph, a sentiment distribution, or an interactive map has been generated. Researchers still need to decide what those outputs mean, how they relate to one another, what interpretive alternatives should be considered, and how an emerging argument should be written. The book therefore presents ISARI as part of the same infrastructure logic that animates the earlier chapters: computational support is valuable not because it displaces human reasoning, but because it may strengthen interpretive rigor when used under conditions of transparency, locality, and methodological accountability.

This role for ISARI places the book in direct conversation with a broader literature on artificial intelligence in education (AIEd). AIEd has increasingly explored AI in relation to tutoring, adaptive support, feedback, writing assistance, and human–AI collaboration (Hwang et al., 2020; Ouyang and Jiao, 2021; Zawacki-Richter et al., 2019). More recent work examines generative AI specifically as a brainstorming partner, writing support system, and complementary collaborator in educational contexts (Jeon and Lee, 2023; Kim et al., 2025; Memmert, 2024; Memmert et al., 2025; Wang, 2025). Yet these conversations have only begun to address the infrastructures required for scholars to think with AI while preserving data privacy, interpretive authority, and institutional trust. The book speaks directly to that gap by treating local, fully offline, open-source, and multimodal AI not as an optional add-on to scholarly work but as part of a future-oriented infrastructure for interpretive and methodological practice (Hwang et al., 2020; Ouyang and Jiao, 2021; Walusimbi et al., 2026).

The chapter proceeds in four parts. First, it presents TESD as the book's central epistemic frame. Second, it clarifies the full range of data handled across the book and provides a comparative table of the software ecosystem. Third, it addresses digital ethnography, ethics, privacy, and the implications of local and multimodal AI for methodological responsibility. Fourth, it explains how the book should be read: not as a set of isolated procedures, but as a system of structured, adaptive, and increasingly integrated workflows that culminate in interpretation, synthesis, and scholarly brainstorming with local generative AI.

Truly Equal Status Design (TESD)

All methodologies contained in this book were designed to analyze information shared by our research participants—via interviews, ethnographic observations, focus groups, essays, open-ended survey responses, social media posts, images, notes, or memos—while relying on data science methodological frameworks. From this perspective, any knowledge to be gained will be a function not only of what people know and choose to share with us, but also of the computational, statistical, and visual analyses that may help uncover structure across our participants' narratives or experiences. In other words, our textual and multimodal databases, which alone may contain thousands of pieces of meaningful information, can be integrated with memos, notes, images, and visual outputs produced using data science, mathematical, and statistical principles that are capable of synthesizing those myriad pieces of information and highlighting structure that is often hidden to the naked eye.

Although we certainly do not believe that true knowledge may only be found by relying on mathematics and its derivatives, we do believe that, when applying the methods discussed in this book, we can reach an understanding of resulting knowledge much more effectively and efficiently than if we do not rely on these methods. From this discussion, it should then be clear that, epistemologically, information provided by a multiplicity of participants may be efficiently structured to form deep and nuanced knowledge when applying data science, data visualization, mathematics, and statistics. Let us simply refer to mathematics here as one form of methodological glue that may help retrieve or build the structure that informs our knowledge-building process.

This discussion, however, does not imply that mathematics plays a more significant role in this process. On the contrary, we believe that the integration of qualitative data sources and quantitative or mathematical analytic strategies must rely on TESD, wherein the balanced interweaving of qualitative data sources and quantitative outputs is vital for researchers to construct deep understandings of the structure and meaning of qualitative and multimodal sources of evidence.

Certainly, this equal status integration is not the novel contribution of this book. On the contrary, TESD is founded on the challenge posed by leaders of the mixed methods designs. Specifically, in Fetters and Freshwater (2015), then new editors of the Journal of Mixed Methods Research (JMMR), the primary outlet worldwide for mixed methodologists, posed what they called the *integration challenge*. In their methodological guidance publication they presented the following equation $1+1=3$ (Fetters and Freshwater, 2015), wherein the 1s may be substituted by qualitative + quantitative and the 3 represents that when truly integrated, the outcome of interweaving qualitative and quantitative will always be greater than the individual components. Guided by this challenge, in 2020 I started working on methodologies that aim to address this integration. For example, in 2023 and 2024, I published methodological guidance papers at JMMR and the International Journal of Qualitative Methods where I stated that rigorous, valid, and potentially useful knowledge is achieved only by integrating the analysis of quantitative outputs with the analysis of qualitative evidence. This analytic strategy implies that if one of these components is conducted poorly, the entirety of the analysis may fall short in (re)constructing the meaning and knowledge embedded in our qualitative evidence (González Canché, 2023a, 2024).

This integration and equal status design notions are important because, in qualitative research, meaning emerges through sustained, context-sensitive engagement with text—be it interviews, observations, reflections, or narratives. When these qualitative datasets are "small," our human brains are capable of detecting nuance, and our analyses may be conducted manually without major barriers or complications. However, as datasets grow in size, complexity, and diversity, researchers increasingly face a core challenge: how to maintain interpretive depth while achieving structural clarity across expansive bodies of narrative and multimodal data. For example, in late 2020, I served as a consultant on a California project where all University of California System students were invited to participate in online surveys. They were asked three open-ended questions regarding how the pandemic impacted their college-continuation decisions or options from social, economic, and family-related perspectives. The responses to these open-ended questions were intended to inform strategies or actionable plans to help students negatively impacted by that generational event. Clearly, the analysis of over 75,000 responses needed to be conducted as quickly and accurately as possible, and qualitative analytic approaches were not feasible from a time-sensitive perspective.

This experience was the trigger for the development of the machine-based text classification frameworks presented in this book. Not only did we need to be time-efficient and effective in the analyses, but we also needed to capture and preserve the voices and concerns of our respondents, whose college going prospects were in peril. This is the main reason why all the analytic tools presented in this book do not seek to *replace qualitative insight with quantification*. Instead, these tools were designed to enhance and extend *qualitative insights*—amplifying interpretive possibilities through transparent, rigorous, and iterative engagement.

Our main premise is that mathematics can be used to help find structure embedded in qualitative, textual, and multimodal data. However, the resulting mathematics-fueled outputs make no intuitive sense to our human brains, thus being useless for building our understanding if the analytic process stops when mathematical outputs are obtained. Our understanding of the structure of qualitative data requires us to go back to the original qualitative or textual data while using the quantitative output as a guide to interpret or add qualitative meaning to those quantitative outputs. This latter step is so important in the interpretative process that, without the constant back-and-forth between quantitative outputs and qualitative input data, arguably no deep meanings or understandings can be built. It is due to this interdependence that, epistemologically, our methodological approach may be referred to as a truly equal-status design.

This interdependence helps researchers maintain clarity and transparency in the rationale driving the interpretation of the data and findings. This added transparency and clarity are important because scholars (Onwuegbuzie, 2000; Johnson and Onwuegbuzie, 2004) have argued that some qualitative researchers do not provide adequate rationale for how they reached their conclusions. Similarly, Constas (1992) stated that qualitative interpretations are not open to public inspection. Although we will not ask researchers who use the methods discussed in this book to provide access to their qualitative input data to all readers due to the need to protect our participants' privacy, as we will discuss in all chapters, our quantitative outputs may be shared without jeopardizing our participant's anonymity or confidentiality while simultaneously enabling readers to interact with those outputs and become an active part of the meaning-building process (González Canché, 2023b, 2024). That is, since the quantitative outputs require solid understandings of the qualitative data, everyone seriously interacting with the resulting studies will be able to follow the rationale used in this analytic endeavor, thus addressing past concerns about the lack of transparency in the analytic process (Constas, 1992; Onwuegbuzie, 2000; Johnson and Onwuegbuzie, 2004).

From our previous discussion, we can infer that the methodological heart of the book is straightforward: knowledge is not produced here by replacing qualitative insight with quantification, but by integrating computational and visual outputs with close interpretive engagement with the original evidence. This commitment is rooted in a **Truly Equal Status Design (TESD)**, which treats qualitative and quantitative elements as interdependent rather than hierarchical. That equal-status logic becomes even more important when evidence extends beyond text into relational, temporal, spatial, affective, and multimodal forms.

Our conceptualization of TESD requires but is not limited to the following four tenets or principles: integration and epistemology, methods as structured, adaptive, and flexible systems, interpretive authority as responsibility and privilege, and the democratization of data science.

Integration and epistemology

Rigorous, valid, and potentially useful knowledge emerges only when quantitative or computational outputs are interpreted in close relation to the qualitative materials from which they are derived. Mathematical structure, machine-learning outputs, network visualizations, automated topic classifications, and spatially contextualized maps do not "speak for themselves." They require researchers to return to the original texts, narratives, interactions, and contexts to understand what those patterns may actually mean. Without that return, the outputs remain at best incomplete and at worst misleading. This is why the book treats quantitative outputs as **maps or guides** rather than as substitutes for interpretation.

TESD emphasizes that the equality at stake is not simply the abstract coexistence of qualitative and quantitative approaches, but their sustained interdependence across the full analytic process. Qualitative evidence is the source of meaning; computational and quantitative procedures provide structure; interactive and dynamic visualizations provide legibility; and interpretive return to the original evidence provides accountability. When these components are integrated carefully, the analytic process becomes both deeper and more inspectable. This matters in relation to long-standing concerns that qualitative interpretations are insufficiently transparent or not open to public inspection (Constas, 1992; Johnson and Onwuegbuzie, 2004; Onwuegbuzie, 2000). The frameworks in this book respond to those concerns not by displacing qualitative reasoning, but by producing quantitative and computational outputs that can be shared, inspected, interacted with, and used to make the rationale for interpretation more explicit.

Within this design logic, ISARI enters at a crucial stage. Once computational outputs are obtained, researchers still face the tasks of comparing interpretations, memoing implications, translating patterns into prose, and deciding how much evidentiary weight a given pattern should carry. ISARI supports that movement. It does not replace the interpretive return to the data; rather, it provides a local environment in which researchers can think with their outputs while preserving both privacy and interpretive authority. In this sense, ISARI extends TESD into the domain of scholarly synthesis. The equal-status logic is preserved: computational outputs guide, original evidence grounds, AI-assisted brainstorming supports, and the researcher remains responsible for the interpretive bridge between them. *Note, however, that the use of ISARI or any other large language model, is optional, for each chapter will provide ample evidence and examples of how to reach integrative meanings without the need to rely on generative or reasoning AI tools.*

Methods as structured, adaptive, and flexible systems

Methods should not be understood only as fixed pipelines. They are better understood as **structured, adaptive, and flexible systems** for generating and validating evidence across modalities. This does not imply looseness or a decline in rigor. Rather, it reflects the fact that in contemporary computational inquiry parts of the process can become iterative and revisable. Theme detection, clustering, feature construction, sentiment identification, visualization choices, and even the relationship between one framework and another may need to be revisited as analysis unfolds

This point is especially important because AI can make methods feel easier than they actually are. A no-code interface or a local brainstorming partner can reduce some barriers to use, but it does not remove the need for interpretive judgment, validation discipline, documentation of choices, or awareness of instability. One of the most important implications of AI for methodology is therefore that methodological responsibility becomes more visible, not less.

Researchers are no longer only carrying out procedures; they are also **orchestrating** interactions among data, models, visualizations, prompts, and interpretations. Expertise increasingly resides not only in execution, but in orchestration. *The book should be read in exactly this way: as an invitation to design transparent, documentable, and revisable workflows rather than to follow rigid pipelines mechanically.*

This shift also changes how validation should be understood. In more traditional computational settings, validation is often discussed in terms of robustness, fit, or alternative specifications. In the methodological ecosystem presented here, validation also involves **triangulation across forms of evidence and across modes of analysis**. Do text-based themes align with network structure? Do temporal patterns align with participants' narratives? Do affective distributions make sense in relation to the classified topics? Do visual outputs remain interpretable when researchers return to the original evidence? And, with ISARI, do locally generated memos and syntheses remain grounded when compared back to the evidence and to the outputs that shaped them? Rigor in this ecosystem therefore depends not on blind adherence to a single pipeline but on disciplined movement across evidence, output, and interpretation.

Interpretive authority as responsibility and privilege

One of the strongest unifying claims of the book is that researchers have both the responsibility and the privilege to remain in complete charge of interpretive authority. This is especially visible in relation to ISARI. Cloud-based AI environments can create the impression that interpretation is partially displaced into opaque infrastructures. By contrast, a local and fully offline environment makes the researcher's role more visible. The researcher selects the brain (ISARI provides six models or brains), frames the prompt, decides which evidence to upload, determines how outputs will be used, and remains fully accountable for the conclusions drawn. This is not a secondary ethical issue. It is a methodological one. The privilege of using AI as a thinking partner is inseparable from the obligation to govern how that partnership is structured and documented.

Since ISARI works locally its use is also not constrained by internet connectivity that may further expand usage gaps, particularly in remote areas (Walusimbi et al., 2026). Access to local brainstorming partners fueled by AI therefore opens access to the use of ISARI as a privilege full with ethical responsibilities to conduct integrative analyses. Notably, since ISARI handles dozens of languages, its use is not constrained to an English-centric focus (Mohammad and Turney, 2013).

TESD and the democratization of data science

The democratization of data science and TESD are inseparable in this book. Researchers should not have to learn to code in order to access rigorous and open analytical methods, but broader access must remain coupled with deeper methodological care. The software environments across the book remove the coding barrier while still keeping the methodological and interpretive demands visible. In this sense, democratization is not only about software availability; it is about creating conditions under which more researchers can enter computational inquiry without surrendering depth, nuance, or interpretive ownership. The inclusion of ISARI strengthens that argument further because it extends democratization beyond classification and visualization into synthesis and writing. Researchers are not merely given tools to compute; they are given infrastructures with which to think.

Textual, Relational, and Multimodal Data

The methodological ecosystem presented here handles a broad range of evidence types. Text remains foundational, but text increasingly appears in relation to network, temporal, spatial, affective, and multimodal data forms. It is therefore most accurate to say that the book handles **textual, relational, and multimodal scholarly evidence**.

Why this broader data ecology matters

This broader definition is not merely descriptive; it is epistemically important. Scholars rarely work with one isolated data format. Interview transcripts may later be coded and modeled relationally. Focus-group exchanges may become temporal interaction networks. Open-ended survey responses may be classified into latent topics and then linked to participant attributes or sentiments. Literature corpora may be mapped through co-authorship or topic structures. Narratives may be geo-contextualized. Screenshots and charts may themselves become evidence that requires interpretation. Scanned pages, illustrations, diagrams, notes, and maps may all become part of the analytic workflow once local multimodal AI is available. The title of the book captures this full scope because the project addresses the analysis of **qualitative, mixed-methods, and multimodal evidence.**

Types of data handled across the book

The chapters in this book collectively handle at least seven broad types of evidence (see Fig. 1.1):

1. **Textual narrative data**: These include interviews, fieldnotes, focus-group transcripts, essays, ethnographic observations, open-ended survey responses, document excerpts, and social media posts. Such materials remain the primary source of information for several frameworks, especially LACOID, MDCOR, and SENA.
2. **Relational data**: These include actor-to-code, actor-to-actor, co-authorship, and participant-to-response structures. Relational representations are central to frameworks such as NAQD, GRATIS, VERITAS, and SENA.
3. **Temporal data**: These include chronologically ordered events, global timestamps, sequential message exchanges, and the evolution of coded responses over time. GRATIS and VERITAS rely especially on these forms of temporal structure..
4. **Affective data**: These include sentiments, emotions, emotional distributions, contradictions across emotional states, and emotional entropy. SENA is particularly designed to engage these data.

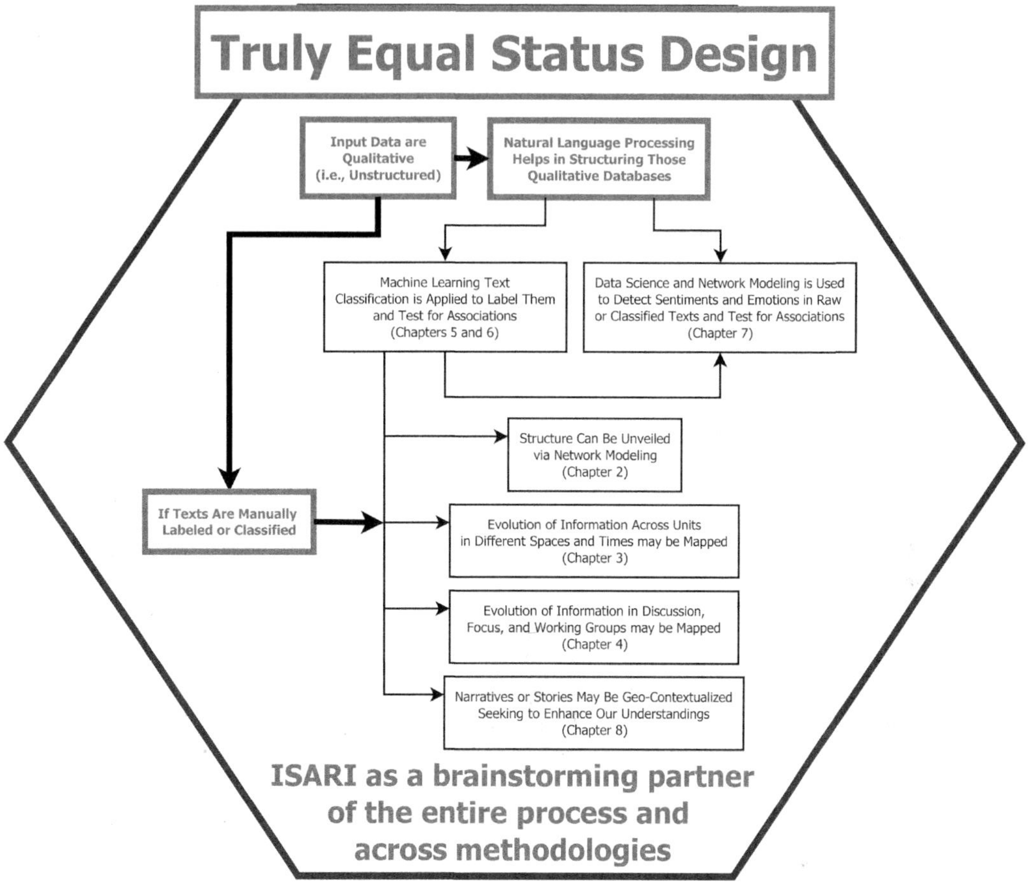

FIG. 1.1 Truly Equal Status Design.

5. **Spatially contextualized data**: These include narratives linked to place, movement, trajectories, location-based contexts, and GIS-supported storytelling. GeoStoryTelling is the framework most directly centered on this kind of evidence (see González Canché et al., 2025 for a quantitative application of GeoStoryTelling).
6. **Documentary and literature-corpus data**: These include article abstracts, research papers, document sets, systematic review corpora, and bibliographic information. MDCOR, LACOID, and SENA are key tools here.
7. **Multimodal and visual analytic data**: These include charts, scanned pages, screenshots, diagrams, maps, illustrations, notes, network graphics, temporal visualizations, and other image-based research artifacts. Through ISARI's multimodal local brains, such materials can now be incorporated into scholarly brainstorming, memoing, comparison, and synthesis.

This fuller evidentiary range validates the book's title. The project is not only about applying data science to qualitative evidence in the narrow sense. It is about a wider ecosystem for the analysis of qualitative, mixed-methods, and multimodal evidence in ways that preserve interpretive depth, methodological transparency, and data stewardship. The central claim is that researchers should be able to move across textual, relational, spatial, affective, and visual forms of evidence without treating any one form as epistemically superior to the others. Instead, the value of the ecosystem lies in the possibility of triangulating across forms of evidence and across analytic modes so that pattern detection, visualization, comparison, and interpretation can strengthen one another.

Comparative overview of the methodological and software ecosystem

To make this ecosystem legible from the outset, Table 1.1 summarizes the Table 1.1 principal evidence types, outputs, and interpretive roles across the software environments discussed in the book. The table is not meant to rank the frameworks or imply that they should always be used in sequence. Rather, it helps readers see how the different chapters address different but complementary parts of the analytic process. Some frameworks are oriented toward the retrieval of latent structure from text. Others map relationships, time, place, or affect. ISARI occupies a distinctive place because it supports the transition from analytic outputs toward memoing, comparison, synthesis, and evidence-grounded writing. In that sense, the table is also a map of how the book moves from extraction and visualization toward interpretation and scholarly reasoning.

The content of Table 1.1 is grounded directly in the methodological descriptions of the frameworks presented in this book. NAQD, GRATIS, VERITAS, LACOID, MDCOR, SENA, and GeoStoryTelling handle different configurations of coded text, temporal sequence, relational structure, sentiment, emotion, place, and multimedia context. ISARI extends that ecosystem by adding local multimodal and integrative AI support for it can handle all analytic outputs discussed in chapters two to eight.

How to read Table 1.1

Table 1.1 should be read as a guide to the functions of the book's software ecosystem rather than as a rigid methodological pipeline. The first column identifies the framework or software

Table 1.1 Methodological ecosystem, evidence types, outputs,and interpretive roles across the book.

Framework / chapter	Main evidence handled	Main outputs	Distinctive contribution	Interpretive role
NAQD (chapter 2)	Coded or classified participant responses; participant roles and group ascriptions	Interactive network visualizations; downloadable relational databases; hypothesis-testing outputs	Maps relational structure across coded qualitative evidence; identifies influential actors, concerns, communities, and shared patterns	Guides researchers back to narratives and tests whether participants' similarities is associated with shared narratives

Framework / chapter	Main evidence handled	Main outputs	Distinctive contribution	Interpretive role
GRATIS (chapter 3)	Textual / qualitative evidence with temporal markers collected across different times and places	Dynamic temporal visualizations	Retrieves and visualizes the chronological evolution of information across participants who were not necessarily co-present	Adds time as a core dimension of qualitative meaning-making
VERITAS (chapter 4)	Real-time interactional data, such as focus-group or group-exchange sequences	Dynamic replay-style visualizations; interaction maps	Models the evolution of interaction and coded contributions in shared real-time spaces	Helps reconstruct how meanings and positions emerge during interaction as well as how knowledge emerged and decayed
LACOID (chapter 5)	Longer texts such as interviews, observations, essays, and extended narrative documents in multiple languages	Topic summaries; coded text outputs; interactive visual summaries	Identifies latent codes in long qualitative materials while reducing aggregation bias	Supports computationally assisted coding while preserving return to original passages and integrates hypothesis testing
MDCOR (chapter 6)	Short texts such as open-ended survey responses or short social media posts in multiple languages	Classified response lists and topic summaries; interactive visual summaries	Closes open-ended responses through machine-driven classification	Makes large-scale short-response datasets analyzable while preserving respondent voice and integrates hypothesis testing
SENA (chapter 7)	Raw or classified texts in multiple languages	Emotion networks; participant-level sentiment distributions; multilingual summaries; interactive visual summaries	Integrates sentiment and emotion analysis with relational thinking and participant-level nuance	Adds affective structure to textual and classified evidence and integrates hypothesis testing
GeoStoryTelling (chapter 8)	Narratives linked to place, movement, trajectories, and multimedia attachments	Interactive maps and multimedia geostories	Integrates spatial context into interpretation and storytelling	Reconnects narrative meaning to place, movement, and context; allows route drawing and distance measurement
ISARI (chapter 9)	Text, notes, memos, coded outputs, charts, diagrams, scanned pages, screenshots, illustrations, maps, temporal visualizations, and other image-based analytic artifacts	Draft memos, prose, structured comparisons, analytic notes, and locally generated multimodal reasoning outputs	Functions as a local, fully offline, open-source, and multimodal brainstorming partner for interpretation, comparison, and writing	Extends the ecosystem from extraction and visualization into synthesis, reasoning, and evidence-grounded scholarly writing while preserving interpretive authority

environment. The second column summarizes the dominant evidence form handled by each framework. The third column indicates the outputs readers can expect, whether those are networks, classifications, maps, emotion distributions, or draft memos. The fourth column identifies the framework's distinctive analytic contribution, and the fifth column clarifies the framework's interpretive role in the broader ecosystem. This final column is especially important because it shows that the frameworks do not compete with one another. They perform different kinds of work and can be combined in complementary ways. ISARI is included in the same table because it belongs to the same ecosystem, but its role is distinctive: it does not replace the other frameworks; it supports the movement from their outputs toward comparison, synthesis, memoing, and evidence-grounded writing.

Isari's Local Brain Ecology

Because ISARI is integral to the architecture of the book, readers also need a preliminary sense of the kinds of local models—or "brains"—through which it operates. ISARI is not anchored in one monolithic model. It is organized as a local brain ecology composed of task-aligned models whose differences matter for speed, interpretive depth, modality, and language coverage. Lightweight multimodal models such as Gemma 3 support quick summaries, memoing, and multilingual everyday prompting. Structured reasoning and programming-oriented tasks can be supported by Qwen2.5-Coder. Concise multilingual explanation and low-resource reasoning can be supported by Phi-4-mini. Deeper reasoning tasks can be handled by DeepSeek-R1 8B in its Qwen3-based distill form. Most importantly for the book's multimodal scope, local vision-capable brains allow researchers to work not only with text but also with scanned pages, charts, diagrams, screenshots, layouts, notes, and other image-based evidence. This brain ecology matters because it transforms local AI from a generic utility into a configurable scholarly infrastructure aligned with different evidentiary tasks.

Digital Ethnography, Data Science, and Ethical Considerations

The relationship between digital ethnography, data science, and ethics is foundational to the book. The methodological ecosystem presented here is designed precisely because digital research settings now generate large and heterogeneous bodies of evidence: written narratives, interactional traces, temporal sequences, visual artifacts, platform-mediated exchanges, and place-linked materials. Computational support can make these materials more tractable, but tractability is not the same as ethical legitimacy. The tools in this book are therefore organized around the principle that computational access must remain compatible with participant protection, transparency of interpretation, and institutional responsibility.

Digital ethnography now includes multimodal traces

Digital ethnography (Pink et al., 2015) increasingly involves not only textual traces but also screenshots, interface captures, visual dashboards, map layers, collaborative traces, replay visuals, and platform-mediated artifacts. In many studies, these materials are not peripheral supplements; they are part of the evidence itself. A methodological ecosystem for digital ethnography can therefore no longer be built as though text were the only relevant format. The frameworks in this book already move in that direction by integrating narrative, relational, temporal, spatial, and affective analysis. ISARI extends that logic by allowing locally controlled engagement with multimodal artifacts as part of the interpretive workflow. This matters especially in fields where screenshots, annotated visualizations, digital documents, and platform traces have become routine evidence for the study of institutions, publics, and culture.

Ethics under local and portable AI

Privacy and confidentiality must be treated not as downstream compliance issues but as part of methodological design itself. A local and fully offline AI environment changes the ethical conditions of possibility for computationally assisted interpretation. Scholars who would hesitate—or be prohibited—from uploading interview excerpts, scanned institutional documents, screenshots, or chart-heavy outputs to external servers can instead work locally under their own control (González Canché, 2025). This does not eliminate ethical responsibility, but it changes the infrastructural baseline. Privacy is no longer treated only as a constraint to be negotiated after the analytic design is chosen; it becomes part of the analytic design from the beginning (González Canché, 2025). This is one of the major reasons ISARI belongs conceptually with the book's other frameworks: locality and full offline execution are methodological as well as technical commitments.

AI and methodological integrity

One of the most important cautions for a methodological book is that AI can make methods feel easier than they actually are. The availability of no-code tools and local brainstorming support should not invite superficial engagement. AI-assisted workflows still require interpretive judgment, validation discipline, documentation of choices, and awareness of instability. Seen in this way, AI does not remove methodological responsibility; it reshapes it. Researchers are increasingly called not only to execute methods, but to orchestrate relationships among data, models, visualizations, prompts, and interpretations. That orchestration is now part of expertise itself. Accessibility and support do not reduce the need for rigor. They increase the importance of explicit documentation, disciplined validation, and careful return to evidence.

Book Plan and Organization

The architecture of the book should be read as a coherent progression across several interrelated movements. First, this opening chapter establishes the epistemic, ethical, and infrastructural foundations of the project: democratizing data science, TESD, multimodal evidence, integration challenges, and the conditions under which computational inquiry can remain aligned with interpretive scholarship. Second, the network modeling frameworks show how coded or interactional qualitative evidence can be rendered relationally and temporally intelligible. Third, the machine-driven text classification frameworks show how raw textual data—whether long interview materials, short open-ended survey responses, or literature corpora—can be organized into patterns, clusters, and themes that would otherwise remain difficult to identify efficiently. Fourth, the integrative extensions add further layers of context and meaning through affective and spatial analysis. Sentiment and emotion become analyzable as structured patterns, and narratives become reconnected to the places and trajectories through which they unfold. Fifth, the ISARI chapter provides the interpretive and synthetic environment through which outputs from the previous frameworks can be compared, memoed, integrated, translated into prose, and used in evidence-grounded scholarly writing. This organization matters because it shows that the earlier chapters are not isolated methods. They are parts of an ecosystem whose ultimate aim is not merely extraction or classification, but better, more transparent, and more responsible interpretation.

Across the chapters, the approaches can be read as moving through different layers of analysis. The early discussion opens a foundation for interpretive data science, which then extends into mapping relational structure and temporal dynamics. From there, attention shifts toward identifying patterns within language, before bringing these elements together across textual, relational, spatial, and affective dimensions. The later chapters turn toward interpretation and synthesis, where these components begin to function as a more integrated analytical system, including support for ongoing scholarly reflection and AI-assisted synthesis without displacing human interpretive authority. This progression offers one way of seeing how the pieces connect, while still allowing each method to stand on its own.

How should one read this book?

The most productive approach is therefore to read each chapter as both a standalone methodological guide and as a node within a broader infrastructure of analysis. Each chapter explains not only how a framework works, but also the kinds of research questions it can help address, the types of evidence it can handle, and the interpretive work it makes possible. Readers may enter the book through the framework most immediately relevant to their data, but the strongest contribution of the book emerges when the chapters are understood relationally. Topic classification can lead into network mapping; network outputs can be temporally or spatially contextualized; sentiment can be related to classified topics or participant positions; and the outputs of all of these frameworks can be brought into ISARI as part of scholarly brainstorming and synthesis. This is why the book should be read not as a catalog of tools, but as a methodological and infrastructural ecology (see Fig. 1.2 and Table 1.1).

Conclusion

This chapter establishes the book as a contribution to the democratization of computational, visual, and AI-supported inquiry for researchers working with textual, relational, spatial, affective, and multimodal evidence. It is grounded in the conviction that qualitative and mixed-methods inquiry must not be displaced by computational tools, but strengthened by them

FIG. 1.2 Democratizing Data Science (DDS). *DDS builds on the idea that qualitative and mixed methods researchers should not have to learn to code to benefit from rigorous, open-source, cost-free software that uses AI, ML, and DSV tools*—just as people do not need to know C++ or TypeScript to benefit from Microsoft Word.

under a Truly Equal Status Design. It also insists that local, open-source, and fully offline infrastructures matter because they make new forms of research support available under conditions of privacy, transparency, and methodological accountability. Within this architecture, ISARI functions as a local interpretive orchestration environment and, when paired with the outputs of the other software tools discussed in this book, extends the ecosystem from extraction and visualization into interpretation, synthesis, and writing. It also makes visible a broader methodological lesson: methods are not merely sequences of fixed steps, but structured, adaptive systems in which researchers coordinate evidence, models, visualizations, prompts, and interpretations. In this framing, the privilege of greater analytic support is inseparable from the responsibility to remain fully accountable for meaning-making. That, ultimately, is the strongest unifying message of the book.

In conclusion, the overarching goal of this book is to broaden participation in the data science movement, particularly by empowering researchers from underserved communities, less-resourced institutions, and methodological traditions that have historically not had broad access to advanced computational tools. From this perspective, the project aims to address the need for innovative frameworks that enable qualitative and mixed-methods researchers to benefit from the insights that can emerge from AI, machine learning, and state-of-the-art interactive visualization methods without any computer-language literacy requirements. DDS, as an equity-fueled methodological movement, seeks to offer true access to rigorous, cutting-edge, and innovative research tools by lifting programming barriers and by making open-source, cost-free, and local infrastructures available for the analysis of multimodal evidence and the strengthening of data storytelling. Fig. 1.1 depicts this theory of change.

The chapters that follow show one possible path through this ecosystem, but they should also be read as an invitation to imagine new combinations, new workflows, and new forms of inquiry under the broader project of democratizing data science. If the opening of this book establishes the possibility of a new scholarly infrastructure, the remaining chapters explore what it can do and how it might continue to grow.

In a nutshell, DDS builds on the idea that *qualitative and mixed methods researchers should not have to learn to code to benefit from rigorous, open-source, cost-free software that uses AI, ML, and DSV tools*—just as people do not need to know C++ or TypeScript to benefit from Microsoft Word. Based on my experience teaching hundreds of qualitative education researchers how to apply interactive visualization and ML tools to qualitative data, it is clear that the applied conceptual understanding of these procedures is not a problem for the overwhelming majority of researchers. The real barrier is the hundreds of R or Python code lines required to apply these concepts to their databases. Naturally, then, by removing the coding proficiency hurdle, the products of this project will empower their research endeavors and help them become active members of and contributors to the applied data science community (see Theory of Change in Fig. 1.1). In addition, and also aligned with the free software movement, all source code and documentation associated with the proposed software application tools are being made publicly available in Code Ocean, "a cloud-based platform focused on computational reproducibility and collaborative research in science, particularly in fields like biology and pharmaceuticals" (Code Ocean, 2025) so that methods researchers can further develop these software tools or build new ones.

In sum, with this book and via DDS we seek to increase the participation of qualitative and mixed-methods researchers in applied data science. This book aims to achieve this goal by offering full and unrestricted access to seven analytic frameworks together with ISARI, a local interpretive orchestration environment that helps researchers work across the outputs generated by those frameworks. Since neither of these methodological frameworks require computer language literacy to generate innovative multimodal and interactive visualizations, apply natural language processing and machine learning, and achieve AI-fuled multimodal integrations, researchers may begin immediately benefiting from the advantages derived from rigurous, transparent, and ethically driven data science tools.

References

Code Ocean. (2025), Available from: https://codeocean.com/about.

Constas, M.A., 1992. Qualitative Analysis as a Public Event: the Documentation of Category Development Procedures. Am. Educ. Res. J. 0002-8312. 29 (2), 253–266. doi:10.3102/00028312029002253.

Fetters, M.D., Freshwater, D., 2015. The 1 + 1 = 3 Integration Challenge. J. Mixed Methods Res. 9 (2), 115–117. doi:10.1177/1558689815581222.

González Canché, M.S., 2023a. Latent Code Identification (LACOID): a Machine Learning-Based Integrative Framework [and Open-Source Software] to Classify Big Textual Data, Rebuild Contextualized/Unaltered Meanings, and Avoid Aggregation Bias. Int. J. Qual. Methods 22. 10.1177/16094069221144940

González Canché, M.S., 2023b. Machine driven classification of open-ended responses (MDCOR): An analytic framework and no-code, free software application to classify longitudinal and cross-sectional text responses in survey and social media research. Expert Syst. Appl. 215, 119265. doi:10.1016/j.eswa.2022.119265.

González Canché, M.S., 2024. Graphical Retrieval and Analysis of Temporal Information Systems (GRATIS): an Integrative Mixed Methodology and Open-Access Software to Analyze the (Non-)Linear Chronological Evolution of Information Embedded in Textual/Qualitative Data. J. Mixed Methods Res. 18 (1), 71–103. doi:10.1177/15586898231166968.

González Canché, M.S., 2025. Ethics and trust in social network analysis: how data science and interactive visualizations impact anonymity, confidentiality, consent, and IRB procedures?. In: Handbook of Social Network Analysis and Education. Edward Elgar Publishing, Cheltenham, UK. Retrieved from Apr 26 doi:10.4337/9781035312191.00013.

González Canché, M.S., Zheng, K., Song, Y., 2025. Standardized testing for diverse talent identification: A framework to address geographical bias in standardized testing and increase diversity in college admissions in the post-affirmative action/race-neutral admissions era. Res. High. Educ. 66, 13. doi:10.1007/s11162-024-09824-4.

Hwang, G., Xie, H., Wah, B.W. and Gašević, D., 2020. Vision, challenges, roles and research issues of Artificial Intelligence in Education. Comput. Educ. Artif. Intell. 1, 100001. doi:10.1016/j.caeai.2020.100001.

Jeon, J. and Lee, S., 2023. Large language models in education: A focus on the complementary relationship between human teachers and ChatGPT. Educ. Inf. Technol. 28 (12), 15873–15892. doi:10.1007/s10639-023-11834-1.

Johnson, R.B., Onwuegbuzie, A.J., 2004. Mixed methods research: a research paradigm whose time has come. Educ. Res. 0013189X. 33 (7), 14–26. doi:10.3102/0013189X033007014.

Kim, J., Yu, S., Detrick, R. and Li, N., 2025. Exploring students' perspectives on Generative AI-assisted academic writing. Educ. Inf. Technol. 30 (1), 1265–1300. doi:10.1007/s10639-024-12878-7.

Memmert, L., 2024. Brainstorming with a generative language model: understanding performance through brainstorming group effects. European Conference on Information Systems. Available from https://www.researchgate.net/publication/379840343_Brainstorming_with_a_Generative_Language_Model_Understanding_Performance_Through_Brainstorming_Group_Effects#fullTextFileContent.

Memmert, L., Cvetkovic, I., Tavanapour, N. and Bittner, E., 2025. Brainstorming with a generative language model: Effect of exposure to AI ideas on brainstorming performance and cognitive load. Bus. Inf. Syst. Eng. doi:10.1007/s12599-025-00974-y.

Mohammad, S.M., Turney, P.D., 2013. Crowdsourcing a word-emotion association lexicon. Comput. Intell. 29 (3), 436–465. doi:10.1111/j.1467-8640.2012.00460.x.

Onwuegbuzie, A.J., Validity and qualitative research: an Oxymoron? Paper presented at the annual meeting of the Association for the Advancement of Educational Research (AAER), 2000.

Ouyang, F. and Jiao, P., 2021. Artificial intelligence in education: The three paradigms. Comput. Educ. Artif. Intell. 2, 100020. doi:10.1016/j.caeai.2021.100020.

Pink, S., Horst, H., Lewis, T., Hjorth, L., Postill, J., 2015. Digital ethnography: Principles and practice, SAGE. https://d1wqtxts1xzle7.cloudfront.net/40273901/71681_Pink_Digital_Ethnography-libre.pdf?1448230959=&response-content-disposition=inline%3B+filename%3DDigital_Ethnography_Principles_and_Pract.pdf&Expires=1777222179&Signature=DnTZNC0lYDcNvdMfES6Ne8Jp9hFBOOgJGexnN8bIysG-GiyTaUD~SsnEbhoKbGI5uS9v4W6E8379k0cUXDnd~AR10tzkmT2NCbCgAtUEuzcfOa61TdLc8Lkv~QGzVijy1in9Za4hPLF4BlaWi4nybnpGMm2K-zHKoZxtnQiJBzCUZSDn5kieGjoT5WMjy1ED1phxd34LvUsoFlLe3EeXBH1EY3i2Z9xFROSDfc~C-nWCdIBI9TeWLyduxV4yp~HBVZQR5L-MxRVC8CFK4l8k6gNotxIynYAFFTFGnVyX69fARI4IiQr8tWeLsDPiCosrl6iVOxiNtrqwgMiecgW33A__&Key-Pair-Id=APKAJLOHF5GGSLRBV4ZA

R-universe. 2025. *R-universe search.* Retrieved July 6, 2025, from: https://r-universe.dev/search

Stallman, R., Initial Announcement - GNU Project. (1983), Available from: https://www.gnu.org/gnu/initial-announcement.html.

Stallman, R. (2005). *Free software, even more important now.* GNU Project. Retrieved June 20, 2025, from: https://www.gnu.org/philosophy/free-software-even-more-important.html

Walusimbi, J., Oguti, A.M., Ssentongo, J.B., Ainebyona, K., 2026. Offline-First LLM Architecture for Adaptive Learning in Low-Connectivity Environments. arXiv preprint. https://arxiv.org/abs/2603.03339. doi:10.48550/arXiv.2603.03339.

Wang, C., 2025. Exploring students' generative AI-assisted writing processes: Perceptions and experiences from native and nonnative English speakers. Tech. Know. Learn. 30 (3), 1825–1846. doi:10.1007/s10758-024-09744-3.

Zawacki-Richter, O., Marín, V.I., Bond, M. and Gouverneur, F., 2019. Systematic review of research on artificial intelligence applications in higher education – where are the educators?. Int. J. Educ. Technol. High. Educ. 16 (1). doi:10.1186/s41239-019-0171-0.

PART II

Mapping Meaning Through Networks: Relational Meaning and Networked Time

2

Network Analysis of Qualitative Data (NAQD)

Introduction

Network analysis is an increasingly important analytical approach utilized to uncover structures in relationships that, due to their complexity and/or dimensionality (i.e., big data or large amounts of information), are hidden at first glance. It follows that qualitatively or quantitatively coded textual information may be conceptualized as relationships between the participants and the codes they provided, which naturally form a relational database. From this perspective, the main argument of this study is that data gathered through qualitative methods via ethnographic approaches, qualitative interviews, or focus groups, and subsequently qualitatively coded (or even quantitatively coded as described in González Canché, 2023a, 2023b), contain mathematical structures that can be retrieved, visualized, or mapped using network analysis methods. When the coded texts include research participants' attributes, the emerging structure may be used to evaluate the degree to which participants' experiences, as captured by their contextualized coded contributions, show differences across groups or whether these experiences are similarly distributed across purposive or analytic samples. Based on these analytical possibilities, the goal of this chapter is to introduce the steps required to harness the *mathematical* power of network analysis in order to uncover structure in qualitatively or quantitatively coded textual content.

The analytic framework discussed in this chapter may be referred to as Network Analysis of Qualitative Data (NAQD). NAQD integrates quantitative, mathematical, and qualitative principles to analyze coded textual data in a way that remains uncommon in qualitative and mixed-methods, and social sciences research. In addition to presenting this analytic framework, and with the goal of democratizing access to data science tools for analyzing qualitative data, we also present a no-code software tool that implements all the elements of NAQD.

The integration of network modeling to analyze coded textual data is becoming more prevalent. For example, commercially available (i.e., for-profit) computer-assisted qualitative data analysis software (CAQDAS) like NVivo, ATLAS.ti, Leximancer, MAXQDA, and QDA Miner all offer network or relational visualization capabilities. Nonetheless, unlike NAQD, none of these software tools offer community detection capabilities, hypothesis testing of similarities across groups, or fully interactive HTML network renderings. The latter, in addition to serving as guides for the interpretation of the resulting network structure, also offer

Data Science, Interactive Visualizations, and Generative AI Tools for the Analysis of Qualitative, Mixed-Methods, and Multimodal Evidence. https://doi.org/10.1016/B978-0-443-21961-0.00009-7

measures of the strength of the relationships among actors and codes, along with information about the total number of contributions made by each actor and their level of influence in this network. Taken together, then, not only is NAQD more versatile in the presentation of the results, but it is also offered completely free of charge and can be modified to accommodate other analytic needs if researchers are familiar with computer and statistical programming. If they are not, note that NAQD has been developed to meet most qualitative and mixed-methods projects' needs, as tested with hundreds of participants who have been trained in this framework since 2015.

The relevance of these outputs extends beyond visualization alone. NAQD's fully interactive HTML renderings, downloadable code-by-code and community-classified databases, and portable document outputs of quadratic assignment procedures are especially valuable because they can circulate into subsequent stages of scholarly interpretation. In the methodological ecosystem developed throughout this book, these outputs are not treated as endpoints but as structured guides that can be further examined in local, fully offline, and privacy-conscious reasoning environments such as ISARI—see chapter 9. There, researchers may upload network renderings, HTML-derived summaries, screenshots, figures, or PDF outputs to support memoing, comparison, synthesis, and evidence-grounded writing. This extension is consistent with the central logic of NAQD itself: network structure is useful not because it replaces qualitative judgment, but because it sharpens and documents the researcher's interpretive return to the evidence.

Network analysis of qualitative data

The analytic framework introduced next builds upon the traditional application of social network analysis in that it highlights structures emerging from participants' interactions or connections within a given group. In this case, a group is conceptualized as individuals participating in an activity that can be clearly delineated with specific boundaries. Following the examples presented in this chapter, these boundaries may be established by participating in the same project as interviewees or by being part of a focus group or research site. In either case, the analyses will be centered on analyzing the text or written data that these participants shared during the qualitative data-gathering process. However, rather than using text mining, text classification techniques, or sentiment analysis of the words used by participants (as is typically done in sentiment and emotion analyses), our NAQD analyses rely on the participants' qualitatively (or quantitatively) coded or labeled transcribed responses.

Logic model guiding NAQD

Fig. 2.1 illustrates the rationale behind the proposed framework. In this figure, the rectangle represents text data (e.g., newspapers, journals, books, social media posts, interviews, observations), which are widely available, particularly in the big data era (González Canché, 2023a). However, these forms of data are considered unstructured because they do not conform to the

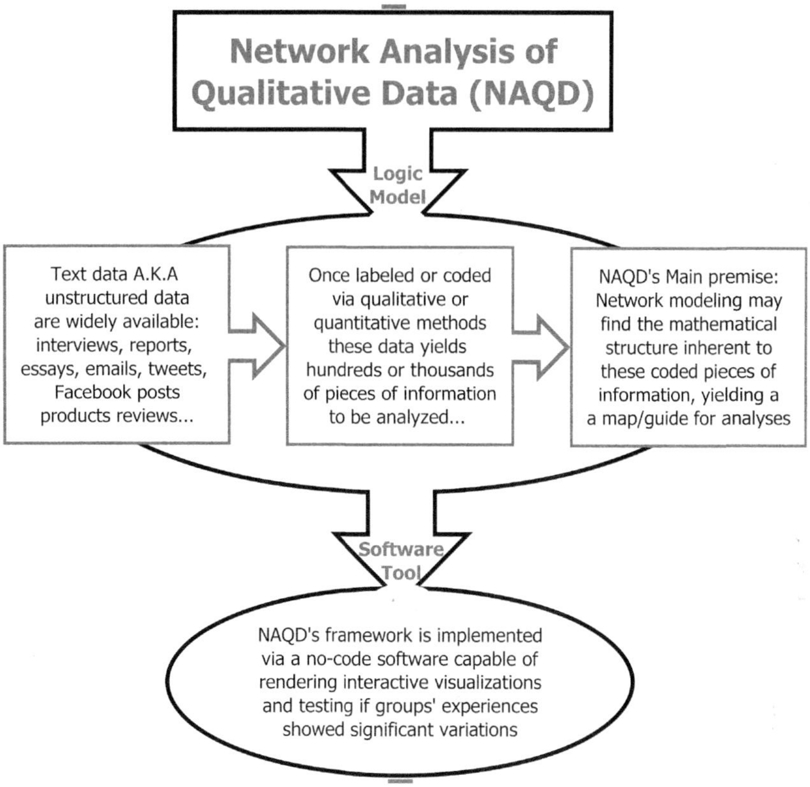

FIG. 2.1 Conceptual Framework.

typical rows-and-columns format (i.e., rectangular form) employed in or required by quantitative analyses or modeling. Typically, in qualitative research, the most common way to *structure these data* involves creating labels or codes for text chunks (such as sentences or paragraphs) that convey meanings relevant to the research questions posed and conceptual frameworks used in a given research project. As demonstrated in González Canché (2023a, 2023b), machine learning text classification can also be utilized to expedite this labeling or coding process. Regardless of how these codes are obtained (qualitatively or quantitatively), the resulting labeled textual data typically consist of hundreds or even thousands of codes, their accompanying text chunks (also known as quotes), as well as attributes of the participants who provided the original unstructured texts.

Labeling or coding process

To briefly illustrate this process, let us present Table 2.1. In that table, we are showing the text responses of participants in a project focused on the challenges and concerns that Minority Serving Institutions (MSIs) across the country face in terms of their production and training of teachers of color. Following Crossley (2010) and Saldaña (2021), these texts may be structured (or

Table 2.1 Short Essays on challenges to train faculty members from minoritized backgrounds.

Actor ID1, Interview 1, Student	**Codes/ Labels**
An accelerated program for our master's. So they're going to go through professional development training with second and third grade teachers.	Professional development
As a result of the training that they are going to be **exposed to, then they will coteach, in a loose sense of the word, with teachers who will go through the training.**	**Hands-on experiences**
They will be in the classrooms with second graders while the third graders are taking what we call third grade gate, that third grade test that they must pass in order to be promoted. …	*Promotion structure*
Actor ID2, Interview 2, Faculty	**Codes/ labels**
I would tell you yes. Jackson State University graduates the majority of the teachers in this metro and beyond area.	*High program impact*
… [Omitted text] …	
You will find them in some of the *largest public school systems and especially Jackson Public Schools. We add some public schools all the time. So, yes, we graduate many, many students for the public schools.*	*High program impact*
Actor ID3, Interview 3, Administrator	**Codes/ labels**
Now with the emphasis on technology and everything that position was repurposed to be the Dean of Undergraduate Students and Cyber Learning.	*Professional development*
So if you were to visit our library you would see that it's no longer traditional as we know.	
The first floor is all about how you use technology so that you can use it and practice to gain hands on experiences. This way when you get to the classroom you will hit the ground running.	**Hands-on experiences**

coded or labeled) so that a word or short phrase is used to capture the meaning of a sentence or paragraph in the original text. More specifically, a code or label in qualitative research is a word, a descriptor, or a short sentence that "symbolically assigns a summative, salient, essence-capturing, and/or evocative attribute for a portion of language-based or visual data" (Saldaña, 2021).

Because the network structure rendered by NAQD is a direct function of the codes assigned to participants' contributions, the rigor of the coding process is methodologically decisive (Charmaz, 2011; González Canché, 2023a; González Canché, 2023b; González Canché, 2026; Saldaña, 2021). The network does not generate meaning independently of coding; rather, it formalizes and visualizes the structure implied by the coding scheme. For that reason, NAQD analyses are strongest when the coding process itself is conducted transparently, systematically, and with clear attention to the conceptual relationships among codes, quotes, and participants. Whether coding is conducted qualitatively, through collaborative interpretive procedures, or quantitatively through machine learning classification (see González Canché, 2024), the analytic quality of the resulting sociograms and comparisons will always depend on the integrity of the coding process that precedes them.

In the example shown in Table 2.1, the highlighted texts and the "codes or labels" to the right configure the label or code summarizing the meaning of the text. For example, the text "The first floor is all about how you use technology so that you can use it and practice to gain hands-on experiences. This way, when you get to the classroom, you will hit the ground running," provided by Actor ID30 is highlighted in yellow, and the label or code assigned to this text was "Hands-on Experiences." Similarly, another text provided by this same participant is, "Now, with the emphasis on technology and everything related to that, it was repurposed to be the Dean of Undergraduate Students and Cyber Learning, which we needed to be trained to use." In this case, this text, labeled as "Professional development," captures the participant's training experiences to develop skills required by the job market. It is worth noting that the selection of "Professional development" in this labeling process was based on our goal to highlight challenges and opportunities that participants expressed regarding the production of teachers of color. As we will further discuss below, these experiences may be grouped based on participants' roles. That is, while there are experiences captured from students' perspectives, there are other perspectives captured by teachers' and administrators' points of view. NAQD was developed to account for potential disparities in their experiences based on diverse roles or participants' attributes.

Continuing with the relevance of keeping track of attributes, one important point to be noted regarding the labeling or coding process is that, as part of this labeling or coding process, we should keep track of who provided us with the original text, including their roles and other attributes. In Table 2.1, we have three actors: "ID1, ID2, and ID3," and it is likely that each actor has attributes that may be useful for us to gain more nuance from our analyses. For example, there could be important variations by gender or by their roles in the college education process. That is, students may be less interested in administrative challenges, whereas administrators may be primarily interested in or provide us with codes related to these types of challenges. As we will further discuss next, NAQD helps us not only to keep track of who mentioned what codes but also to statistically test whether differences across groups of participants, as captured by their attributes, may also be observed during the analytic process. We will further elaborate on these integrative statistical comparisons below.

Returning to Fig. 2.1, note that once all textual data have been labeled, the resulting relationships, along with keeping track of participants' attributes of interest, result in datasets that are difficult to analyze manually. That is, these resulting relationships entail keeping track of the providers of texts, labels or codes, the text content or quotes, as well as their attributes, all of which may result in hundreds and typically thousands of pieces of information that need to be synthesized to capture commonalities and patterns. From this perspective, although we are not used to referring to qualitatively (or quantitatively) coded texts as big data, these resulting labeled or coded datasets translate into important analytical challenges. Not only do we need to deal with this multiplicity of information and meanings, typically manually, but while doing so, we should try to be as transparent and rigorous as possible. This latter point has been a theme of concern for qualitative methodologists. For example, Attride-Stirling, 2001 has argued for greater disclosure of how qualitative researchers conduct their analyses and reach their conclusions.

To address this analytical challenge, NAQD's main premise is that network modeling or relational analyses are useful methodological tools that may be used to find the mathematical structure embedded and inherent in qualitatively or quantitatively coded databases. That is, as further discussed in this chapter, once our textual resulting datasets have been labeled, we can link these labels to the participants who provided us with those quotes while also being able to consider these participants' attributes, which may be useful in highlighting the resulting structure. From this perspective, then, NAQD's interactive outputs may serve as a guide or map to inform our analyses. These maps provide more transparency to the analytic process, as they reveal the structure of the most prevalent topics or codes, as well as who the most important providers of these codes were. More specifically, NAQD is especially well suited for research questions concerned with how perspectives, concerns, and experiences are structured across groups within a sample, and with whether those structures are similar or meaningfully different as a function of participants' roles or attributes (González Canché, 2026; González Canché, 2022d,). In other words, NAQD may also help us unveil how or if participants' attributes or roles may help explain their textual contributions to our research projects. Moreover, these techniques may also serve to test whether certain groups of individuals in our samples had similar or different experiences, as reflected by the distribution of their coded texts. By rendering structure through interactive, inspectable, and shareable outputs while still requiring return to the original evidence, NAQD contributes to the broader goal of opening the analytic black box of qualitative research Attride-Stirling, 2001; Constas, 1992; Johnson and Onwuegbuzie, 2004; Onwuegbuzie and Leech, 2007) without reducing qualitative interpretation to mere quantification.

Despite the timeliness and relevance of uncovering the structure of these labeled datasets, the implementation of NAQD poses significant statistical and computer programming challenges. The data science and feature engineering procedures required to implement the analytic processes serve as a major deterrent to the use of NAQD by most qualitative and mixed-methods researchers. With the goal of democratizing access to the benefits of data science (see also González Canché, 2023), all backend processes required to implement NAQD have been packaged into a user-friendly software tool available for both Mac and Windows operating systems. In short, NAQD can be implemented using a no-code (i.e., absolutely no programming requirements) software tool that is also offered at no monetary cost.

##NAQD and Mixed-methods Social Network Analysis

Recent scholarship in mixed methods social network analysis has emphasized that network approaches are most useful when they move beyond the descriptive mapping of ties and toward deeper integration between structure, process, and meaning (Edwards, 2010; Froehlich et al., 2020; Fuhse and Mützel, 2011). Relatedly, qualitative social network analysis has argued that network research becomes especially valuable when it helps researchers access the meanings actors attach to positions, relations, and collective configurations rather than reducing social life to abstract structure alone (Luxton and Sbicca, 2021). NAQD is directly aligned with these calls. It uses qualitatively or quantitatively coded textual data as the basis for network construction, not to displace interpretive inquiry, but to strengthen it by making patterns, commonalities, and differences across groups more visible, more inspectable, and more analytically testable. In this sense, NAQD contributes to a broader movement in qualitative and mixed-methods research that seeks not simply to visualize relationships, but to integrate network structure with the substantive meanings embedded in participants' narratives (González Canché, 2026).

NAQD also occupies a distinctive position among other network-based approaches to discourse and process data. Epistemic Network Analysis (ENA), Ordered Network Analysis (ONA), and Transition Network Analysis (TNA) offer important ways to model relationships among codes, concepts, or temporally ordered states (Saqr et al., 2025; Shaffer et al., 2016; Tan et al., 2023). NAQD differs in a fundamental respect: it is built on two-mode affiliation networks that directly connect participants to the codes applied to their contributions. This design makes NAQD particularly well suited for research questions centered on how perspectives, concerns, and experiences are distributed and structured across participant groups, rather than only on relationships among codes themselves.

Background: currently available applications to map coded textual data

As briefly mentioned above, there are currently qualitative software applications that provide researchers with the ability to map their coded textual databases. Accordingly, the goal of this section is to describe these applications and highlight how NAQD is similar to or different from those software tools. To achieve this, we will briefly describe some of the most popular and influential software tools.

Atlas.ti networks

Atlas.ti is a commercial software tool that aids in the process of labeling textual data while maintaining the resulting wealth of information, making it available for researchers or analysts to retrieve, analyze, and ultimately craft their findings and conclusions. Once these labels or codes have been completed, Atlas.ti allows users to *manually* build a visualization with codes, families (i.e., codes grouped into categories), and quote elements. In this visualization-building process, the connecting lines may be assigned different meanings by the researchers—that is, a connecting line itself is given a qualifying label.

Note that during this process, the emphasis is placed on codes, quotes, and connecting lines manually linked by researchers, rather than on allowing readers to see who (i.e., participants) mentioned a set of codes—wherein linking elements capture the participants' contributions, similar to the relations shown in Table 2.1. Although not necessary, during this visualization-building or crafting process, researchers may also add a "document" to the resulting visualization or map. Considering that documents typically represent individuals (i.e., research participants and interviewees), the inclusion of a document represents adding some of the participants that researchers consider important actors or units in the resulting visualization network or sociogram (González Canché, 2019). Finally, researchers may also incorporate research memos (notes) and groups (or families of codes) into this visualization. An example of this visualization exercise may be found on the Atlas.ti website by following this link: https://doc.atlasti.com/ManualWin.v9/MainConcepts/EntityTypeNetworks.html. Due to copyright issues, we are not presenting this visual, but it is fair to state that this crafting visualization is not much different from using Microsoft Power Point or the Dia software (see http://live.gnome.org/Dia) to manually build a conceptual map that researchers argue best captures the story they want to share.

In other words, the resulting Atlas.ti Networks visualization serves to highlight what researchers considered valuable in explaining their findings. This is in contrast to the goal of NAQD, where we seek to highlight the structure emerging among hundreds or thousands of relationships resulting from linking participants to the codes and the texts they provided to our research endeavors. From this view, NAQD is similar to the subsequent visualization approaches we present, all of which aim to highlight the resulting structure emerging from connections like those discussed above, where actors are connected to codes and codes are connected to quotes (see González Canché, 2019; M.S., 2022d). In Atlas.ti, the resulting network may highlight these actor-to-code-to-quote relationships, but when elements are added to the Atlas.ti plotting canvas, their connections are not automatically added; instead, it is the researcher who "manually" includes them.

Nvivo project maps

Similar to Atlas.ti's networks, users may select codes, files, memos (research notes that may help researchers make sense of their thought process), and themes (or families that group a set of codes) to include in the proposed network diagram or representation. A salient difference between Atlas.ti and NVivo is that, in the latter, the connections are retrieved from the relationships observed in the data. To illustrate this process, let us build from our example shown in Table 2.1. Assume we add the code "Professional development" to the NVivo Project Maps visualization canvas. If only this element is added, no connections are drawn. However, if Actor ID1 is added to this map, NVivo will then automatically draw a connecting line between the "Professional development" code and Actor ID1. On the other hand, in Atlas.ti, the researcher would need to draw that connecting line manually. In short, when elements are connected—that is, when codes are assigned to a theme or family (using NVivo language)—and these elements are added to the map, their connecting lines are automatically added to such a visual map. This implies that, if all codes and actors are added to the map, the result will be an aggregated depiction of the complete set of connections. NAQD builds from this rationale, but it does not ask researchers to add certain elements and exclude others. Instead, all codes and actors participating in the project are added automatically, situating their positions in the resulting sociogram based on measures of relevance or influence. These measures are referred to as centrality measures, and we will describe them in depth in our methods section.

Leximancer concept maps

Concept maps in Leximancer contain the themes (which we have been referring to as labels or codes) that this software detected via unsupervised text classification, along with the most highly correlated words associated with those themes. The resulting concept map is represented in a network form. The size of the nodes (represented as circles or bubbles) is based on their frequency in the transcripts or texts analyzed. Different from Atlas.ti's Networks and NVivo's Project Maps, these concept maps are automatically generated as a result of the unsupervised text classification and represent an aggregated description of the most frequent connections of themes and cooccurring words in these texts or quotes. Based on the unsupervised nature of this classification, all sentences across the corpora (all collections of texts) are labeled into one of the resulting themes (i.e., codes, as we have described them above). A commonality of Leximancer, Atlas.ti, and NVivo is that, unlike NAQD, their resulting visualizations have no measures of centrality that may help us detect who the most important actors (i.e., providers of labels) or labels/codes are in the resulting sociogram. Arguably, this lack of centrality or relevance identification obscures or limits the researchers' thought process in highlighting or retrieving the structure hidden in hundreds or thousands of connections.

MAXQDA maxmaps

MAXMaps follows the same rationale as NVivo's Project Maps. Researchers may select codes, documents, memos and quotes and categories (themes, families) that should be added to the plotting canvas. Once selected elements are added, lines among such elements are also automatically drawn. This follows the rationale employed by NAQD. Nonetheless, different from NAQD, users may also link objects that are not connected following the rationale shown in Table 2.1 (i.e., $actors \rightarrow quotes \rightarrow codes$). That is, researchers may, like with Atlas.ti, idiosyncratically craft their visualizations as an analytic tool based on their analytic needs or goals. More to the point, the connecting lines may be formatted to allow researchers to highlight attributes they consider relevant. More information about this crafting process can be found in this link https://www.maxqda.com/help-mx20/maxmaps/what-does-maxmaps-do.

Another visualization tool available in MAXQDA is called code cooccurrence. This tool highlights the number of documents (interview transcripts or essays, for example) that contain the same pair of codes. Researchers may decide to display these frequencies as another strategy to highlight structure in qualitative data. This is also an analytic strategy implemented in MAXQDA by describing the frequency with which a given actor provided text that was labeled under a given code, as well as the degree to which these relationships were more or less prevalent in this actor's discourse or contributions to the resulting network.

Qualitative Data Analysis (QDA) miner's link analysis

Finally, the last CAQDAS we have found to integrate network visualization tools is QDA Miner. Link analysis allows the visualization of connections between codes and between codes and cases (i.e., research participants). The goal of this visualization is to highlight patterns and structures of cooccurrences. The resulting visual applies a layout that places codes that are more densely connected together and codes that are less prevalent in the working space as peripheral—that is, as far from the center of the plot as possible, which is one of the properties implemented by NAQD. The closeness of codes based on the prevalence of cooccurrence may also be represented in a dendrogram and in a proximity plot. These visualizations are possible given that a version of Table 2.1, as shown in this study, may be transformed into a matrix, and with this matrix, traditional classification and correlation approaches may be implemented. Indeed, in our methods section, we discuss the required transformations needed to test for similarities or differences in our participants' experiences, as implemented by NAQD. Finally, note once more that in the QDA Miner Link Analysis, no centrality or relevance information is considered when creating these visuals, a common trend across all the software tools described.

This brief review of currently available software that employs network analysis principles to visualize relationships obtained from textual data serves to highlight the contribution of NAQD to these visualization capabilities by incorporating attributes into the resulting visualization, as well as offering the possibility to evaluate whether certain groups are more concerned about certain topics than others.

Conceptual and methodological frameworks

Let us start this section by asking: why are we relying on network analysis and network visualizations in the analysis of qualitative data? To address this question, let us express that when we move from participants' qualitative contributions to data frames, matrices, and then to sociograms (or the network representations of networks or matrices), we gain access to a comprehensive vantage point of analysis (Crossley, 2010). Network analyses enable researchers to detect or uncover active or central topics (i.e., labels or codes) via measures of influence. When these measures of centrality or influence are presented in tabular form, the identification of central topics and actors becomes possible. However, when these same measures of influence are presented in visual form, much more nuanced understandings of the overarching structure embedded in qualitative or coded textual data may be gained.

For example, the resulting visualizations may reflect areas of overlap or consensus among our research participants regarding their expressed points of view or experiences—as captured by the codes or labels they provided us with. These visualizations may also show areas where disagreements or discrepancies were more prevalent, as expressed by the mathematical structure uncovered by network analyses and visualizations. This uncovering exercise is similar to what Alexander et al. (2019) referred to as *hot zones* in their discussion of their proposed *visual replay methodology*. In our case, these *hot zones* would be represented at the center of the sociograms (i.e., network visualizations) and will tend to be larger in size. By the same token, the network visualization techniques rendered by NAQD also allow researchers to identify peripheral conversations or points of discussion (i.e., *cold zones*). These cold zones will be represented at the periphery of the sociograms and will tend to have smaller sizes.

Overall, although most current applications of network visualizations discussed in the previous section do not rely on network analysis principles to identify relevance like NAQD, the possibility of visualizing patterns (or telling stories with these maps of relationships) is the reason why network visualizations have become prevalent in CAQDAS. Sociograms are tools that allow for a comprehensive depiction of the entire web of relationships, thus going beyond the mere quantification and visualization of qualitatively coded content (Crossley, 2010; González Canché, 2019; M.S., 2022d). Nonetheless, we should always remember that the resulting sociograms are just a tool and not the end goal of NAQD. That is, researchers, as experts, should interpret these maps, and during this interpretative process, other questions may emerge that may lead researchers to go back to the field or to ask more questions of their research participants.

Network modeling

Network analysis and modeling are particularly useful ways to "deal with complex systems in the real world" (Mitchell, 2006, p. 1199), such as the set of relations that form the structure of qualitatively coded textual content. In the context of NAQD, network modeling is particularly useful in helping researchers make sense of multiple sets of relationships emerging as a function of not only linking participants to the codes or labels they provided us with, but also indirectly connecting participants to others precisely because they shared a similar set of codes or labels. Conceptually speaking, then, two participants that shared the same codes or labels may have also shared similar experiences regarding the topic of study or the research goal under analysis. Pragmatically, two actors will be indirectly connected to each other given their commonalities in providing us with similar textual content that was eventually coded or labeled when sharing their experiences.

Operationally, network thinking emphasizes relations among specific events, including connections among individuals and their actions or among broader categories, labels, or codes, as defined by researchers (González Canché, 2019; Maroulis et al., 2010). Formally, networks are a collection of *potentially interactive units* (Csárdi, 2006; Kolaczyk and Csárdi, 2014; Mitchell, 2006; Wasserman and Faust, 1994). These units are typically referred to as *nodes* or *vertices* (e.g., actors, participants, or entities that participants may interact with or be ascribed to), and the connections resulting from their interactions are referred to as *edges* or *links* (Wasserman and Faust, 1994). When the units configuring a given network are of the same *class* and hierarchy (e.g., students interacting with other students), they form a one-mode network. When the units configuring the network are different (e.g., professors ascribed to their employing universities or ascribed to a particular set of beliefs), they form *two-mode networks.*

NAQD relies on two-mode networks given our focus on the analysis of participants' contributions of *codes.* In this case, this contribution is operationalized asinformation, or the first mode, and participants are the providers of such information, or our second mode. Following (Crossley, 2010), although network analyses can be focused on individuals within the network, we can also observe groups of participants who clustered around certain topics or viewpoints during their interviews (referred to as "hot zones" above, see Alexander et al., 2019).

Highlighting structure

NAQD applies graph theory (Biggs et al., 1986) and interactive visualizations to highlight the mathematical structure inherent in coded pieces of textual information, which will eventually be used as a map or guide for analyses. The integration of graph theory and interactive visualization enables the simultaneous depiction of all contributions taking place in our research projects. However, when considering participants' attributes, the resulting visualizations may be disaggregated by their attributes, which could potentially ease the discovery of patterns or structures that may be masked by aggregated analyses. From this perspective, our description of centrality measures that capture participants' and codes' centrality or influence may be computed at the aggregate level, but disaggregated analyses by groups are also possible with NAQD. With respect to the latter, NAQD was programmed to test whether groups of participants shared similar "concerns" or "experiences" as reflected in their codes, or whether those concerns or experiences differed across groups. In sum, NAQD highlights structure via measures of influence, and when group information is provided, it tests whether certain groups are more similar than others via quadratic assignment procedures (QAP), as we will illustrate below.

In addition, NAQD offers the possibility of detecting communities of actors as a function of having more densely connected relationships among a group of codes than what would be expected to be observed by chance alone. We will discuss this in a subsection called "community detection".

From authors and codes to networks: Network operationalization

To explain the transformation from authors' texts to codes and their relationship, let us go back to Table 2.1. As can be seen in that table, textual datasets may be represented by collections of texts where we can identify the providers of those texts, as well as attributes, such as the roles the actors have in the research setting. In our working example, we have four groups of participants: students, administrators, faculty, and faculty with administrative roles. We will further elaborate on how we programmed NAQD to use this information. For now, let us discuss how we can move from authors and codes to networks.

The information presented in Table 2.1 includes actors who shared their experiences, which were transcribed into texts. These texts were then labeled or coded to generate codes or labels summarizing the content (or quotes). When we initially described Table 2.1, we demonstrated these relationships to illustrate the coding or labeling process. In our current discussion, we elaborate on how to operationalize these relationships to construct networks.

Generally speaking, a network or graph may be represented asa

$$G=(V,E), \tag{1}$$

where V stands for the units configuring the network and E are the relationships observed or with the potential to be realized amongst those units. As described above, for NAQD the units configuring this network are *actors* and the *codes* these actors provided us with, as a result of the coding or labeling or coding of their texts or information. From this view then, we can capture these relationships as

$$actors \rightarrow codes \tag{2}$$

More generally, we can standardize this relational thinking by substituting each element with a unique identifier (or ID as referred to this generic identifier from now on). By referring back to Table 2.1, we can observe the following relationships.

$$\text{ID1} \rightarrow \text{Professional Development} \tag{3a}$$

$$\text{ID1} \rightarrow \text{Hands} - \text{on Experiences} \tag{3b}$$

$$\text{ID1} \rightarrow \text{Promotion Strcuture} \tag{3c}$$

$$\text{ID2} \rightarrow \text{High Program Impact} \tag{3d}$$

$$\text{ID2} \rightarrow \text{High Program Impact} \tag{3e}$$

$$\text{ID3} \rightarrow \text{Professional Development} \tag{3f}$$

$$\text{ID3} \rightarrow \text{Hands} - \text{on Experiences} \tag{3g}$$

As can be seen from these relationships, there are cases where the same participant mentioned the same code twice, such ID2 whose contributions were classified under "High Program Impact." Note that, in this example, we omitted some text between these codes to illustrate this repetition. With this information in mind, ID2's contributions may be represented as:

$$\text{ID2} \overset{2}{\rightarrow} \text{High Program Impact}, \tag{4}$$

Where the number above the arrow indicates how many times this participant's textual contributions were classified under high program impact. In other instances, different actors may have provided text labeled under the same codes, such as actors ID1 and ID3, who both provided texts labeled under "Professional development" and "Hands-on experiences."

In terms of data structure, Eqs. (3a) to (3g) represent lists of connections, which are also called edgelists in network modeling (Kolaczyk and Csárdi, 2014; González Canché, 2019), and these relationships can be stored as shown in Table 2.2. This is the structure that NAQD was programmed to read and transform into networks. Table 2.2 shows actors, the codes retrieved from their texts, as well as attributes like the role played in the network or research setting, and other contextual attributes such as the type of institution where these roles were observed. Note that with this information, each set of relationships may be limited to an attribute, which may be retrieved from the roles or the school type in Table 2.2. In our example, we relied on role as the attribute of interest to test for similarities or variations in the distribution of codes, as we will further explain later.

Table 2.2 Text structure network analysis of qualitative data.

Actor	Code	Content	Role	Gender
A1	Campus environment	understanding the ...	Student	Man
A1	Campus environment	significant is the ...	Student	Man
A1	Campus environment	You may be familiar ...	Student	Man
A1	Campus environment	while it has great ...	Student	Man
A2	Campus environment	One of the problems ...	Faculty	Woman
A2	Campus environment	So they're, some ...	Faculty	Woman
A2	Campus environment	So I think state ...	Faculty	Woman
A2	Campus environment	My thing is having ...	Faculty	Woman
A3	Recruitment	impaction issue ...	Administrator	Man
A3	Recruitment	SB1440, are you ...	Administrator	Man
A3	Recruitment	So that means that ...	Administrator	Man
A3	Recruitment	enrolment caps ...	Administrator	Man
A3	Recruitment	student demographic ...	Administrator	Man
A4	Student needs	drive us to learn ...	Faculty and Administrator	Woman
A4	Cultural relevance	majority of our ...	Faculty and Administrator	Woman
A4	Resources	manage our resources ...	Faculty and Administrator	Woman
A4	Resources	colleges and departments ...	Faculty and Administrator	Woman
A4	Resources	programs and colleges ...	Faculty and Administrator	Woman

Centrality or relevance of units

As described above, an important contribution (and point of departure) of NAQD with respect to other software tools is its detection of structure in our coded/labeled datasets. That is, NAQD was programmed to automatically detect actors' and codes' relevance/influence based on social network analysis measures (Freeman, 1978). Specifically, NAQD computes three measures of centrality that unveil the mathematical structure in the network through actors' and codes' positions and paths or connections (Borgatti, 2006). These three centrality measures are degree, betweenness, and eigenvector, as we describe next.

First, let us note that the information presented in Table 2.1 can be transformed into an edgelist, as represented in Fig. 2.2. This list of connections can then be uploaded to the NAQD user interface (UI) to obtain the network visualization. When more cases are present (say at least 10 human actors), comparison analyses may be conducted. For this brief example, we used the information shown in Table 2.1 as a "toy complete network" to form an edgelist and then visualize it as a sociogram using NAQD to elaborate on the meaning and usefulness of centrality measures to highlight structure.

Degree centrality

This form of centrality consists of measuring the number and frequency of connections that a given actor has in the network (Borgatti, 2006). In terms of the *number* of connections, NAQD

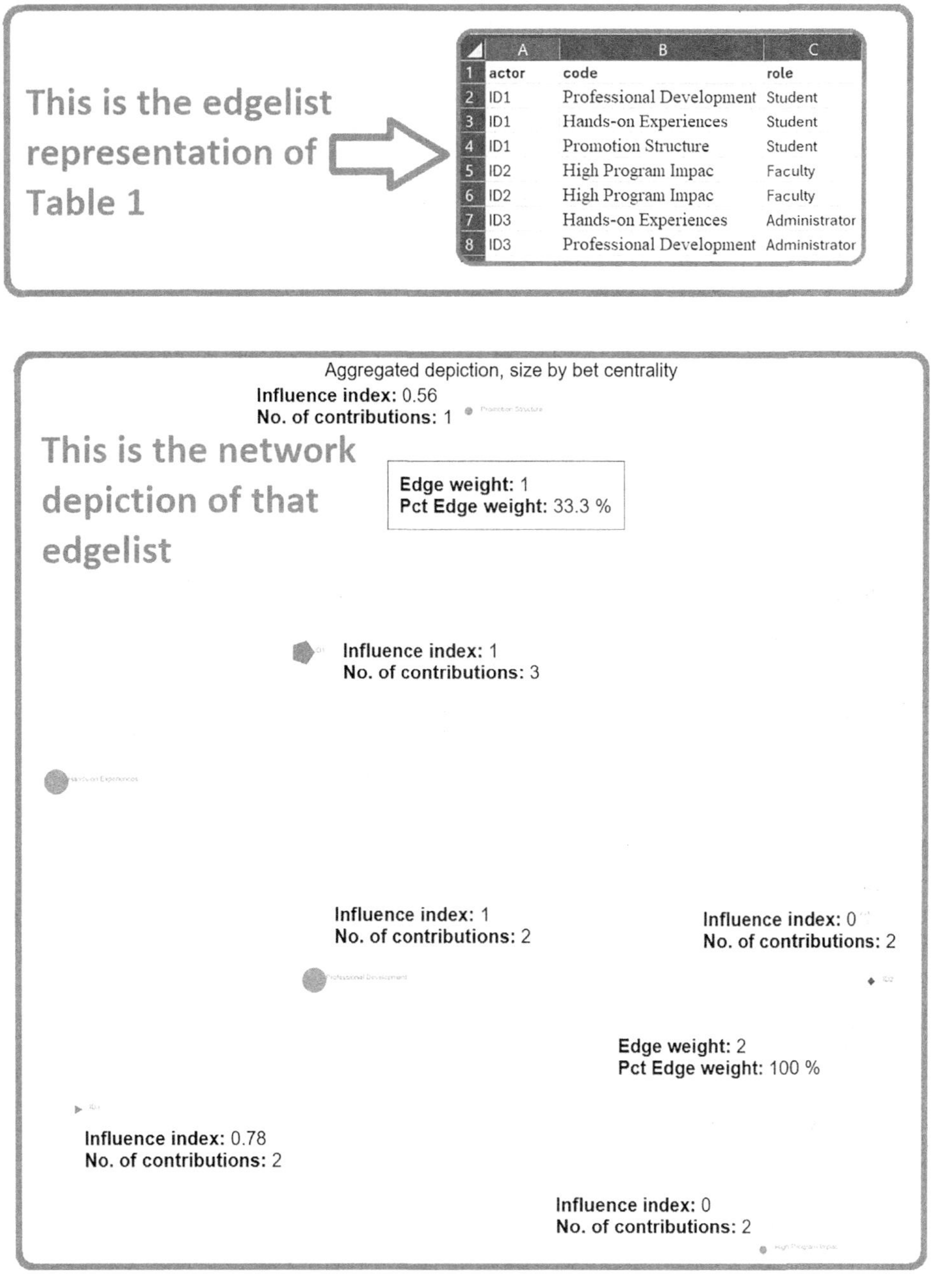

FIG. 2.2 From Edgelists to Sociograms.

computes degree centrality by counting the number of times an actor provided information that was eventually classified as a code—similar to the aggregated connections depicted in Eq. (4) discussed above. Pragmatically, when we click on an actor, a dialog box will display "No. of contributions: X." In this case, X is a place holder counts the number of pieces of text or quotes labeled under a set of codes. In other words, under "No. of contributions," the resulting HTML output will show the total number of codes provided by that actor. When we click on a code, this number will count how many times a given code appeared in that network. Specifically, if a human actor displays "No. of contributions: 15," this means that such an actor contributed 15 codes to the network. But if a code, let us say "Professional development," displays "No. of contributions: 97," this means that there were 97 text chunks or codes classified as "Professional development."

In Fig. 2.2, we can see that when we clicked on actor ID1 (i.e., the pink pentagon), the number of contributions was 3. This can be corroborated in both the edgelist shown in Fig. 2.2 and in the information presented in Table 2.1. In the case of the code "Professional development" in orange, we can see that this number is 2, and based on the edgelist information and Table 2.1, we know that this number 2 comes from actor ID1 and ID3 having mentioned text that was classified under "Professional development."

This brief discussion implies that degree centrality, in its role as a counting device at the actor or code levels, does not reflect the relevance of that code for a given actor in particular. To address this potential downside, we programmed NAQD to also include code relevance or *frequency* per actor by clicking on the line(s) connecting each actor with their codes. In the case of actor ID2 (blue diamond in Fig. 2.2), we know that this actor only had two contributions, and both were labeled as "High Program Impact." In this case, when clicking on their connecting line, we can see that a dialog box indicates that actor ID2 mentioned this code twice. This same dialog box also shows that this relationship represents 100% of all their contributions in the network—see "Pct Edge weight: 100%" in the resulting HTML sociogram in Fig. 2.2. In other words, not only does NAQD show the number of times this code was provided by this actor, but also the percentage of this actor's total number of contributions that this relationship represents. Had actor ID2 provided 10 codes in total instead, and two of these codes were classified as "High Program Impact," then a click on this same connecting line would read "Pct Edge weight: 20%," as this actor's total number of contributions would have been 10, with 2 of them (or 20%) being captured by this relationship alone. These "Pct Edge weight" estimates then allow us to further contextualize the relevance of each code for each actor when clicking on their connecting line, and the relevance of the actor or code for the network when clicking on that actor or code.

Betweenness centrality

This measure goes beyond frequency counts. Instead, it detects the extent to which a given actor or code falls in between other actors or codes (Wasserman and Faust, 1994) thus effectively serving as a communication bridge in the system to ease the flow of information. In other words, betweenness centrality highlights *both* the participants who served as bridges of information *and* the codes that connected the individuals in this network. NAQD uses betweenness centrality as an attribute to add size to each actor or code. To the extent that the size of a node (actor or code) is larger than other actors or codes, this indicates that such an actor or code is serving as a relevant bridge of information. Based on these myriad relationships, researchers can observe which actors and codes were the most influential by serving as bridges of information in the whole network.

Returning to Fig. 2.2, we can see a clear pattern. Actors and codes that do not fall between other units have the smallest sizes. For example, actors ID2 and ID3 as well as the codes "High Program Impact" and "Promotion Structure," have betweenness values of zero. On the other hand, ID1 falls between the codes "Promotion Structure," "Hands-on Experience," and "Professional Development." Since these latter two codes are also connected to actor ID3, these codes have high betweenness centrality values.

Eigenvector centrality

This measure appears as the "Influence Index" in a dialog box when an actor or code is clicked in the NAQD HTML output (see Fig. 2.2). Eigenvector centrality measures the overall influence of a given code or participant as a function of the relevance of its connections (Borgatti, 2006; Freeman, 1978). Its Influence Index is standardized to range between 0 and 1, with 1 indicating the maximum level of influence, and it highlights individuals or codes that are central as a function of being connected with individuals and/or codes that are actively connected in the network (Borgatti, 2006; Freeman, 1978). Hence, an actor is central in eigenvector centrality if such an actor provided codes or established connections with other actors who were also active or central in the network (influence begets influence).

Given that, mathematically, the eigenvector centrality of a unit i is proportional to the sum of centralities of the units j to whom unit i is connected (Borgatti, 2006; Freeman, 1978), the level of influence of each actor and/or code may serve to further contextualize their contributions and strengthen our understandings. For example, in Table 2.1 and Fig. 2.2 we can note that actor ID1 was the most active contributor in the network with 3 codes in total. Two of these codes were also present in actor's ID3 connections. With these pieces of information, we can infer that actors connected with the codes provided by the most active contributor of the network (i.e., actor ID1) will increase their relevance or influence in the network. In this virtuous cycle, since actor ID3 shared the two most shared codes (same ones as actor ID1), this actor will have a higher level of influence than actor ID2, despite the fact that both have the same number of codes, but these codes in the latter case were not shared by anyone else in this mini network example shown in Table 2.1 and Fig. 2.2.

More specifically, note that ID3 with two codes shared, has an influence index of 0.78 (maximum value is 1), whereas actor ID2 has an influence index of zero. This property is also observed in the codes. For example, the code "Promotion Structure," which was only shared once (by the most active actor), has an influence index of 0.56, whereas the code "High Program Impact," which was shared twice but by an actor without power or influence, has an influence index of zero. These findings clearly indicate, then, that more than frequency, what matters in codes and actors' influence is considering the relevance of their connections—that is, my influence is a function of the influence of my connections. Empirically, as shown in Fig. 2.2, when clicking on an actor or code, we will note that a dialog box will read "Influence index: Y," and when the value of Y reaches 1, this indicates that such a unit is the most influential from an eigenvector centrality perspective.

This finalizes our discussion of these centrality measures. Our description of the findings will further elaborate on these measures of centrality and their role in highlighting structure across each of the resulting networks. This is important because, if we disaggregate the analyses by actors' attributes, we will obtain one sociogram per group.

Community detection

NAQD offers the possibility of identifying clusters or groups of individuals belonging to a given community as a function of being more densely connected to a set of codes than to others. More technically, NAQD implements a spin-glass model and simulated annealing to detect communities, or a set of nodes with many edges inside the community and few edges outside it (Kolaczyk and Csárdi, 2014; Reichardt and Bornholdt, 2006). Following the structure of the network built from edgelists as described in this chapter, the spin-glass model allows links between actors and nodes to be weighted—that is, to account for the number of times each actor mentioned texts that were labeled under a given code.

Procedurally, the spin-glass model implements simulations to iteratively identify the optimal number of communities to be observed in our networks by comparing the resulting structures of communities against graphs or networks of the same size and with similar density levels in their connections (Reichardt and Bornholdt, 2006).

Note that this community detection function is only implemented if researchers do not select a column in their datasets that groups participants by color. That is, in looking at Table 2.2, we can see that there is a column called "Gender." If, instead of allowing the spin-glass model to detect communities, we prefer to use gender to add colors to the actors in our HTML visualization, we simply need to select the column "Gender" in section D of the NAQD UI (we will detail this UI later in this study). Alternatively, the column "Role" could also be used for this purpose. Notably, if comparative analyses are preferred, we could use the column "Gender" to test the hypothesis that women and men have similar distributions of their contributions to the network, as we describe in the next subsection.

Finally, note that if we allow NAQD to automatically identify communities via spin-glass, we can then download a dataset containing the community to which each of our actors was ascribed. To interpret the meaning of these communities, we can observe the resulting maps and identify the sets of codes to which actors classified in the same community (i.e., same colors) were ascribed. That is, these communities will reveal that a group of individuals consistently mentioned the same issues or experiences during their discourses or participation in our research projects.

Testing for similarities or differences in experiences based on attributes

NAQD enables testing the hypothesis that some networks are more similar than others as a function of participants' roles or attributes. For example, in our working example, we have four groups: faculty, administrators, students, and faculty and administrators. This latter group, classified as "faculty and administrators," is composed of faculty members who are also involved in administrative functions. Accordingly, their concerns or experiences should, in theory, be more closely aligned with the concerns raised by faculty and administrators than with the concerns raised by students. This hypothesis may be tested by relying on the QAP discussed next.

The QAP is a nonparametric analytic procedure that does not rely on normality assumptions and does not assume independence. QAP enables testing for statistical significance using network (e.g., matrix) data (Whitbred, 2011) and takes two main forms: correlation and regression analyses. In the former, QAP measures the extent to which two networks are statistically associated. This translates into measuring whether every corresponding dyad or pair of connections in two matrices tend to vary in the same direction (positive correlation with a maximum value of 1), the opposite direction (negative correlation with a minimum value of −1), or whether they are independent (i.e., a correlation of 0). In our example, we can ask whether *faculty* tended to express similar concerns regarding the formation of teachers of color than *students*.

When more than two groups are present, NAQD tests if similar concerns are expressed across all groups dyadically. That is, these analyses translate into testing all combinations across groups. To conduct these comparisons NAQD relies on combinations in the form: $C(n,r)=\frac{n!}{(n-r)!r!}$, where n is the number of groups and r is the number of cases compared at a time. That is, with four groups being compared two at a time, we have a total of six QAP tests to be computed by NAQD automatically.

Procedurally speaking, QAP can be classified under the random permutation test family, which is sometimes referred to as a randomization test (Phipson and Smyth, 2010). Random permutation tests begin by recording a given test statistic of interest across two mutually exclusive and exhaustive groups. These statistics can include mean differences, correlation coefficients, or regression coefficients, for example. In each test, one set of observed estimates is recorded, and then thousands of estimates are obtained by shuffling all the values of the networks randomly X number of times—NAQD employed 50,000 permutations. Each time this random shuffle or permutation occurs, the statistic obtained is also recorded. This procedure enables the recreation of a sampling distribution wherein the coefficients obtained through a random process should have a mean of zero, given the lack of inherent structure upon which the data-generating process rendering these coefficients was founded (as shown in the histograms located in Fig. 2.7, to be discussed in our application section). Once this sampling distribution based on randomly shuffled estimates is obtained, analysts can compare the number of times randomly generated estimates were greater in magnitude than the estimates obtained with the observed data.

Using a correlation coefficient as an example, NAQD estimates the proportion of times that the randomly generated correlation coefficients between two networks are larger than the correlation obtained with the observed data. Once more, the foundational rationale behind this test is that randomly obtained coefficients should, on average, be distributed around zero. Hence, if randomly generated correlation coefficients exceed the magnitude of the correlation coefficient obtained with observed (unshuffled) data, then we may conclude that our observed correlation is not better than what we would expect to see by chance alone. If the proportion of times that randomly generated correlation coefficients are greater in absolute magnitude than our coefficient obtained with actual data approximates zero, then one can conclude that our observed correlations are better than chance by 1 minus the proportion of times randomly generated coefficients were greater. Specifically, assume that 0.005 is the proportion of times out of 50,000 permutations that the random coefficients surpassed the magnitude of our correlation coefficient obtained with actual data. Then we can conclude that 99.5% of the time, we would expect our observed results to hold true if we had access to other networks.

The QAP correlation implemented by NAQD begins by estimating a Pearson correlation coefficient across the corresponding cells of two matrices (Phipson and Smyth, 2010). This requires that the units in the networks are the same and follow the same order across matrices. For example, the representation shown in Fig. 2.3 has three units. Note that these matrices contain the same units and follow the same order (U1, U2, and U3)—a process that is computed automatically by the NAQD software. The process takes the dyadic correlation among U*i* and U*j* in matrices 1 and 2 to assess whether a given dyadic intersection in one network (i.e., faculty experiences) is of a similar magnitude in the other network (i.e., students' concerns). The second step is depicted in the shuffled section of Fig. 2.3, where all rows and cells of the matrix are randomly permuted. Each time this permutation happens, its corresponding Pearson coefficient is recorded.

NAQD was programmed to run these random shifts 50,000 times. Moreover, by default, the diagonal of the matrix is ignored, as it represents self-selection within each matrix—a default procedure implemented in relational modeling (González Canché, 2026). (Although NAQD does not implement regression, but QAP correlations, the regression-based form follows a similar rationale in terms of calculating R-square and regression coefficients, across intersecting cells of the matrices (Whitbred, 2011). After each permutation, the resulting correlation coefficient is stored to test the proportion of times these estimates were larger than those obtained with the observed data.

	Observed Matrix 1			Observed Matrix 2		
	U1	U2	U3	U1	U2	U3
U1	2	1	5	4	2	5
U2	1	4	0	0	5	0
U3	1	0	6	2	0	7
	Observed Matrix 1			Shuffled Matrix 2		
U1	2	1	5	?	?	?
U2	1	4	0	?	?	?
U3	1	0	6	?	?	?

FIG. 2.3 Matrix Manipulation Behind Quadratic Assignment Procedures.

FIG. 2.4 Projected one-mode matrices.Based on this discussion, the matrices built from qualitative data analyzed here are the one-mode representation of the qualitative actors and codes network. All one-mode transformations are shown in Fig. 2.4. As can be seen in that figure, each matrix accounts for a total of 26 codes contained in each of the resulting subnetworks (one per group). In sum, this transformation indicates that the dimensions of the complete code matrices are 26 by 26 codes obtained from participants interviews and in total there were four matrices, one per each group of interest.

Feature engineering

Recall that the networks handled by NAQD are two-mode networks because we have actors (one mode) providing us with codes (second mode). In this respect, it is worth noting that QAP requires symmetric matrices or networks, which can be achieved through a simple transformation. In his seminal duality study, Breiger (1974) demonstrated that two-mode networks can be represented or projected as one-mode matrices via matrix algebra.

This needed transformation is obtained as follows. First the two-mode network needs to be represented as a two-mode matrix, wherein the number of rows counts the number of actors and the number of columns contains the number of codes. This process renders a set of matrices in the form $[n,m]$ where n and m may be of different lengths. Taking one matrix at the time, the transformation or one-mode projection as illustrated by (Breiger, 1974) is accomplished as follows. Assume

$$A=[n,m],\text{thus } A^T\%*\%A=[\boldsymbol{m},n]\%*\%[n,\boldsymbol{m}]=\mathrm{A}_{one-mode}=[m,m] \tag{5}$$

This transformation shown in Eq. (5) is possible given that the resulting dimensions of $A_{one-mode}$ in this process is the outer (shown in bold font) of the matrix multiplication (see Breiger, 1974). In our case, these one-mode projected matrices will contain four code by code networks, one for each of the four groups we have in our dataset.

Application

In this section, we present a reproducible application in which we describe all the steps and the data provided with the software tool (also downloadable here: https://cutt.ly/ErutuQYC) so that researchers can interact first-hand with NAQD.

Data

The edge list analyzed contained 887 rows (for an example see Table 2.2, and the link https://cutt.ly/ZriTKHXu to access the data:). These data included 26 participants representing four roles: administrator, faculty, students, and faculty/administrators. In addition, the edge list contains participants' sex: women and men. As can be seen in Table 2.3, these groups based on roles are mutually exclusive and exhaustive; that is, no participant may be both a faculty member and a student, for example. Table 2.3 also shows how active each actor was in providing texts that were eventually labeled in codes. Finally, from Table 2.3, we can see that of the 887 codes obtained, 341 (or 38.44%) were provided by faculty alone. The second most active group was students, with 221 codes provided by them (or 24.92%). Although this descriptive information is useful for gaining a perspective on participation levels individually and by groups, this information does not tell us anything about the potential statistical significance of these relationships.

Table 2.4 shows the code distribution by role. In this table, we can see all 26 codes contained in the network. Although outside the scope of our chapter, let us briefly note that the qualitative codes were created from the written information contained in a set of 26 interviews and followed a rigorous qualitative coding process (see Maxwell, 2019 for an ample discussion on qualitative research design). No differentiation of participants' roles was performed when the codes were generated. After the complete transcription of the interviews, open coding (Creswell and Creswell, 2018; Lewis, 2015) was conducted in NVivo (Charmaz, 2011) followed by process coding (Bogdan and Biklen, 2007) to highlight practices, interpretations, and activities. The codes identified were compared and refined through two additional rounds of review and deliberation. Redundant codes were collapsed to generate larger categories.

In Table 2.4, we can see that there are some instances where no codes were provided by a given group. For example, faculty and administrators did not mention text labeled under "Community Service." Similarly, students did not mention text labeled as "Cost of Program" and "Retention." Finally, administrators did not mention "Cohort Model" and "Experience being a minority teacher." In cases like this, since NAQD needs squared matrices, as discussed above, these lack of mentions will still be kept across all projected matrices, but they will be populated by zeroes, which does not alter the structure of these relationships and thus does not pose any threats to the integrity of the data.

Research questions

In this section, we present examples of research questions that can be addressed with NAQD.

Table 2.3 Code-by-Code Transformation of Table 13.

ID	Administrator	Faculty	Faculty/Admin	Student	
ID1	21	0	0	0	
ID2	0	35	0	0	
ID3	0	4	0	0	
ID4	0	0	0	37	
ID5	0	0	0	34	
ID6	0	0	45	0	
ID7	0	54	0	0	
ID8	26	0	0	0	
ID9	0	22	0	0	
ID10	35	0	0	0	
ID11	0	28	0	0	
ID12	0	35	0	0	
ID13	0	0	66	0	
ID14	0	0	0	22	
ID15	0	21	0	0	
ID16	0	0	55	0	
ID17	0	0	0	16	
ID18	0	0	0	56	
ID19	0	0	0	24	
ID20	56	0	0	0	
ID21	0	0	0	32	
ID22	0	25	0	0	
ID23	0	0	21	0	
ID24	0	15	0	0	
ID25	0	41	0	0	
ID26	0	61	0	0	**Total Contributions**
Sum codes	**138**	**341**	**187**	**221**	**887**

1. How do faculty, administrators, and students conceptualize and prioritize the most important issues affecting the landscape of teacher education programs at MSIs?
2. How does the emphasis on these issues vary when qualitatively generated data are analyzed based on the roles and functions that respondents hold in an MSI (i.e., student, faculty, administrator)?
3. Is there evidence to suggest that participants prioritize certain qualitatively coded issues more than others based on their roles and functions within an MSI? If so, what are the implications of these differences and similarities?

Table 2.4 Code distribution by roles.

Code	Administrator	Faculty	Faculty/Admin	Student
Campus Environment	**16**	22	**27**	9
Challenges	**22**	**35**	**26**	18
Classroom Environment	1	12	8	12
Clinical Work or Student Teaching	5	**35**	8	**37**
Cohort Model	0	10	4	1
Community Service	2	2	0	2
Cost of Program	0	1	6	0
Cultural Relevance	6	17	4	5
Curriculum	2	19	4	4
Experience attending MSI	1	1	2	1
Experience being a minority student	4	4	3	4
Experience being a minority teacher	0	6	2	9
Faculty	2	5	15	5
Future Plans	1	1	7	7
Inspiration	1	13	3	**34**
Job Counseling	3	14	4	4
Mentorship	7	9	1	8
Opportunities	3	10	4	8
Pedagogy	2	**32**	4	10
Recruitment	8	7	8	7
Reflection	4	6	1	3
Resources	14	5	14	5
Retention	2	5	4	0
Standardized Testing	5	16	2	2
Student Needs	5	8	6	5
Support System	**22**	**46**	**20**	**21**
Sum codes	**138**	**341**	**187**	**221**

Findings

NAQD data analyses may be separated into two parts: The first shows the sociogram representation of participants and the codes they are associated with. The second part of the data analysis may also be separated into two parts. Each consists of accounting for participants' attributes, and with this information, we can see if the resulting disaggregated networks present differences in their structure. This is a more qualitative test, as we will be observing the resulting networks. As part of this second set of analyses, we can also test the hypothesis that some networks are more similar than others as a function of the participants' roles.

Network analysis of connections (sociograms)

The two-mode sociograms resulting from these networks were analyzed following central actors or players' principles. In these analyses, the color, size, and position of participants and their respective qualitatively-generated codes reveal the structure resulting from observing the density of these connections. As stated above, these analyses may be represented in two contexts. The first is the aggregate representation of all these connections, which addresses the first research question, and the second is the disaggregated analyses by actors' role (in this case). The latter is important, as it is likely that concerns expressed by participants may vary as a function of their role in their respective colleges and universities.

Fig. 2.5 shows the aggregate network representation of all 887 connections across 26 interviewees and their 26 codes. Note that the interactive version of this figure can be found at https://cutt.ly/8rikqLil. As stated above, in this representation, the width of the links represents a measure of the strength of the relationship between actors and codes. For example, taking ID13 and the code "Campus Environment" indicates that this actor mentioned text classified under this code a total of 15 times, which represented 22.7% of this actor's contributions. By clicking on the node ID13 we can also see that he is a man who served as a *faculty member with administrative* roles on his campus.

In terms of codes' positions, Fig. 2.5 also shows that the most relevant codes (i.e., hot zones of discussion) were "Challenges," "Support System," "Campus Environment," and "Clinical Work or Student Teaching." Here, we can also see that ID26, ID16, ID20, and ID13 were central actors in this network, but only the latter was classified in another community, as can be seen by its pink color. Notably, four more of these actors in pink (i.e., the same community as ID13) were also faculty (ID15), faculty and administrators (ID10, ID8), and faculty and administrators (ID6). This community algorithm indicates that these actors may be concerned with similar, if not the same, challenges or experiences. A complete mixed-methods analysis would elaborate more deeply on the meaning of these community detections by elaborating on the codes these actors were concerned about.

As stated above, these networks also place codes that are not relevant at the periphery of the visualization (i.e., cold zones of discussion). For example, "Retention," "Experience attending MSI," "Experience being a minority teacher," "Experience being a minority student," "Mentorship," and a few other codes were noncentral issues or points of discussion, as only a few actors brought up these issues in their interviews. Notably, we can easily identify the actors who mentioned these issues in case we want to follow up with them—see actors ID4, ID5, ID22 who mentioned "Experience being a minority teacher," or actors ID13 and ID15 who mentioned "Retention" issues, for example.

Disaggregated analyses of sociograms

Fig. 2.6 is configured by four sociograms, each corresponding to a central player analysis of these two-mode networks. These sociograms resemble the aggregated depiction, except for the fact that only members of each of the four groups are analyzed, along with their measures of

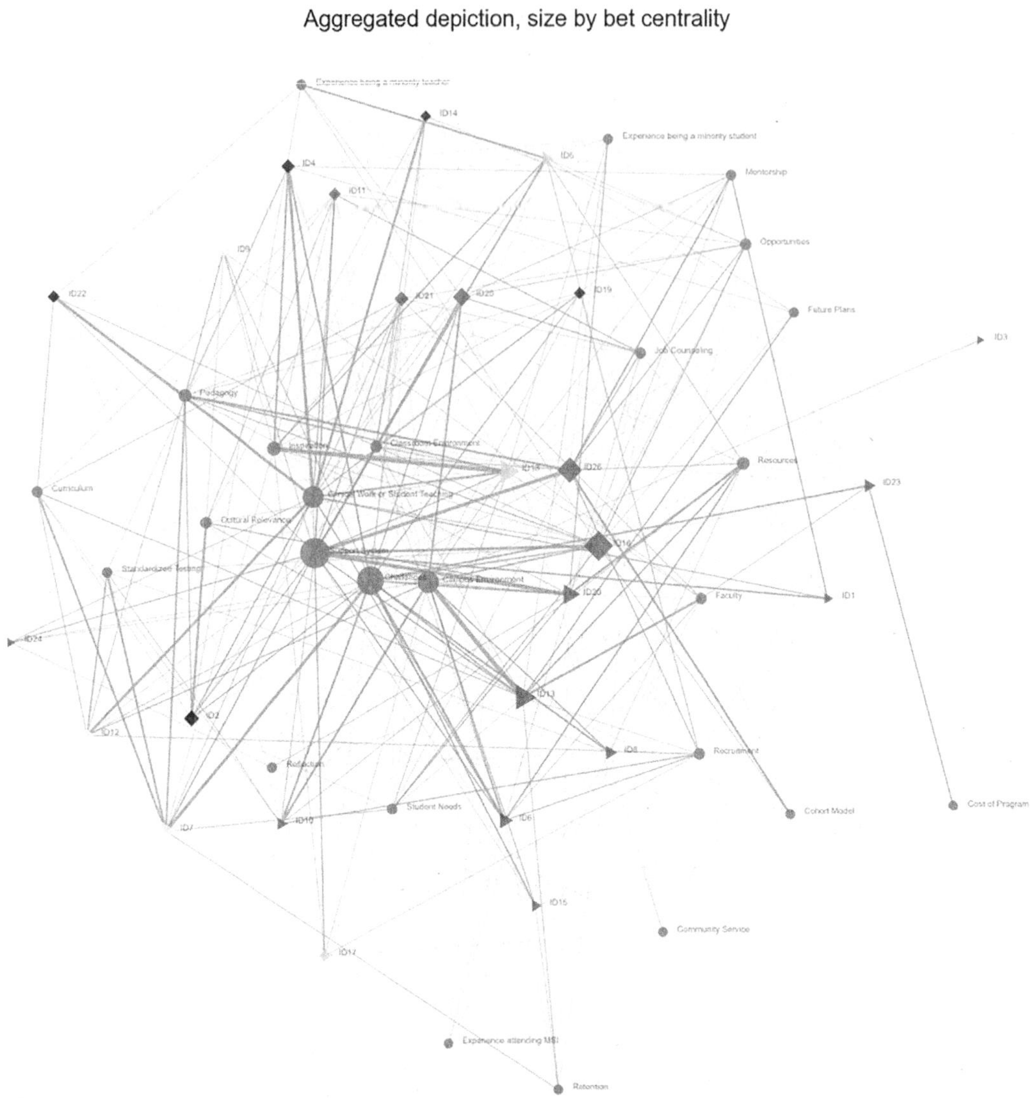

FIG. 2.5 Sociogram of Aggregate Connections (see interactive version here: https://cutt.ly/8rikqLil).

centrality. As discussed earlier in the chapter, eigenvector centrality shows that a given unit in the network will be relevant to the extent that such a unit is connected to other relevant units in the network. Betweenness centrality highlights the extent to which a given unit falls between other units or serves as a bridge between actors or units in this case. This last notion is important, as it highlights both interviewees who served as bridges of information and codes that connected individuals in this network. Eigenvector centrality also highlights individuals who were central as a function of providing codes that were also provided by other active individuals

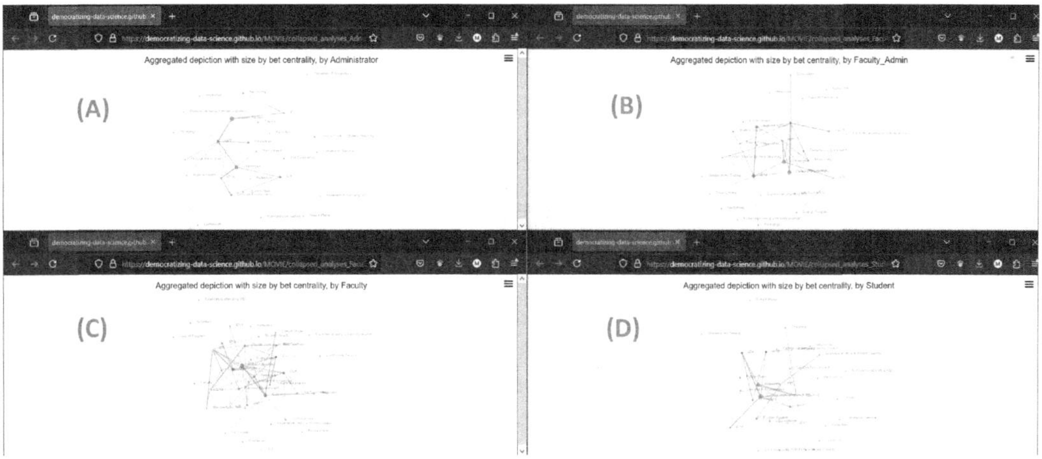

FIG. 2.6 DisaggregatedSociograms. HMTL access to (A) https://cutt.ly/GricQWKw, HMTL access to (B) https://cutt.ly/DriOLXrJ. HMTL access to (C) https://cutt.ly/9ricUqEW. HMTL access to (D) https://cutt.ly/zricl5pb.

in the network. Once more, degree centrality aids in both counting contributions and assessing how important a relationship with a code is for a given actor.

Fig. 2.6(A) includes only the relational information provided by administrators. The interactive version of this figure can be found at https://cutt.ly/GricQWKw. An advantage of these disaggregated analyses is the added clarity in highlighting relevant codes. For example, in Fig. 2.6(A), we can see that the two most important codes were "Support System" and "Challenges." On the other hand, it seems that, for administrators, what happens inside the classroom does not carry much weight. That is, only two individuals, ID1 and ID10, mentioned "Classroom Environment" and "Curriculum," respectively.

In the analysis of faculty and administrators, we can see that "Support System" and "Challenges" were also important topics. However, in addition, this group mentioned "Campus Environment" as another relevant topic. In this case, "Curriculum" remained a peripheral topic, but "Classroom Environment" was no longer peripheral. The sociogram for faculty is shown in Fig. 2.6(C), with its interactive version available at https://cutt.ly/9ricUqEW. For this group, the most important code was "Support System," but in this case, "Curriculum" became one of the most important codes as well. Moreover, "Pedagogy" emerged as another important code. For faculty, the two most peripheral issues were "Experience attending MSI" and "Future Plans," both with only one mention each. Finally, the sociogram for students is shown in Fig. 2.6(D), with its HTML version available here: https://cutt.ly/zricl5pb. In this case, the codes that emerged as the most relevant were "Clinical Work or Student Teaching" and "Inspiration." Similar to their faculty counterparts, "Experience attending MSI" was a peripheral code, but these peripheral codes also included "Standardized Testing" and "Cohort Model," both with one mention each.

Testing for differences across groups

Although we can start to see certain patterns and similarities across groups, at this point, we still have no formal way to address our third research question, which asks whether we have any evidence to suggest that some groups of participants *prioritize* (or are concerned with) certain qualitatively coded issues more than others based on their roles and functions held in an MSI. To address this question, we rely on QAP correlations.

As stated above, NAQD will automatically conduct dyadic analyses across all groups. In this case, since we have four groups, the resulting number of comparisons will be six. The results are shown in portable document format (PDF available here: https://cutt.ly/ZriYlnEv). In QAP, significant results indicate that the two matrices or networks are more correlated than what can be expected to happen by chance alone. This means that the textual contributions that were eventually coded or labeled are shared more frequently than by chance alone by actors who belong to the same group. When no statistically significant correlations are found, the conclusion is that, for those groups of participants, there is a disconnect among their experiences or concerns.

Fig. 2.7 contains all QAP correlations. For context recall that the qualitatively coded dataset was part of a project sponsored by the Kellogg Foundation (see González Canché, 2019). This

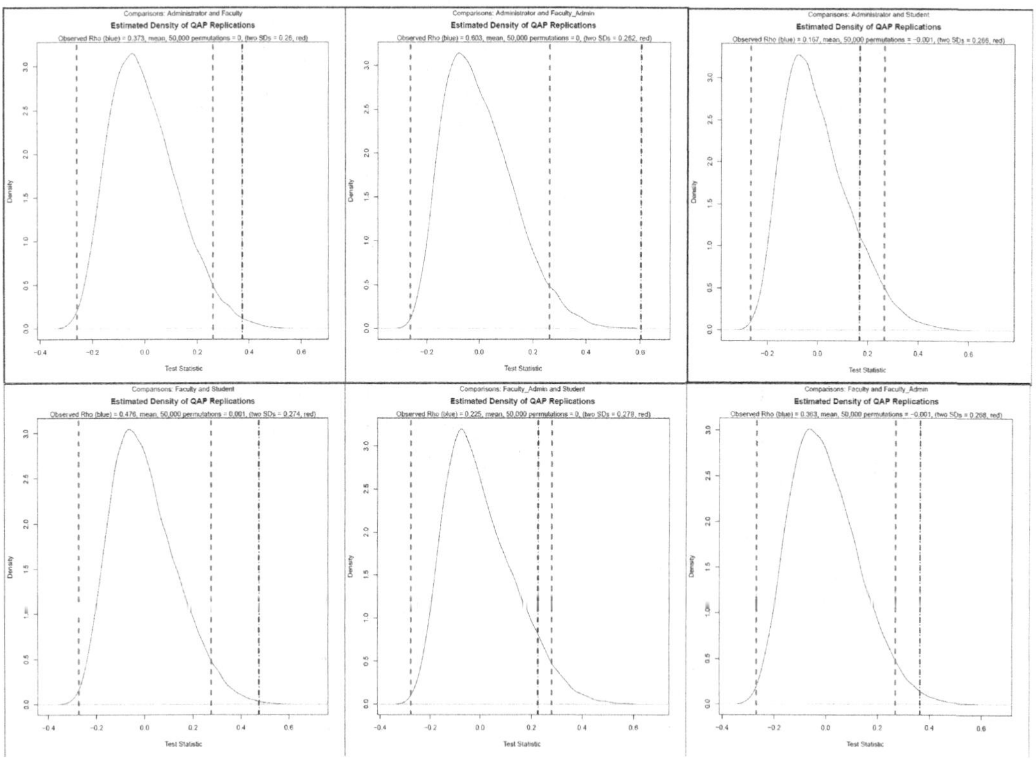

FIG. 2.7 QAP outputs.

project focused on the challenges and concerns that MSIs across the country face in terms of their production and training of teachers of color. This dataset, in addition to the $Actor_i \rightarrow Code_j$ relationships, includes the role of each respondent (or the role they hold in this research): student, administrator, faculty, or faculty-administrator. Given that each of these roles is expected to translate into perceived challenges and concerns that may be correlated across groups with different magnitudes (e.g., faculty members' and students' concerns or perceived challenges may be more correlated than students' and administrators' concerns/challenges, for example), QAP tested network correlations across all groups, rendering six comparisons:

1. Administrators and faculty (rho = 0.373***)
2. Administrators and faculty-administrators (rho = 0.603***)
3. Administrators and students (rho = 0.167)
4. Faculty and faculty-administrators (rho = 0.363***)
5. Faculty and students (rho = 0.476***)
6. Faculty-administrators and students (rho = 0.225)

The observed network correlations (rho) indicated that the concerns of *students and administrators,* as well as *students and faculty/administrators,* are not shared at a statistically significant level. That is, these groups expressed challenges or experiences that were different at a network structural level. The integrative framework behind NAQD now requires a qualitative analysis to guide how or why this may be the case by observing the content of the qualitative codes. Hence, "qualitative + quantitative = more than the individual components" (Fetters and Freshwater, 2015, p. 116). Note also that the example database includes the content of the quote and one individual attribute of the actors (gender), which may be useful in better understanding the meaning of those discrepancies in experiences and perceived challenges.

The remainder of the comparisons indicated that administrators, faculty, and faculty with administrative functions statistically shared similar concerns or experiences, with the experiences of administrators and faculty-administrators being the most alike (rho = 0.603***). Finally, students and faculty also had similar experiences, with the second-highest correlation coefficient found across tests (rho = 0.476***). This means that faculty, by being employed in a college or university, do share concerns with administrators; however, by being active in the teaching and learning process, faculty also shared concerns with students. This is interesting because students and administrators, by themselves, do not share the same or similar concerns, as found in their respective correlations. As stated above, these outputs are a guide to highlight the structure embedded in this qualitative dataset. Within the broader analytic architecture of this book, such guides can also be used as prompts for local scholarly reasoning in ISARI, where researchers may compare alternative readings of the network structure, memo possible interpretations, and begin translating those patterns into evidence-grounded prose while retaining full interpretive authority. However, even an integrative or brainstorming exercise is implemented as part of the analytic ecosystem, a qualitative analysis is necessary to better understand how these lack of correlations are substantively different in our participants' experiences. In this respect, although the sociograms are useful in starting to guide this qualitative analysis, the understanding of the quotes associated with the codes may be particularly relevant in this integrative process, wherein qualitative and quantitative results may lead to deeper and more nuanced understandings and stronger storytelling. ISARI can also help researchers to brainstorm how to best integrate our participants' qualitative stories with their resulting positions in the network.

NAQD outputs and software architecture

In this section, we will discuss all the outputs generated by the NAQD UI, as well as describe in detail all its components.

NAQD outputs

NAQD is a user-friendly standalone implementation available on Mac and Windows operating systems that produces the following four outputs:

1. NAQD produces a downloadable database with results of community allocation using spin-glass modeling. When analysts do not include a column containing an attribute for color, the resulting visualization will add the spin-glass communities as colors to identify patterns. This database also offers actors' roles (e.g., student) and other personal (e.g., gender) and nonpersonal attributes (e.g., community assigned) that may strengthen the sense-making process based on subsequent qualitative analyses. That is, not only will these analyses include community allocations, but authors may also further analyze whether participants with certain attributes are more prone to be located in some communities instead of others (either qualitatively or quantitatively—via chi-squared analyses, for example).

The structure of this database is shown in Fig. 2.8. To access this database, follow this link: https://cutt.ly/ZriTKHXu. This figure contains the ID names, the community ascribed, information about how the color is assigned (in this case, the color shown in our sociograms is based on the spin-glass community), and the shape of their nodes (in this case, we used gender). In addition, this table contains a cross-tabulation to start gauging any potential relationship. Based on this cross-tabulation, a preliminary analysis of these relationships shows that most students (5 of 7) were classified in the sixth community detected. Faculty administrators and administrators were classified in the first and second communities, and faculty were spread out into communities 1, 2, 3, 5, and 6. The latter distribution helps explain why faculty were able to connect with both students as well as administrators and faculty/administrators as well. (The cross-tabulation is not generated by NAQD; we conducted it manually by loading this data in R and applying the following code: table(a$community, a$role), where "a" is the database shown in Fig. 2.8.)

2. If actors' attributes are added, as explained earlier, NAQD will also generate a downloadable PDF that includes all combinations of QAP. This file can be accessed here: https://cutt.ly/ZriYlnEv. These combinations will comprehensively test whether actors

	A	B	C	D	E
1	**names**	**communit**	**role**	**color**	**shape**
2	ID1	2	Administrator	color-based_on_community_assigned	man
3	ID8	1	Administrator	color-based_on_community_assigned	man
4	ID10	1	Administrator	color-based_on_community_assigned	man
5	ID20	2	Administrator	color-based_on_community_assigned	man
6	ID2	3	Faculty	color-based_on_community_assigned	woman
7	ID3	1	Faculty	color-based_on_community_assigned	man
8	ID7	5	Faculty	color-based_on_community_assigned	man
9	ID9	5	Faculty	color-based_on_community_assigned	woman
10	ID11	2	Faculty	color-based_on_community_assigned	woman
11	ID26	2	Faculty	color-based_on_community_assigned	woman
12	ID12	5	Faculty	color-based_on_community_assigned	woman
13	ID15	1	Faculty	color-based_on_community_assigned	man
14	ID22	6	Faculty	color-based_on_community_assigned	woman
15	ID24	2	Faculty	color-based_on_community_assigned	man
16	ID25	2	Faculty	color-based_on_community_assigned	woman
17	ID6	1	Faculty_Admin	color-based_on_community_assigned	man
18	ID13	1	Faculty_Admin	color-based_on_community_assigned	man
19	ID16	2	Faculty_Admin	color-based_on_community_assigned	woman
20	ID23	1	Faculty_Admin	color-based_on_community_assigned	man
21	ID4	6	Student	color-based_on_community_assigned	woman
22	ID5	4	Student	color-based_on_community_assigned	man
23	ID14	6	Student	color-based_on_community_assigned	woman
24	ID17	2	Student	color-based_on_community_assigned	woman
25	ID18	6	Student	color-based_on_community_assigned	man
26	ID19	6	Student	color-based_on_community_assigned	woman
27	ID21	6	Student	color-based_on_community_assigned	woman
28					

Cross-Tabulation

Communities	Administrator	Faculty	Faculty_Admin	Student
1	2	2	3	0
2	2	4	1	1
3	0	1	0	0
4	0	0	0	1
5	0	3	0	0
6	0	1	0	5

FIG. 2.8 Classified database with initial cross-tabulations.

belonging to certain groups (or performing certain roles) share statistically significant similarities regarding their qualitatively gathered and preprocessed information. This PDF output matches Fig. 2.7 and is the result of advanced matrix transformations and management (see Fig. 2.4 as well). Moreover, these analyses in the PDF should be added to mixed-methods studies as outputs to illustrate the findings of a NAQD study.

3. NAQD will also produce an aggregated network depiction of actors and codes. The colors represented in this depiction are, by default, based on community allocation, as depicted in the first output described above. However, adding flexibility to this analytic framework, researchers can easily select actors' attributes instead, which may help them better understand these relationships. Moreover, this HTML network rendering is fully interactive, offering measures of the strength of the relationships among actors and codes, along with information about the total number of contributions made by each actor and their level of influence in this network. For example, when researchers click on a given actor (or institution, as in the case of the original study), a measure of influence will be displayed. This measure ranges from 0 to 1, with 1 being the highest eigenvalue centrality, indicating that this actor was providing codes that other influential actors deemed relevant (Borgatti, 2006). In addition, the "No. of contributions" information shows how many codes this actor provided during the interview process—also referred to as degree centrality. This same information box will reflect the role of this actor and the personal and nonpersonal attributes they may have. In addition, the size of each node is weighted by betweenness centrality, or the relevance that actors and codes had in connecting actors to other actors and codes to other codes through actors (Borgatti, 2006). These three centrality measures are estimated at the actor or code level, thus aiding in the identification of key elements in the network (Borgatti, 2006; González Canché, 2019).

The HTML rendering will also provide information when clicking on a line connecting actors to codes. In this case, NAQD was programmed to display information about how relevant a given code was for an actor. That is, we will be able to assess the percentage with respect to the total number of contributions that a specific code was provided by each actor. For example, assume *actor A* provided 10 codes in total, and 8 of them were classified as *code j*. In addition to the HTML output indicating width differences based on the prevalence of these connections, when clicking on the line, they will see that for actor A, this code represented 80% of her contributions.

4. Finally, if actors' roles are selected, NQDA will automatically render a disaggregated set of network depictions of actors and their codes. Once more, these networks will be fully interactive and will also display centrality levels and connection relevance as previously described. Note that all these HTML outputs are stored locally as standalone files that can be shared with colleagues or posted online—just as we did in this study. Because interactive network visualizations may reveal structural clues that could weaken anonymity in small, distinctive, or contextually sensitive samples, the online sharing of HTML outputs should always be treated as optional and governed by the same principles of confidentiality, data minimization, and participant protection that guided the original qualitative design (González Canché, 2025a). In this respect, it is always recommended that, if posting the results online, all data protection measures and the anonymity of the participants are ensured as much as possible. Recall that posting the analyses online is completely optional, and we should always prioritize the safety and trust of our participants (see González Canché, 2025b for a discussion on how qualitative data science may be jeopardizing trust in the social sciences).

Software access and architecture

Mac users can download NAQD here: https://cutt.ly/unZDUMq, and Windows users can download it here: https://cutt.ly/KnL9frz. To execute the application for the first time on a computer, the following steps need to be performed:

1. Extract the zip folder
2. Open the folder starting with "NAQD-." It contains two subfolders: NAQD-Win and NAQD-Mac, for Windows and Mac operating systems, respectively
3. Double-click on the file called "NAQD-Mac" or "NAQD-Win"

The first time NAQD is executed, you may be required to select the menu "View" and "Force Reload" to generate all paths to this newly extracted folder. If, after forcing a reload, NAQD does not display as shown here https://cutt.ly/JnJO0MH, close the window and execute NAQD again; if needed, force reload once more when the software is launched.

NAQD is housed in a secure server location, and all executable files are virus-free but may require administrative permission to be deployed. This is particularly true for Mac users, as they need to go to systems security and select to execute NAQD, even if this software was not developed by Apple. Mac users are also recommended (and in some instances they do need) to place the NAQD software (and all software products discussed in this book) in their applications folder for newer versions can only read and write files from this location. Finally, each of these executable files for Mac and Windows operating systems contains all the source code required to modify and/or expand the application capabilities in the file called "app.R."

NAQD is also supported through a public GitHub repository that centralizes software files, tutorials, example data, and backend materials (access here: https://cutt.ly/Pr5i05ch). This strengthens the framework's reproducibility and transparency by allowing researchers not only to replicate the procedures described in this chapter, but also to inspect, preserve, and extend the software environment over time.

Note that NAQD has also been peer-reviewed by Code-Ocean, a multisoftware platform that seeks to ensure quality, reproducibility, and transparency in scientific research. To view NAQD in Code Ocean, follow this link: https://cutt.ly/2riYDEhH. In sum, all outputs and supplementary materials are available in the centralized repository: https://github.com/democratizing-data-science/NAQD/tree/main.

NAQD user interface and architecture

Fig. 2.9 shows NAQD's UI. This UI is configured into four parts. In the first two sections, we load our data or load the data included in the distribution of NAQD for replication purposes. The third section includes the only required steps, which are the selection of actors and codes. The fourth section is optional, but it is where we can select the column in our datasets that will be used for comparison analyses. This is achieved using the "role" column. Populating this column enables the comparative analyses, disaggregating the visuals by attributes, as well as conducting the QAP correlation analyses.

In the following lines, we describe these steps in more detail.

NAQD execution requires the following steps:

Load your data or select the example provided

1. NAQD describes the example that comes preloaded.
2. When you load your database in comma-separated values (CSV) format, this data will be displayed in the main panel.
3. Data access
4. To upload a database, select "Click to browse" and search for a file stored locally (see Fig. 2.10(A)).
5. Alternatively, you can load the example data provided by selecting "Interview or multiple focus groups example" under Section A and clicking "Click to load data example shown below" (see Fig. 2.10(B)).
6. Select the "actor" column (or provider of the code)

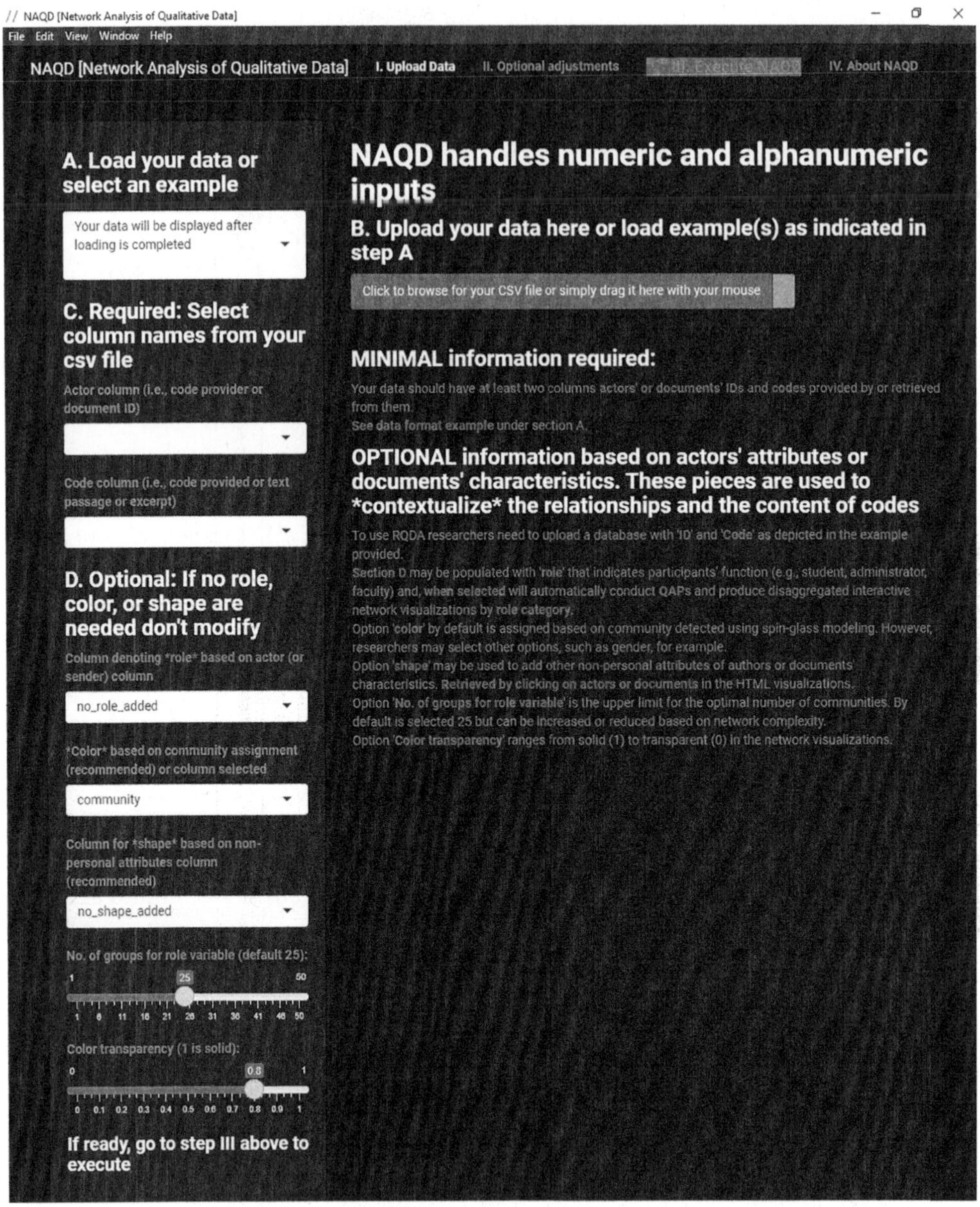

FIG. 2.9 NAQD User Interface.

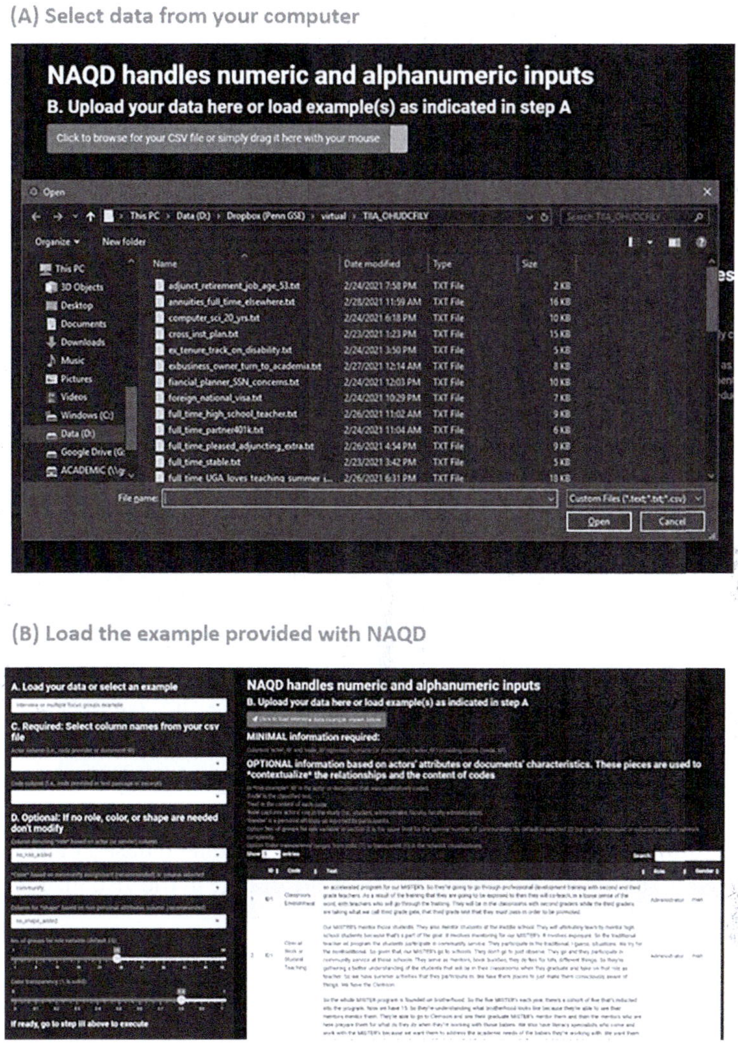

FIG. 2.10 Data Handled by NAQD and Loading Process.

7. Select the “Code“ column
8. OPTIONAL (These pieces of information aid in the contextualization of information generation or allow the execution of QAPs.):

Column denoting *role*: It indicates specific roles of the actors (e.g., students, faculty, administrators). If selected, all combinations will automatically be tested using QAPs, and network visualizations will be produced. It is recommended to have at least three actors per role to conduct QAPs.

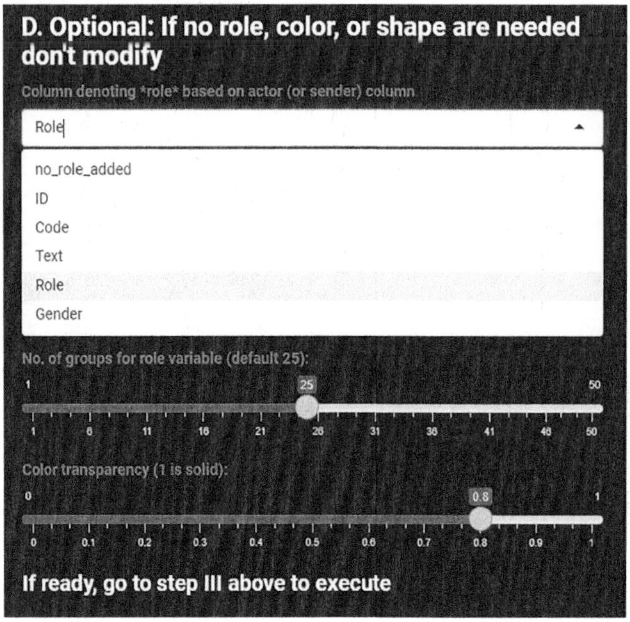

Column for ***color***. By default, the colors assigned are calculated based on ***community*** attribution using spin-glass modeling. All codes are orange in the visualization (this color can be changed in the optional adjustments tab).

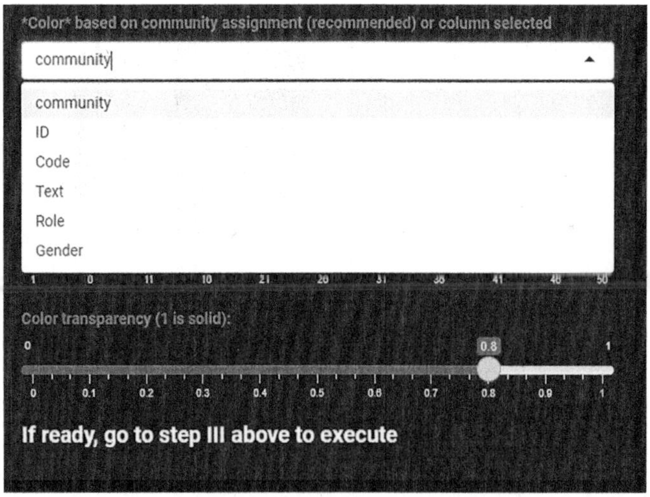

Column denoting ***shape*** that may further help in the contextualization of the information provided by actors. This attribute may be personal (e.g., gender) or nonpersonal (e.g., field of study). In addition to being displayed when clicking on an actor, this attribute will be reflected in the shape of the nodes in the network. All nodes are displayed as circles in the visualization (this shape can be changed in the optional adjustments tab).

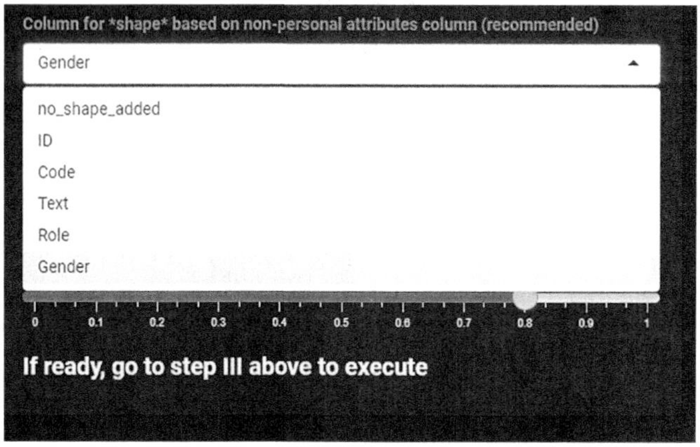

Researchers can modify the ***optimal number of communities*** to detect. The recommended number is 25 (Csárdi, 2006). This number does not mean that exactly 25 communities will be detected, but rather that the maximum may be 25.

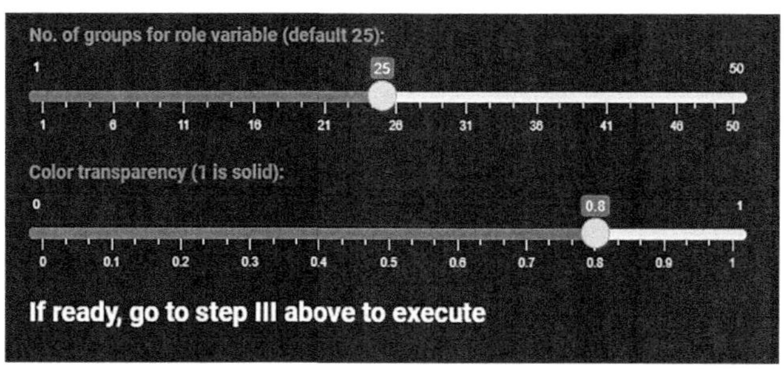

You can also select the ***transparency*** level of the colors to be plotted, with 1 being solid and 0 being transparent. The recommended value is 0.8.

9. When ready, click "Execute NAQD" on the top panel to obtain all four outputs.

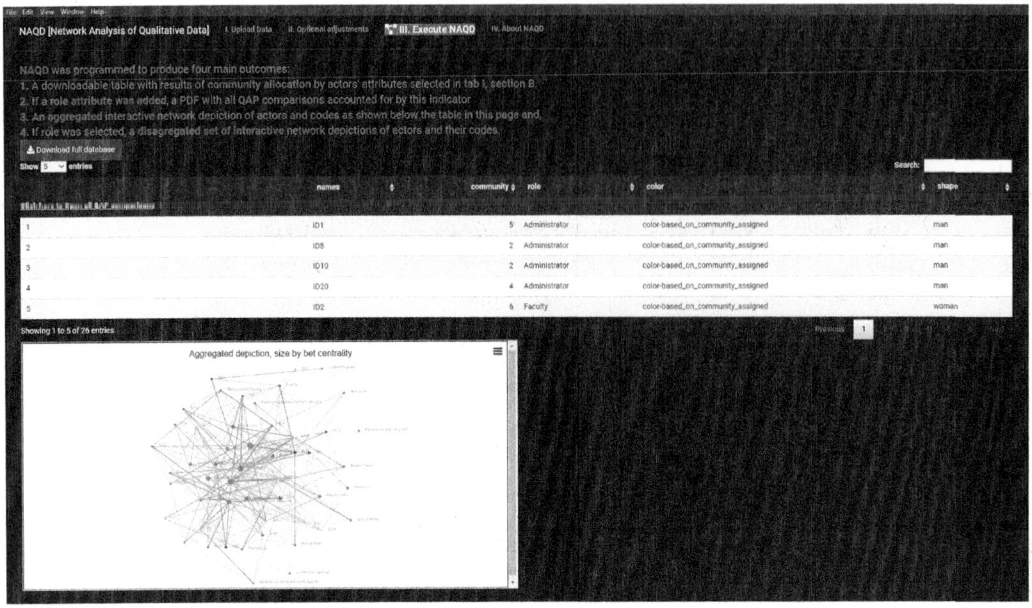

a. The database displayed can be downloaded in CSV format.
b. If a role is selected, a PDF will be generated containing all combinations based on roles by clicking "Click here to open all QAP comparisons" at the top of the table.

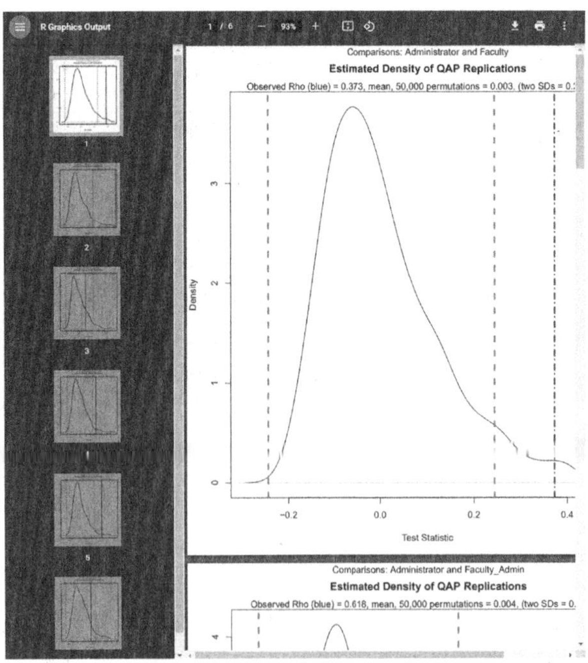

a. In addition to the aggregated network depiction shown here, this network will be automatically launched in your browser window with the address indicating the local folder where this file (or these files) is stored. Moreover, all *n* disaggregated network depictions by role will also be launched.

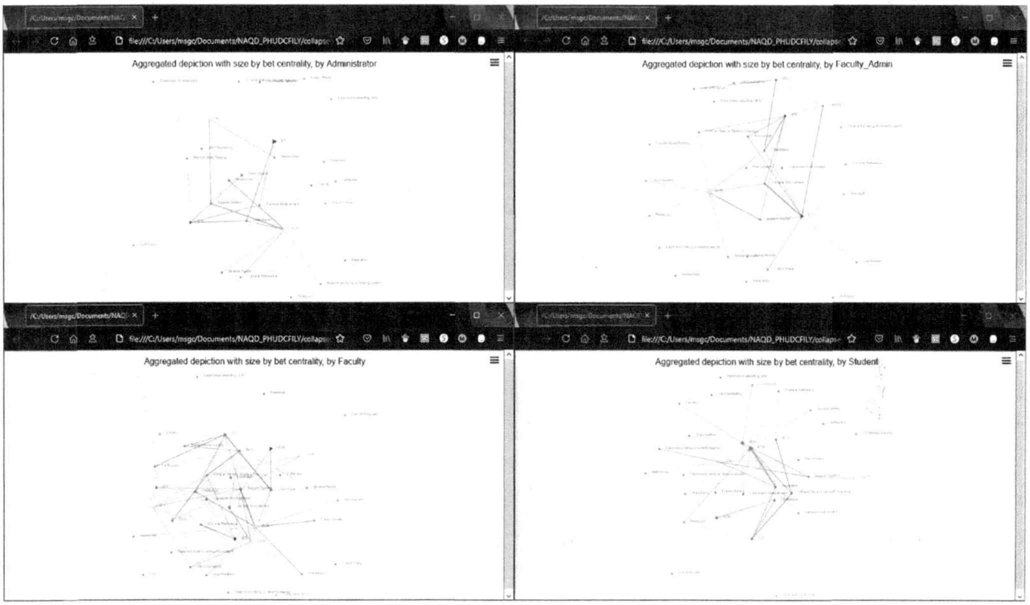

All HTML files can be easily distributed as standalone files. Their display does not require an internet connection.

Integrating NAQD Outputs into Local Scholarly Brainstorming

One of the strengths of NAQD is that its outputs are portable across analytic stages. The fully interactive HTML renderings, downloadable databases, and PDF summaries of quadratic assignment procedures are useful not only for identifying structure, but also for supporting subsequent interpretive work. In the broader ecosystem presented in this book, NAQD outputs can be brought into ISARI, a local, fully offline, open-source, and multimodal brainstorming environment designed to support scholarly memoing, comparison, synthesis, and writing.

This possibility matters because network visualizations often reveal patterns whose substantive meaning still requires careful interpretation. For example, researchers may wish to ask why certain actors occupy central or peripheral positions, what explains the clustering of specific codes, how disaggregated sociograms differ across roles, or how QAP results should be translated into narrative argument. Because ISARI can work with textual notes, screenshots, figures, charts, scanned pages, and other image-based analytic artifacts, researchers can upload

NAQD outputs in multiple forms—for instance, screenshots of the HTML sociograms, exported figures, the PDF discussing quadratic assignment results, or even notes derived from the interactive network renderings—and use them as the basis for local scholarly brainstorming.

This integrative workflow should not be understood as a substitute for methodological rigor. On the contrary, it extends NAQD's central logic. The outputs remain guides, not conclusions. ISARI can help researchers compare interpretations, generate analytic memos, identify possible explanatory contrasts, and begin drafting evidence-grounded prose, but the responsibility for deciding what the patterns mean remains with the researcher. In that sense, NAQD and ISARI are methodologically aligned: both are designed to strengthen interpretive transparency without displacing interpretive authority.

Closing thoughts

NAQD is an integrative analytic framework and software tool designed to unveil the structure embedded in qualitative data. Although all the steps described in this study may be replicated with textual data that has been coded or labeled via machine learning approaches, our example focuses on qualitatively coded information. At this point, please note once more that the NAQD outputs represent a step forward toward the consolidation of our understanding. As such, these outputs are not the end product but a guide that will hopefully serve to strengthen our storytelling and reach deeper and more nuanced understandings.

Similar applications of NAQD can be found in GRATIS (González Canché, 2023d) wherein, in addition to relying on network analyses, we can incorporate time to observe the evolution of these contributions. From this perspective, while NAQD does not incorporate time in the modeling strategy, this is not necessarily considered a limitation of the methodological framework but rather a function of the types of questions and analyses that we may ask. That is, some studies may benefit from the implementation of GRATIS, whereas others may require the use of NAQD, or vice versa. Moreover, some studies may benefit from the integration of both GRATIS and NAQD to strengthen the storytelling, address the research questions, and meet the study's purpose.

To conclude, we reiterate that complete NAQD analyses (and GRATIS analyses) will benefit most from using the outputs discussed herein as guides, which can be complemented and strengthened through a qualitative understanding of the meaning of the codes being plotted. In this respect, it is vital to note that if the coding or labeling process (or the data collection itself) was not conducted with rigorous methodologies, the resulting outputs obtained with NAQD would still reflect this lack of quality, potentially leading to biased or incorrect conclusions. It is only when the entirety of the process is conducted with quality assurance that research projects employing NAQD will yield useful and potentially relevant results. Our main goal in offering NAQD is to provide qualitative and mixed-methods researchers with a framework and tool that is both easy to use and capable of strengthening the analytic process, thereby helping to open the analytic black box typical of most qualitative studies.

References

Alexander, E., Eppler, M.J., Bresciani, S., 2019. Visual replay methodology: a mixed methods approach for group discussion analysis. J. Mixed Methods Res. 13 (1), 33–51. doi:10.1177/1558689816664479.

Attride-Stirling, J., 2001. Thematic networks: an analytic tool for qualitative research. Qual. Res. 1 (3), 385–405. doi:10.1177/146879410100100307.

Biggs, N., Lloyd, E.K., Wilson, R.J., 1986. Graph Theory, 1736-1936, Oxford University Press.

Bogdan, R., Biklen, S., 2007. Qualitative Research for Education: An Introduction to Theory and Practice, Allyn and Bacon.

Borgatti, S.P., 2006. Identifying sets of key players in a social network. Comput. Math. Organ. Theory. 15729346. 12 (1), 21–34. doi:10.1007/s10588-006-7084-x.

Breiger, R.L., 1974. The duality of persons and groups. Soc. Forces. 0037-7732. 53 (2), 181–190. doi:10.1093/sf/53.2.181.

Charmaz, K., 2011. Grounded theory methods in social justice research. The Sage Handbook of Qualitative Research Vol. 4 (CA: SAGE: Thousand Oaks); 359–380.

Constas, M.A., 1992. Qualitative analysis as a public event: The documentation of category development procedures. Am. Educ. Res. J. 29 (2), 253–266. doi:10.3102/00028312029002253. 0002-8312.

Creswell, J.W., Creswell, J.D., 2018. 5th, Thousand Oaks, CA, SAGE Publications.

Crossley, N., 2010. The social world of the network: combining qualitative and quantitative elements in social network analysis. Sociologica 4 (1), 1–34.

Csárdi, N., 2006. The igraph software package for complex network research. InterJournal, Complex Systems, 1695, 1–9.

Edwards, G., 2010. Mixed-method approaches to social network analysis. NCRM Resource Repository. Available from https://eprints.ncrm.ac.uk/id/eprint/842.

Fetters M.D. and Freshwater D., 2015. The 1 + 1 = 3 Integration Challenge. J. Mix. Methods Res. 9 (2), 115–117. doi:10.1177/1558689815581222.

Freeman, L.C., 1978. Centrality in social networks conceptual clarification. Soc. Netw. 03788733. 1 (3), 215–239. doi:10.1016/0378-8733(78)90021-7.

Froehlich, D.E., Van Waes, S. and Schäfer, H., 2020. Linking quantitative and qualitative network approaches: A review of mixed methods social network analysis in education research. Rev. Res. Educ. 44 (1), 244–268. doi:10.3102/0091732X20903311.

Fuhse, J. and Mützel, S., 2011. Tackling connections, structure, and meaning in networks: quantitative and qualitative methods in sociological network research. Qual. Quant. 45 (5), 1067–1089. doi:10.1007/s11135-011-9492-3.

González Canché, M.S., 2019. Geographical, Statistical, and Qualitative Network Analysis: A Multifaceted Method-Bridging Tool to Reveal and Model Meaningful Structures in Education Research. Springer Science and Business Media LLC, pp. 535–634. doi:10.1007/978-3-030-03457-3_12.

González Canché, M.S., 2022d. Network analysis of qualitative data: an integrative software application to visualize and assess similarities in participants' qualitative contributions. J. Mixed Methods Res. 16 (3), 373–377. doi:10.1177/15586898211051584.

González Canché, M.S., 2023a. Latent code identification (LACOID): a machine learning-based integrative framework [and open-source software] to classify big textual data, rebuild contextualized/unaltered meanings, and avoid aggregation bias. Int. J. Qual. Methods 22, 16094069221144940.

González Canché, M.S., 2023b. Machine driven classification of open-ended responses (MDCOR): an analytic framework and no-code, free software application to classify longitudinal and cross-sectional text responses in survey and social media research. Expert Syst. Appl. 215, 119265.

González Canché, M.S., 2023d. Graphical retrieval and analysis of temporal information systems (GRATIS): an integrative mixed methodology and open-access software to analyze the (non-)linear chronological evolution of information embedded in textual/qualitative data. J. Mixed Methods Res. Original work published 2024. 18(1), 71–103. 10.1177/15586898231166968.

González Canché, M.S., 2023. Spatial Socio-econometric Modeling (SSEM: A Low-Code Toolkit for Spatial Data Science and Interactive Visualizations Using R Springer Texts in Social Sciences. Springer International Publishing. 10.1007/978-3-031-24857-3.

González Canché, M.S., 2024. Graphical retrieval and analysis of temporal information systems (GRATIS): An integrative mixed methodology and open-access software to analyze the (non-)linear chronological evolution of information embedded in textual/qualitative data. J. Mix. Methods Res. 18 (1), 71–103. doi:10.1177/15586898231166968.

González Canché, M.S., 2025a. Ethics and trust in social network analysis: how data science and interactive visualizations impact anonymity, confidentiality, consent, and IRB procedures? In: Handbook of Social Network Analysis and Education (pp. 114–131), Edward Elgar Publishing. doi:10.4337/9781035312191.00013.

González Canché, M.S., 2025b. Network analysis of qualitative data: an integrative software application to visualize and assess similarities in participants. Qual. Contrib. doi:10.24433/CO.8492516.v1.

González Canché, M.S., 2026. Network analysis of qualitative data (NAQD): An analytic framework and software for qualitative data science visualization and hypothesis testing. Int. J. Qual. Methods. 25. doi:10.1177/16094069261439942.

Johnson R.B. and Onwuegbuzie A.J.. 2004. Mixed methods research: A research paradigm whose time has come. Educ. Res. 33 (7), 14–26. doi:10.3102/0013189X033007014.

Kolaczyk, E.D., Csárdi, G., 2014. Statistical Analysis of Network Data with R. In: Use R!, Vol. 65. Springer, New York.doi:10.1007/978-1-4939-0983-4.

Lewis, S., 2015. Qualitative inquiry and research design: choosing among five approaches. Health Promot. Pract. 1524-8399. 16 (4), 473–475. doi:10.1177/1524839915580941.

Luxton I. and Sbicca J., 2021. Mapping movements: a call for qualitative social network analysis. Qual. Res. 21 (2), 161–180. doi:10.1177/1468794120927678.

Maroulis, S., Guimerà, R., Petry, H., Stringer, M.J., Gomez, L.M., Amaral, L.A.N., Wilensky, U., 2010. Complex systems view of educational policy research. Science. 0036-8075. 330 (6000), 38–39. doi:10.1126/science.1195153.

Maxwell, J., 2019. Qualitative Research Design SAGE Research Methods Foundations, Sage Publications, Sage Publications.

Mitchell, M., 2006. Complex systems: network thinking. Artif. Intell. 00043702. 170 (18), 1194–1212. doi:10.1016/j.artint.2006.10.002.

Onwuegbuzie A.J. and Leech N.L., 2007. Validity and qualitative research: An oxymoron? Qual. Quant. 41 (2), 233–249. doi:10.1007/s11135-006-9000-3.

Phipson, B., Smyth, G.K., 2010. Permutation P-values should never be zero: calculating exact P-values when permutations are randomly drawn. Stat. Appl. Genet. Mol. Biol. 15446115. 9 (1). doi:10.2202/1544-6115.1585.

Reichardt, J., Bornholdt, S., 2006. Statistical mechanics of community detection. Phys. Rev. E. 1539-3755. 74 (1). doi:10.1103/physreve.74.016110.

Saldaña, J. (2021). The Coding Manual for Qualitative Researchers, UK: SAGE Publications Ltd..

Saqr, M., López-Pernas, S., Törmänen, T., Kaliisa, R., Misiejuk, K., Tikka, S., 2025.

Transition network analysis: A novel framework for modeling, visualizing, and identifying the temporal patterns of learners and learning processes. In: Proceedings of the 15th international learning analytics and knowledge conference, (pp. 351-361).

Shaffer D.W., Collier W. and Ruis A.R., 2016. A tutorial on epistemic network analysis: Analyzing the structure of connections in cognitive, social, and interaction data. Learn. Anal. 3 (3), 9–45. doi:10.18608/jla.2016.33.3.

Tan, Y., Ruis, A.R., Marquart, C., Cai, Z., Knowles, M.A., Shaffer, D.W., 2023. Ordered Network Analysis. In: Damşa, C., Barany, A. (Eds), Advances in Quantitative Ethnography. ICQE 2022. Communications in Computer and Information Science, vol 1785. Springer, Cham. doi:10.1007/978-3-031-31726-2_8.

Wasserman, S., Faust, K., 1994. Social Network Analysis: Methods and Applications, Cambridge University Press.

Whitbred, R., 2011. Quadratic assignment procedure (QAP). In: Encyclopedia of Social Networks. SAGE Publications, Inc. pp. 733–734.

3

Graphical Retrieval and Analysis of Temporal Information Systems (GRATIS)

Introduction

Most mixed methods and qualitative methodologies available today capture a snapshot of the world, thus forfeiting the nuances gained by being able to observe and analyze the chronological evolution of processes, events, and information that shape qualitative research findings and conclusions. From this perspective, the incorporation of *time* as part of our integrative mixed methods frameworks, may enhance our perceptions of the social world and strengthen our level of understanding of the processes and the consequences of change or continuity in social settings (Adam, 1990; Neale, 2015a, 2019a). Accordingly, guided by the idea that *a short video can reveal much more than a single photo*, and considering that *all qualitative datasets inherently contain temporal elements* that can be used to recreate the specific contexts (These contexts include the content of participants' most recent and their previous most recent contribution. That is, assume that at time t, a participant mentioned X, and that at the previous time (or t–1) that same participant mentioned Y. The simultaneous retrieval of X and Y enables us to contextualize not only the content of X and Y, but also how each participants' contributions evolved over time. As further explained below, by default our software maps the two most recent contributions provided by each participant, but analysts may select up participants' 10 most recent contributions and this can be achieved with a single click as described in the user interface section (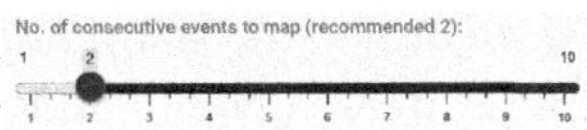

). In addition, we may add personal (i.e., marital status, sexual orientation) and non-personal attributes (i.e., tone of delivery—upset, sad, happy, employment status) of our participants, which may be useful to better understand their particular standing on a given topic, as we detailed in the Background and Procedural Diagram section below.) during which participants provided their insights and contributions, the development of methodologies capable of mapping these dynamic processes, represents a relevant contribution to fully integrative mixed methods research. This relevance is justified given that, despite the availability of temporal elements and the analytic benefits that the incorporation of time represents to illuminate processes, trends, and dynamics in our studies (Artale et al., 2007a; Neale, 2015a), these temporal elements remain very underutilized or foregone in qualitative and mixed methods analyses.

The main reasons behind this underuse of temporal elements in qualitative datasets are: (1) the complexity of accounting for the multiplicity of elements evolving simultaneously, (2) the assumption that the incorporation of time as part of the analyses requires long periods of time,

Data Science, Interactive Visualizations, and Generative AI Tools for the Analysis of Qualitative, Mixed-Methods, and Multimodal Evidence. https://doi.org/10.1016/B978-0-443-21961-0.00002-4

when in reality all qualitative and textual datasets have an inherent time component (Adam, 1990; Neale, 2019a) that can be mapped, as we discuss below, and (3) the lack of user-friendly software tools capable of effortlessly visualizing the emergence and decay of these chronological elements in qualitative data. Regarding the latter point, as we further illustrate in our "Background" section below, even the most recent advancements in tools for visualizing qualitative data, such as those depicted in the computer-assisted qualitative data analysis software (CAQDAS) Networking Project, do not yet handle temporal information as part of their methodologies. Specifically, although NVivo, ATLAS.ti, Leximancer, MAXqda, and QDA Miner do offer network or relational visualization capabilities (like the capabilities employed in our framework), such software tools **"stop at presenting an aggregated depiction of all connections,"** instead of mapping the evolution of information, processes, and knowledge construction over time. Our proposed methodology, and its accompanying software, were designed to capture the evolution of these temporal elements following participants' discourses linearly—that is, chronologically as they spoke or constructed their dialogs—or non-linearly, which allows their discourses to evolve organically and even "jump" back and forth in the spatial-*temporal narrative continuum* of their speeches or episodic memories, as we further elaborate below.

In this study, we offer an integrative mixed methodology that relies on data science, dynamic network visualization, and mining/retrieval tools typically employed in quantitative research to provide a means to analyze the dynamic processes that shape qualitative and mixed methods research findings and conclusions. Accordingly, based on the fact that *all qualitative datasets inherently contain temporal elements* (these temporal elements may be chronological or linear, following our participants' discourses or speeches, or non-linear, allowing for a more nuanced and likely more accurate description of their experiences through time; please see the section called "Non-linear Temporal Dimensions in Qualitative Research" below) that can be used (1) to recreate the specific context during which participants provided the insights that eventually informed or will inform research findings and conclusions, and (2) to capture their experiences through time (Adam, 1990; Neale, 2019a), the purpose of this paper is to contribute to the mixed methods research literature by introducing *Graphical Retrieval and Analysis of Temporal Information Systems* (GRATIS), an analytic framework and research methodology, and its corresponding open-access software application to visualize and analyze the chronological evolution of processes, events, and information taking place in qualitative and mixed methods research settings. GRATIS's reliance on data science tools also serves to combat the assumption that longitudinal qualitative studies require large timespans. We showcase how all qualitatively- or machine-learning-coded textual data, even those with short timespans (i.e., essays, interviews, social media posts), may be analyzed with no extra data preparation (i.e., encoding or feature engineering), rendering fully integrative/interactive HTML outputs. More specifically, GRATIS enables the analysis of coded textual data without requiring additional feature engineering beyond the coding structure itself, which is one of the reasons it is especially useful for researchers seeking to move efficiently from coded evidence to dynamic temporal interpretation.

Throughout this chapter, coded data should therefore be understood broadly as including qualitatively coded, machine-learning-coded, and computer-assisted coded textual data (see

González Canché, 2022c, 2022d, 2023a, 2023b). GRATIS was designed to remain flexible across these coding pipelines, provided that actors, codes, and their associated textual content can be linked in an edgelist structure. Note also that GRATIS may be applied to all labeled or coded data, even if this labeling process did not consider the *time element* as part of the coding strategy. This means, then, that *previously coded qualitative data* may be re-analyzed with GRATIS, which may lead to new insights and even to different takeaways.

This portability of outputs matters because the value of temporal visualization does not end when a dynamic map has been rendered. Once the evolution of information has been made visible, researchers still need to decide what those trajectories mean, how specific turning points should be interpreted, which contrasts deserve further memoing, and how temporal patterns should be translated into narrative argument. In the broader architecture of this book, GRATIS outputs can therefore be understood as structurally rich guides that may also be brought into ISARI, a local, fully offline, open-source, and multimodal brainstorming environment designed to support scholarly comparison, memoing, synthesis, and evidence-grounded writing. In this way, GRATIS and ISARI are methodologically aligned: GRATIS makes temporal structure visible; ISARI helps researchers think with that structure while preserving interpretive authority and data sovereignty.

This proposed analytic framework responds to the integration challenge wherein "qualitative + quantitative = more than the individual components" (Fetters and Freshwater, 2015a, p. 116). By relying on the analytic power and flexibility of temporal dynamic visualizations and data retrieval, we aim to take data integration to a "new level ..., where visual synergy gains enable a 'value-added' [and fully integrated] research outcome" (Alexander et al., 2019a, p. 34). Procedurally, as shown in Fig. 3.1 (and fully described in the "Procedural Diagram" section), this analytic framework follows an exploratory sequential analytic design (Fetters and Freshwater, 2015b, p. 209) where the qualitative phase is followed by the quantitative one, and it concludes with the integration of the understandings gained through both phases.

The paper is structured as follows. We present a background section that reviews visualization strategies currently available in commercial software to highlight structure in qualitatively coded data, relying on *static network modeling* techniques. This background discussion enables us to start introducing the rationale and data formats employed by GRATIS. Subsequently, we present the conceptual foundation of GRATIS: temporal information systems (TIS) (Artale et al., 2007a) and discuss how TIS is operationalized via temporal network analysis methods (Butts et al., 2020a; E. D. Kolaczyk and G.Csárdi, 2014; M. Mitchell, 2006; Lazega et al., 1995). The integration of these concepts and methods enables GRATIS to map and visualize the simultaneous evolution of information provided by research participants, which, in turn, may strengthen our substantive understanding of the topic of study. In the next section, titled "*Procedural Diagram, Coding Approaches, and Data Formats*," we bring together all concepts and methods discussed so far to form a coherent and comprehensive depiction of GRATIS. Subsequently, we introduce the software application (freely available for Mac and Windows platforms) that implements GRATIS and conclude with a discussion of next steps and the integration and contribution to mixed methods research.

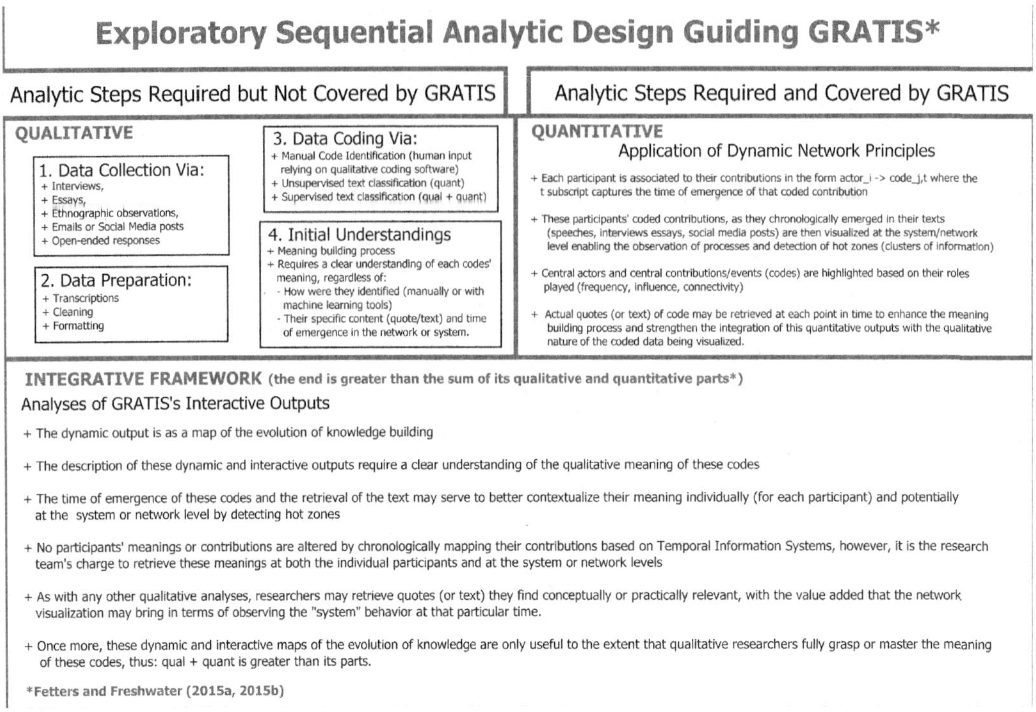

FIG. 3.1 Procedural Diagram.

Background

As briefly discussed above, even the newest advancements in *qualitative data visualization* that aim to highlight the structure embedded in *coded or labeled textual data* only present an aggregated snapshot of such structure, rather than the chronological progression (linear or nonlinear (Linear refers to presenting the evolution of coded information as provided by participants speeches or discourses. Nonlinear accounts for changes in temporality that allows participants' discourses jump back and forth in time following their storytelling, similar to thematic analyses (V. Braun and V.Clarke, 2006). An important takeaway from this opening paragraph is that *coding or labeling textual data* is fundamental for these visualizations to work. Accordingly, although a comprehensive description of coding data, either qualitatively or quantitatively, is beyond the scope of this study (due to space limitations), before presenting how currently available software tools are visualizing these coded texts, let us describe how this coding process should look using a short illustration.

Coding or labeling textual data

A code in qualitative research is a word, descriptor, or label that "symbolically assigns a summative, salient, essence-capturing, and/or evocative attribute for a portion of language-based or visual data" (Saldaña, 2013a, p. 3). Language-based or visual data imply that a code may be added to textual data as well as to video or photographs. For clarity, in this study, we focus on coding textual data, but to the extent that visual data may be translated into text, we could assign codes or labels to such data and apply GRATIS to the resulting codes. Based on this description, coding may be understood as the process of fracturing textual data (Denzin and Strauss, 1988) into *labeled* categories or *codes* to ease their comparison and aid our sense-making efforts (Miles and Huberman, 1994a) when we put these data back together for the purpose of pattern detection, explanation, or theorization (Charmaz, 2014a; Saldaña, 2013a; González Canché, 2022b, 2023a, 2023b; Elliott, 2018).

With this brief description in mind, note that Saldaña (2013a) documented 32 qualitative coding methods, which, due to space limitations, cannot be covered in this study. Nonetheless, to be as clear and as precise as possible, we next describe the underlying process researchers would need to follow to label or code textual datasets using qualitative and quantitative methodologies (via supervised and unsupervised machine learning). In short, although the following examples are general or broad by design, they seek to set a common ground for our subsequent presentation of GRATIS.

General coding processes

Fig. 3.1 shows the methodological steps upon which GRATIS relies. Although the coding of text data is required for GRATIS to work, such a coding process *is not implemented* by the software accompanying this study (Software that classifies qualitative data may be accessed at González Canché (2022c, 2022d, 2023a, 2023b). However, note that GRATIS handles qualitatively and quantitatively coded texts, as we illustrate in the following paragraphs.

As indicated in Fig. 3.1, after data collection has concluded and texts are retrieved or transcribed from interviews, essays, social media posts, websites, or open-ended responses, for example, we may proceed with the coding or labeling process of such databases. For the sake of clarity, let us assume that our textual database is configured by texts retrieved from three participants (A, B, and C), as shown in Table 3.1. Assume also that each participant wrote a short essay on their reasons for enrolling in a graduate data science seminar—we provide access to the complete version of this dataset as part of our distribution of the GRATIS software (i.e., the software comes preloaded with this dataset). However, the examples discussed in this study are based on a subset of this essay dataset containing only participants A, B, and C. This database version is available in (González Canché, 2023c) and is provided to ease the full replication of our analyses.

Qualitative coding example

Following Crossley (2010a) and Saldaña (2013a), these texts may be represented (and coded) as shown in Table 3.1. In this representation, we see that the essays have been coded or labeled so that a short word or sentence is used to summarize the highlighted text (this highlighted text is also referred to as a quote). At this point in the coding process, each quote is assigned

Table 3.1 Short essays on the reasons to enroll in a data science seminar.

Actor A, Essay 1	Codes/Labels
I am focusing on an institutional analysis of nonprofit intermediary college access programs using neo-institutional and organizational theories—which is the organizational data I am using for the SSNA exercises.	Utility_Applicability
I'd like to use SSNA to study the access organization's partnerships and reach within the U.S.	Research_&_Dissertation
Actor B, Essay 2	**Codes/Labels**
I hope to become more comfortable with quantitative methods for research and to become more proficient using R and other statistical software packages.	Utility_Applicability
I also aim to develop a better understanding of the possible methods for modeling social networks and the *potential uses of these models for demonstrating impacts and results to third-party stakeholders such as campus administrators, legislators, and the general public.*	**Skillset_Building** *Utility_Applicability*
Finally, I believe that this course will aid the research that I am conducting for my assistantship	Research_&_Dissertation
Actor C, Essay 3	**Codes/Labels**
I am taking this course because I am interested in learning how to analyze networks using R. I took a SNA class in the fall and learned how to analyze everything in UCINET, but *I would like to learn how to expedite network analysis when using more than one network via R code.*	**Skillset_Building** *Utility_Applicability*
I hope to be able to learn the code and stat packages necessary in order to be able to perform basic descriptive network analyses in R (centrality, density, etc), cluster analyses in R, analyzing/visualizing both two mode and one mode networks in R, key player analysis in R, and ERGMs in R.	Skillset_Building

(or linked) to a label, and such a label and quote may both be linked or connected to the person (or unit) providing that text. These connections or links may be expressed in a tabular form, as shown in Table 3.2. All qualitative coding software and natural language processing approaches have an export function to save these lists of codes as Microsoft Word or Excel files that resemble Table 3.2 and/or Fig. 3.2, wherein each actor is linked to the code provided as part of their discourse (speech or response), and each code is linked to its corresponding quote. Accordingly, each actor can be linked to each of their quotes, and the emergence of these quotes may be retrieved based on the linear chronological evolution (i.e., moving from top to bottom of each participant's discourse). If, instead of following this linear evolution, researchers have a theoretically or empirically relevant reason to re-order these codes (and their respective quotes) following a predetermined analytic rule or strategy—such as showing all instances across participants that deal with *theme A* first, and after exhausting this theme, proceeding to present the second set of codes or labels associated with *theme B* (see Braun and Clarke (2006))—then they should rearrange or re-order these codes and quotes accordingly. In our conceptual section, we also showcase how researchers may follow non-linear representations based on other

Table 3.2 Lists of actor-to-codes/labels connections and codes' textual content.

t	Provider	Code/Label	Quote/Text Content
1	Actor A	Utility_ Applicability	I am focusing on an institutional analysis of nonprofit intermediary college access programs using neo-institutional and organizational theories—which is the organizational data I am using for the SSNA exercises.
2	Actor A	Research_&_ Dissertation	I'd like to use SSNA to study the access organization's partnerships and reach within the US.
1	Actor B	Utility_ Applicability	I hope to become more comfortable with quantitative methods for research and to become more proficient using R and other statistical software packages.
2	Actor B	Skillset_ Building	I also aim to develop a better understanding of the possible methods for modeling social networks
3	Actor B	Utility_ Applicability	Potential uses of these models for demonstrating impacts and results to third-party stakeholders such as campus administrators, legislators, and the general public
4	Actor B	Research_&_ Dissertation	I believe that this course will aid the research that I am conducting for my assistantship.
1	Actor C	Skillset_ Building	I am taking this course because I am interested in learning how to analyze networks using R.
2	Actor C	Utility_ Applicability	I would like to learn how to expedite network analysis when using more than one network via R code.
3	Actor C	Skillset_ Building	I hope to be able to learn the code and stat packages necessary in order to be able to perform basic descriptive network analyses in R (centrality, density, etc), cluster analyses in R, analyzing/visualizing both two mode and one mode networks in R, key player analysis in R, and ERGMs in R.

temporally associated meanings that emerge from participants' discourses—see "Non-linear Temporal Dimensions in Qualitative Research" below.

Unsupervised quantitative coding example

Fig. 3.2 shows the output resulting from applying *unsupervised text classification* following natural language processing and machine learning methods via Latent Dirichlet Allocation and Topic Modeling (see González Canché, 2022c, 2022d, 2023a, 2023b for software access and examples to apply these machine-learned codes or labels). This unsupervised text classification was applied to these same three essays contained in Table 3.1. In this classification method, analysts may decide to mirror line-by-line coding (Saldaña, 2013a) where the classification will be applied to each of the sentences configuring our texts, or to paragraphs or to complete texts (see González Canché, 2022c, 2023a). Concisely, after thousands of learning iterations, each sentence will be classified into *n* number of latent codes (or labels) and the learning process link each of these codes to the text that is most likely to represent that latent code based on word to code and text

Most representative texts per latent code and coherence assessments

code	text	relative_text_contribution	relative_group_fit	quote
V3	Actor A.doc.1	1	1	I am focusing on an institutional analysis of nonprofit intermediary college access programs using neo-institutional and organizational theories - which is the organizational data I am using for the SSNA exercises.
V1	Actor A.doc.2	0.899	0.87	I'd like to use SSNA to study the access organization's partnerships and reach within the U.S.
V1	Actor B.doc.1	0.942	0.87	I hope to become more comfortable with quantitative methods for research and to become more proficient using R and other statistical software packages.
V1	Actor B.doc.2	0.973	0.87	I also aim to develop a better understanding of the possible methods for modeling social networks.
V2	Actor B.doc.3	0.961	0.868	And the potential uses of these models for demonstrating impacts and results to third-party stakeholders such as campus administrators, legislators, and the general public.
V1	Actor B.doc.4	0.883	0.87	Finally, I believe that this course will aid the research that I am conducting for my assistantship.
V1	Actor C.doc.1	0.979	0.87	I am taking this course because I am interested in learning how to analyze networks using R.
V2	Actor C.doc.2	1	0.868	I would like to learn how to expedite network analysis when using more than one network via R code.
V1	Actor C.doc.3	1	0.87	I hope to be able to learn the code and stat packages necessary in order to be able to perform basic descriptive network analyses in R (centrality, density, etc), cluster analyses in R, analyzing/visualizing both two mode and one mode networks in R, key player analysis in R, and ERGMs in R.

Showing 1 to 9 of 9 entries

FIG. 3.2 Output of code detection based on machine-learning text classification.

to code probabilistic distributions (T. L. Griffiths and M.Steyvers, 2004). Note-that in the output shown in Fig. 3.2, the latent codes are simply referred to as V1, V2, V3. This is different from the qualitative codes shown in Tables 3.1 and 3.2. This means that, with this machine learned output, researchers now have to read the content (quotes) and assign a meaningful label, descriptor, or code (Saldaña, 2013a) to such latent codes (González Canché, 2022c, 2022d, 2023a, 2023b). A recent description of this labeling process can be found in Chang et al. (2021a), González Canché (2022c, 2022d, 2023a, 2023b), Ho et al. (2021a), and Poth et al. (2021a) (Note also that although the number of latent codes per actor shown in Fig. 3.2 was the same as those found relying on the manual coding approach shown in Table 3.2, this match may not necessarily be the case when applied to other collection of texts—in natural language processing this resulting number would depend on how the input textual data are decomposed for the topic modeling analyses (i.e., by sentences, paragraphs, or documents.)). Finally, note that the ordering of these latent codes and texts followed the linear evolution based on each participants' speeches or discourses. As before, we may retain this order or update it based on another criterion.

Before moving to the next subsection, let us share that the software Leximancer conducts unsupervised text classification, but instead of labeling the codes as V1, V2, ..., V*n,* it names those codes (called themes in this software) with the most frequent word in a given sentence (or sentences, based on users' analytic decisions). This "theme" labeling process can be qualitatively analyzed, similar to the output shown in Fig. 3.2, where the most representative texts of latent codes are displayed (see https://doc.leximancer.com/doc/LeximancerManual.pdf).

Supervised quantitative coding example

Supervised text classification involves the combination of both qualitative and quantitative coding or classification methods. For example, as demonstrated in software like QDA Miner and its required WordStat add-on module, researchers can load previously coded texts (likely following qualitative coding methods) into software like QDA Miner and then apply machine learning-based supervised classification to compare text-to-code probabilistic distributions and classify those texts (i.e., quotes) as being more likely to represent a given pre-determined label or code. When these pre-determined codes are assigned meaningful labels, the supervised classification process stops at this point; researchers do not need to revisit the learned classes or codes to assign meaningful labels, as those labels would have been created *a priori*. Specifically, as outlined in the QDA Miner User Guide (p. 242) (see https://provalisresearch.com/uploads/QDA-Miner-5-User-Guide-V1.2.pdf), each text (quote) is assigned to one of the pre-coded labels based on the similarity of word distributions within the text. In the example shown in Fig. 3.3, the "Economy" label was assigned to the penultimate row in this figure, with the bold and blue font indicating a 93.7% probability that this text "correctly" belongs to the label "Economy." Similarly, since all previous texts were identified as belonging to the "Politic" code with high probabilities, the probabilities of those texts belonging to the "Economy" code in the column labeled "Economy" are zero or very close to zero.

After this supervised classification is completed, we will once more have access to the quote, the label (or code), and the participant who provided such a quote or text. This information can then be exported to build a dataset like the one shown in Table 3.2, and with that information, we can apply the GRATIS procedures that we will discuss below, also following linear or nonlinear temporal elements.

Visualization tools for coded data

Before presenting the GRATIS procedures, let us describe how currently available software tools use the elements shown in Table 3.2 to render network-based visualizations.

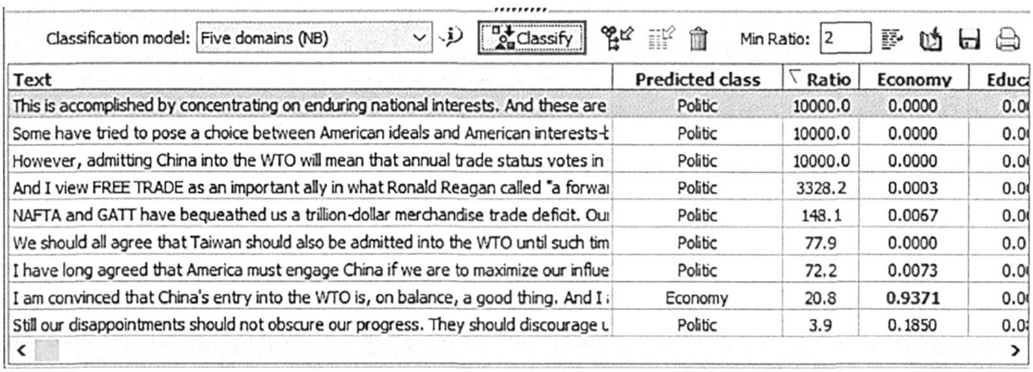

FIG. 3.3 Example of Supervised Learning in QDA Miner.

Atlas.ti networks

The networks tool in Atlas.ti allows users to manually build a visualization with codes and quote elements, such as those contained in Table 3.2 and Fig. 3.2. In this building process, the connecting lines may be assigned a meaning by the researcher—that is, a connecting line itself is given a qualifying label. During this process, researchers may also add a "document" to the resulting visual. Considering that documents typically represent individuals (i.e., research participants or interviewees), the inclusion of a document signifies adding the participants that researchers consider as important actors or units in the resulting network. In addition, researchers may incorporate research memos (notes) and groups (or families of codes) into this visualization (see https://doc.atlasti.com/ManualWin.v9/MainConcepts/EntityTypeNetworks.html). The resulting visualization serves to highlight what researchers consider valuable in explaining their findings. This approach differs from the subsequent visualization methods we present, which aim to emphasize the resulting structure emerging from connections, such as those discussed above, where actors are connected to codes and codes are connected to quotes (see González Canché, 2022a). In Atlas.ti, the resulting network may highlight these actor-to-code-to-quote relationships, but when elements are added to the Atlas.ti plotting canvas, their connections are not automatically generated; instead, it is the researcher who "manually" includes them.

Nvivo project maps

Similar to Atlas.ti's networks, users may select codes, files, memos, and themes to include in the proposed network diagram or representation. A salient difference between Atlas.ti and NVivo is that, in the latter, the connections are retrieved from the relationships observed in the data. To illustrate this process, let us build from our example shown in Table 3.2. Assume we add the code "Utility_Applicability" to the visualization canvas. If only this element is added, no connections are drawn. However, if Actor A is added to this map, NVivo will then automatically draw a connecting line between the "Utility_Applicability" code and Actor A. On the other hand, in Atlas.ti, the researcher would need to draw that connecting line manually. In short, when elements are connected—that is, when codes are assigned to a theme or family (using NVivo language)—and these elements are added to the map, their connecting lines are automatically added to such a visual map. This implies that, if all codes and actors are added to the map, the result will be an aggregated depiction of the complete set of connections, but no temporal elements are captured in the resulting network visualization (see https://help-nv11.qsrinternational.com/desktop/concepts/About_project_maps.htm).

Leximancer concept maps

Concept maps in Leximancer contain the themes (which we have been referring to as labels or codes) that this software detected via unsupervised text classification, along with the most highly correlated words associated with those themes. This concept map is represented in a network form. The size of the nodes (represented as circles or bubbles) is based on their frequency in the transcripts or texts analyzed. Different from Atlas.ti's and NVivo's visuals, these concept maps are automatically generated as a result of the unsupervised text classification and represent an aggregated description of the most frequent connections of themes and configuring words (or concepts, as called in Leximancer, even though they are single words) co-occurring in these texts or quotes. Based on the unsupervised nature of this classification, all sentences across the corpora (all collections of texts) are labeled into one of the resulting themes (i.e., codes, as we have described them above). Finally, similar to all our previous discussions of Atlas.ti and NVivo, there are no temporal elements being mapped in these outputs. Moreover, these "themes," their quotes, and the units that provided those codes may be exported in a tabular form that resembles Table 3.2 and Fig. 3.2.

MAXQDA maxmaps

MAXMaps follows the same rationale as NVivo's project maps. Researchers may select codes, documents, memos and quotes and categories (themes, families) that should be added to the plotting canvas. Once connected elements are added, lines among such connected elements are also automatically drawn. Nonetheless, users may also link objects that are not connected following the rationale shown in Table 3.2 (i.e., actors → codes → quotes). That is, researchers may craft their visualizations as an analytic tool. More to the point, the connecting lines may be formatted to allow researchers to highlight attributes they consider relevant.

Another visualization tool available in MAXQDA is called code cooccurrence. This tool highlights the number of documents (interview transcripts or essays, for example) that contain the same pair of codes. Researchers may decide to display these frequencies as another strategy to highlight structure in qualitative data. Again, the end result may be an aggregated depiction of all selected elements to be visualized in a network diagram (see https://www.maxqda.com/help-mx20/maxmaps/what-does-maxmaps-do).

QDA miner's link analysis

Finally, the last CAQDAS we have found to integrate network visualization tools is QDA Miner. Link analysis allows the visualization of connections between codes and between codes and cases (i.e., research participants). The goal of this visualization is to highlight patterns and structures of cooccurrences. The resulting visual applies a layout that places codes that are more densely connected together and codes that are less prevalent in the working space as peripheral—that is, as far from the center of the plot as possible, which is one of the properties implemented by GRATIS. The closeness of codes based on the prevalence of co-occurrence may also be represented in a dendrogram and in a proximity plot. These visualizations are possible given that a version of Table 3.2, as shown in this study, may be transformed into a matrix, and with this matrix, traditional classification and correlation approaches may be implemented. Indeed, in Table 3.3, we discuss this resulting transformation as implemented by GRATIS. Finally, note once more that in the QDA Miner Link Analysis, no temporal information is considered when creating these visuals.

Network analysis of qualitative data

A special mention in these visualization tools corresponds to the network analysis of qualitative data (NAQD). This is a special mention because the software that implements NAQD (see

Table 3.3 Aggregated matrix representation of participants' contributions.

	Utility_ Applicability	**Research_&_Dissertation**	**Skillset_Building**
Actor A	1	1	0
Actor B	2	1	1
Actor C	1	0	2
Time 1	**Utility_ Applicability**	**Skillset_Building**	**Research_&_ Dissertation**
Actor A	1	0	0
Actor B	1	0	0
Actor C	0	1	0
Time 2	**Utility_ Applicability**	**Skillset_Building**	**Research_&_ Dissertation**
Actor A	0	0	1
Actor B	0	1	0
Actor C	1	0	0
Time 3	**Utility_ Applicability**	**Skillset_Building**	**Research_&_ Dissertation**
Actor A	0	0	0
Actor B	1	0	0
Actor C	0	1	0
Time 4	**Utility_ Applicability**	**Skillset_Building**	**Research_&_ Dissertation**
Actor A	0	0	0
Actor B	0	0	1
Actor C	0	0	0

González Canché, 2022a) was designed to (1) highlight the structure embedded in coded textual data by plotting all the resulting connections and (2) statistically test, via quadratic assignment procedures (random permutation tests, see (Rubin and Canché, 2019) whether certain groups are more concerned about certain topics than others. Like GRATIS, NAQD has not integrated text coding or labeling capabilities, but researchers may rely on machine learning procedures to code these texts free of charge (see González Canché, 2022c, 2022d, 2023a, 2023b) or may rely on a Microsoft Word macro that serves as CAQDAS to label these texts in a similar fashion to NVivo or Atlas.ti, for example. (Access to and instructions to use this Microsoft Word macro are available here: https://github.com/democratizing-data-science/word_as_a_coding_tool.). Once those labels are created, qualitatively or quantitatively via machine-driven text classification, researchers may apply NAQD—and GRATIS, as further discussed below.

This brief review of currently available software that employs network analysis principles to visualize relationships serves to highlight the contribution of GRATIS to these visualization capabilities by incorporating temporal attributes into the resulting maps.

Why rely on network visualizations?

When we move from participants' qualitative contributions to data frames, or to matrices, and then to sociograms (or the network representations of matrices), we gain access to a comprehensive vantage point of analysis (Crossley, 2010a). Network analyses enable researchers to highlight active or central topics via measures of influence, as well as areas of overlap or consensus, or even areas where disagreements are more prevalent among research participants [similar to what Alexander et al. (2019a), referred to as *hot zones* in the visual replay methodology (VRM) discussed below]. These visualization techniques also allow researchers to identify peripheral conversations or points of discussion (i.e., *cold zones*). These are the reasons why network analyses have become prevalent in CAQDAS as tools that allow for a comprehensive depiction of the entire web of relationships, thus going beyond the mere quantification and visualization of qualitatively coded content (Crossley, 2010a; González Canché, 2022a).

A common theme observed in our review, and arguably the main limitation of the current application of network analysis of qualitative data, however, is the aggregated nature of the resulting visualizations. The final outputs depict the summation of participants' contributions (i.e., actor-to-code relationships) that took place in a given qualitative research setting, while completely omitting the temporal (linear or non-linear) evolution of these contributions and the textual content of each of these contributions. This is an important limitation because, although the chronological evolution of participants' reasons may provide further insights into their priorities and interests, this information is currently forfeited in available CAQDAS.

To be more specific about how currently available CAQDAS aggregates network analyses, let us observe Table 3.2. In that table, the chronological information is captured with a column t, where $t=1$ represents the first contribution of each actor as shown in Table 3.1, $t=2$, their second contribution, and so on. Although conceptually straightforward, the modeling of these evolving relationships require dynamic network representations as we discuss below.

Current approaches forgo temporal information by simply counting the number of times an actor mentions a given code. Let us illustrate this using Actor B's coded essay in Table 3.2:

$$\text{ActorB} \xrightarrow{2} \text{Utility\_Applicability} \tag{1a}$$

$$\text{ActorB} \xrightarrow{1} \text{Skillset\_Building} \tag{1b}$$

$$\text{ActorB} \xrightarrow{1} \text{Research\_\&\_Dissertation} \tag{1c}$$

Where the arrows indicate that Actor B provided those codes, the number on top of the arrow counts how many times Actor B provided those codes throughout her/his essay. In this case, Actor B provided two pieces of text (quotes) classified or coded as Utility_Applicability. The other two codes provided by this actor were Skillset_Building and Research_&_Dissertation, with a frequency of one in each instance. This depiction, although useful in summarizing code relevance at the actor level, fails to include temporal and contextual components, such as the order in which actors provided these codes (and the textual content of these codes) throughout their essays. Since this order of emergence in participants' discourses (and their content) may indicate participants' priorities, for example, aggregated representations forego the nuances gained by analyzing the chronological evolution of information and processes, which inherently limits the extent of our understanding, even though that information exists *in all coded textual data* and can be retrieved and mapped.

In continuing with our discussion, note that when all the connections represented in Table 3.2 are taken together and aggregated, they may be transformed into a *weighted* and *directed* matrix representation contained in Table 3.3. Table 3.3 is a weighted and directional matrix because the cells may take values greater than 1, with higher values indicating a higher frequency or relevance of a given code for a given participant in the network. The directionality indicates that actors (in the rows of the matrix) are sending connections to the columns.

Finally, one can plot these relationships in a sociogram form (plot of connections) to capture a final snapshot of all these relationships as shown in Fig. 3.4. The sociogram shown in Fig. 3.4 enables the observation of actors' and codes'/labels' relevance (Crossley, 2010a). Actors A and C were placed at the periphery (i.e., corners) of the network, whereas Actor B was at the center of the plot based on her heightened contribution level in this small example. In terms of code relevance, we see that Utility_Applicability was the most central (or at least the most frequent) reason for these participants' decisions to enroll in this seminar. Again, although these features are important, this depiction alone falls short in capturing time-based processes and in retrieving the meaning (i.e., quote or text as discussed in our coding section above) of each $actor \rightarrow code$ relationship as they evolved.

Moving beyond aggregated depictions

(Alexander et al., 2019a) introduced the VRM, a graphic approach to analyze *processes* "by integrating the accuracy of visually elicited quantitative counts of contributions [codes or labels as discussed above] and the thoughtfulness of qualitative reflections" (Alexander et al., 2019a, p. 36). This focus on processes and dynamics allows researchers to observe and analyze the contextualized situations that shape qualitative research findings and conclusions (Alexander et al., 2019a).

It is the integration of time into the analytic process that allows researchers relying on GRATIS to address questions such as "How did the contributions of participants evolve?" and "What codes or labels did they prioritize in the evolution of their discourses, speeches, essays, or social media posts?" By being able to visualize the evolution of information, we can also observe the clustering of topics/issues/subject areas, or "hot zones" (Alexander et al., 2019a), which capture trends around which the majority of participants' contributions converged. Similarly, we can also capture "cold regions," where the opposite was true—that is, by observing contributions that were provided by a few or just one participant at a given time.

In addition to enabling the observation of the evolution of information or knowledge as an output to be analyzed by researchers, the insights gained with GRATIS via replay and recording of time-evolving visual textual data may become interactively intersubjective (Alexander et al., p. 46). That is, researchers may offer participants the opportunity to interpret whether the

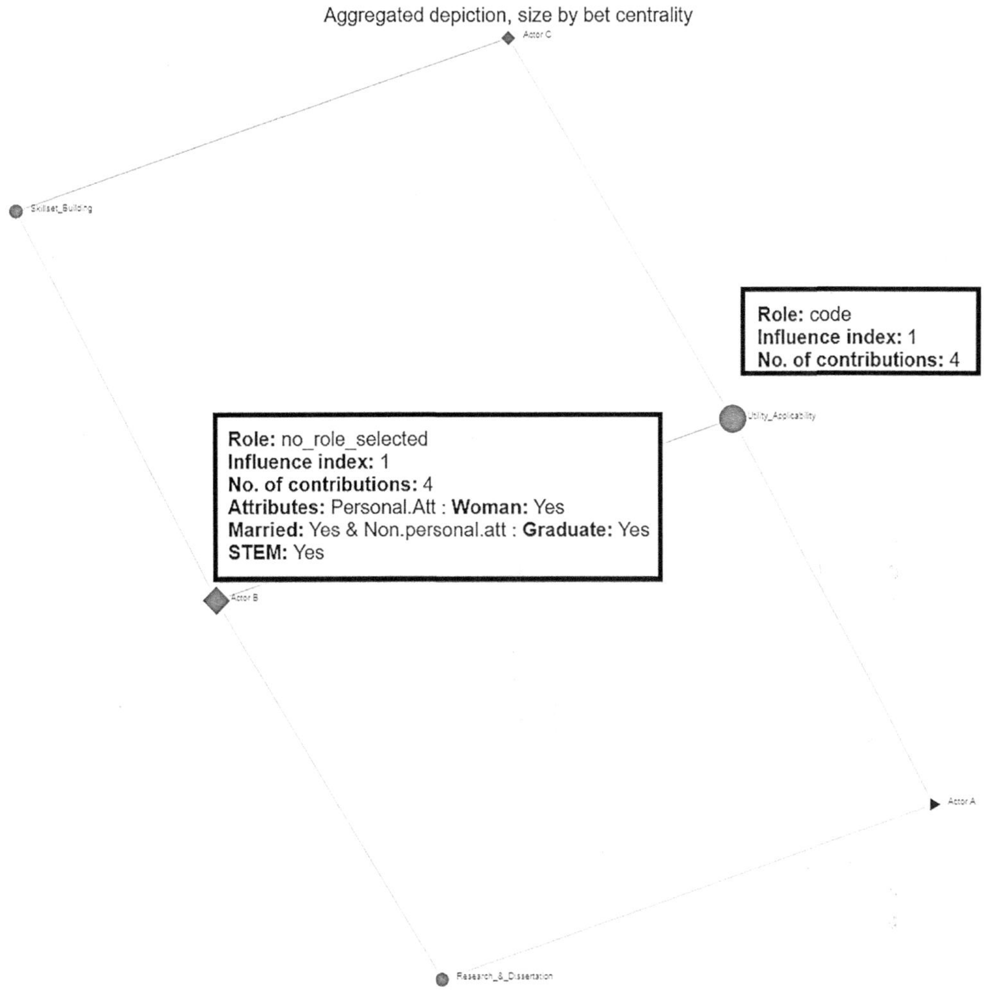

FIG. 3.4 Aggregated Network Representation.

researcher or research team accurately captured their individual insights. Moreover, these participants may also be offered the opportunity to observe how their insights collectively compare with those provided by other research participants—i.e., *interactive reviewability* (Alexander et al., 2019a). This process may help both empower research participants and assess the quality of the analyses and, perhaps more importantly, the transparency and accuracy of researchers' interpretations.

Procedurally, the implementation of GRATIS requires the iterative creation of multiple matrices, as opposed to one aggregated matrix, as shown in Table 3.3. Specifically, this iterative process proceeds as follows: in going back to Table 3.2, we can observe that the first contributions of Actors A and B were Utility_Applicability, whereas for Actor C, it was Skillset_ Building. To account for their temporal elements, these relationships take the forms:

$$\text{ActorA} \rightarrow \text{Utility\_Applicability}_1 \tag{2a}$$

$$\text{ActorB} \rightarrow \text{Utility\_Applicability}_1 \tag{2b}$$

$$\text{ActorC} \rightarrow \text{Skillset\_Building}_1 \tag{2c}$$

In this case, the subscript 1 at the end of the code represents the chronological appearance of this code or label in the actor's discourse. For example, the subscript 3 in the next equation.

$$\text{ActorC} \rightarrow \text{Skillset\_Building}_3 \tag{3}$$

Indicates that for Actor C, Skillset_Building also represented his/her third contribution (hence the value 3 in column *t* in Table 3.2). From a dynamic perspective, there will be an Nth number of matrices to be plotted as sociograms, where N is the highest number of contributions made by a given participant. In returning to the matrix representation of these connections throughout time, we will have the following matrix for the first contribution.

The major difference between this matrix and the final aggregated version discussed above (see Table 3.3) is that this is a binary matrix, where there is only the presence or absence of connections at time *t*. Accordingly, note that since no one provided information labeled as Research_&_Dissertation at time 1, this code (column) contains only zeroes.

The second contribution is represented as follows.

The matrix generated at time 3 shows that, as of this time, Actor A no longer makes any contributions to the network.

Finally, by the fourth and final observation in this small example, only Actor B remained active in the system/network.

It is through these temporally disaggregated matrices that GRATIS enables the observation of the emergence and decay of these *chronological contributions* in research findings. Moreover, the reliance on data mining/retrieval and visualization using interactive HTML outputs enables the use of the quote (i.e., *textual content*) of each *actor to code relationships.* The content of the column "Quote/Text Content" in Table 3.2 qualifies each actor-to-code relationship and changes at each point in time.

As an attribute of that relationship, researchers may retrieve this content with a single click on the line connecting actors to their codes at time *t*. For example, Fig. 3.5 shows the GRATIS output at time 1, resulting from clicking on each of the participants' connections, which matches our original data. Note that although GRATIS's main contribution is the incorporation of time, it also automatically generates an aggregated depiction of the network, similar to currently available CAQDAS, given its remarkable summary properties.

Having discussed how GRATIS contributes to currently available visualization and analytic practices, the next section examines TIS and network analyses as the conceptual and methodological engines fueling GRATIS. Subsequently, we describe the software application and discuss how GRATIS may be applied, relying on the nonlinear temporal evolution of discourses.

Temporal information systems and network modeling

As discussed in the introduction, GRATIS, as an analytic framework, is founded on the notion that observing and analyzing the evolution of trends, dynamics, and changes in our studies through the lens of time (Artale et al., 2007a) may uniquely enhance our perceptions of the social world and strengthen our understanding of such social events (Adam, 1990; Neale, 2015a, 2019a). Despite these analytic benefits, most currently available data models and database systems are designed to capture a snapshot of the world, which results in analyses that provide limited and partial views of actual circumstances, thus falling short in capturing the dynamics of the world or how it evolves as time passes (Artale et al., 2007a, p. 6). Though these shortcomings can impact both qualitative and quantitative studies, the following discussion focuses on qualitative and mixed methods research.

TIS

Although all qualitative datasets contain hundreds or thousands of temporally contextualized pieces of information (Neale, 2015a, 2019a) and these temporal elements are fundamental to contextualizing and informing research findings and conclusions, qualitative analyses rarely convey a clear depiction of the temporal or social context in which ideas emerged or were exchanged. To formally incorporate time-based information, GRATIS relies on TIS (Artale et al., 2007a) as its guiding framework.

TIS treats the recording and easy retrieval of knowledge over time as offering the possibility of depicting the emergence, maintenance, and decay of information and ideas as provided by research participants. The first step in capturing these social dynamics consists of systematically recording the evolution of events as they emerge (J. Chomicki and D. Toman, 1998; Theodoulidis et al., 1991a). In this respect, a salient attribute of GRATIS as an analytic framework and software is that most, if not all, qualitatively coded or machine learning-categorized datasets can be analyzed with GRATIS *with minimal or no extra data cleaning*. That is, as depicted in our background section, GRATIS simply requires each participant's contributions to be stored following the chronological emergence of information they provided, as shown in Table 3.2 and Fig. 3.2.

Linear and nonlinear global timestamps

Our focus on the chronological emergence of events allows GRATIS to standardize the evolution of information across all participants under *global timestamps* (Artale et al., 2007a; Jensen et al., 1998b) rather than at specific time points that are unique to each recorded event and may obstruct the compatibility of simultaneous analyses. For example, if we aim to observe how the evolution of information took place across 40 individual interviews and rely on actual time points (rather than global timestamps, as proposed by GRATIS), it would be impossible to map the simultaneous generation of information *unless* the 40 interviews were conducted simultaneously by 40 interviewers in different spaces. (In the case of these 40 interviews, the data collection timespan was six months, resulting in over 40 hours of transcribed text.) GRATIS overcomes this challenge by assigning global timestamps under a chronological approach. Here, events are recorded under the umbrella of "first, second, third," and so on as part of an emergence of contributions (see Table 3.2 under column t). (As further described, the ordering may follow chronologically linear time evolutions as shared by our participants. Nonetheless, researchers may re-order these codes to capture non-linear temporal elements, but this latter decision must follow clear analytic standards, and research participants should be consulted to corroborate whether this rearrangement captured their experiences (interactive reviewability). In addition to ensuring compatibility, as discussed next, this approach allows for the detection of general trends (or hot zones, as depicted by Alexander et al., 2019a), while also allowing for a focus on specific cases or actors, hence meeting the three modeling requirements of TIS: orthogonality, reducibility, and compatibility (Artale et al., 2007a, p. 13).

In addition to this linear chronological coding timespan approach, which follows the "top-to-bottom" paragraph order of participants' interviews or essays, for example (as depicted in Tables 3.1 and 3.2), researchers may build a nonlinear storytelling timespan. The latter focuses on the time period covered by our participants' storytelling, wherein the analytic focus is placed on their experiences and episodic memories. If, for example, we want to understand the strategies that adjunct faculty members have used to save for retirement, the coding timespan may cover each faculty member's trajectories in academia, which may range from a few months to decades of seniority and involve the development of nonlinear, chronologically-based themes focused on participants' experiences (V. Braun and V.Clarke, 2006). In the following lines, we elaborate further on this important topic.

Nonlinear temporal dimensions in qualitative research

Temporal dimensions in qualitative research are not constrained to linearity in the form time 0, time 1, ..., time n. Instead, these temporal dimensions depend on the organic evolution of the discourse(s) as expressed by our research participants. Guided by complex systems thinking (Maroulis et al., 2010a; Mitchell, 2006), when discourses are operationalized as a network, they become dynamic, nonlinear, and flexible. That is, what our participants shared regarding "time X," may have prompted them to share more information regarding "time $X-n$," or "time $X+n$." Here, X is a point in time, and n represents time units before or after such point in time took place. The n time units may be hours, days, years, or decades, for example, before or after a given point in "time X." For instance, let us assume that a participant shared with us the following:

"I finished college in 1997. Who knew that 10 years *later*, in 2007, I would start my graduate coursework in a different country and that these graduate school years would be extremely challenging in many aspects of my life? This experience was in stark contrast to my middle school dreams of studying abroad. The real world did not match my expectations."

Here, this participant offered information initially situated in 1997, but then there was a jump in time ten years into the future (time $X+n$), when more information about graduate education was offered. Also note that in this same example, our research participant jumped back to her high school years, which following this no-linear rationale may be expressed as "time $X-n$."

As analysts interested in capturing evolution of information and/or events, we could do so by respecting or following participants' speech as they shared it with us (i.e., verbatim from a discourse generation point of view), or we could try to capture the emergence of their speech following their nonlinear storytelling. This nonlinear time-coding scheme is more challenging/complex but may result in a perhaps more accurate timeline depiction of their experiences. If a temporally linear depiction suffices, then our examples shown in Tables 3.1 and 3.2 and Fig. 3.2, are enough to capture our participants' discourses evolution. However, if we instead seek to offer an analysis of discourse evolution that follows a more nuanced description of storytellings or events, we must rearrange the labels based on *how we perceived that our participants' stories evolved*. Specifically, going back to our example regarding "time X" discourses that prompted "time $X-n$" stories, we could then present "time $X-n$" information and after exhausting this part of our participant's discourse, move to presenting coded information pertaining to "time X" and then "time $X+n$" as we illustrate in Table 3.4.

When this particular rearrangement, shown in Table 3.4, is conducted, the input or assessment of our participants becomes more important than ever, as we would need to ask them *whether our interpretations of their nonlinear timelines* accurately capture their experiences. Here, the notion of *interactive reviewability*—where participants may be asked to interact with the resulting evolution of their experiences—is even more relevant than in the linear or "top-to-bottom" depictions, as it helps validate our analytic decisions. If our proposed resulting timeline is inaccurate from our participants' perspectives, we (researchers) should then rearrange the labels until they are satisfied with the temporality to be plotted by GRATIS. When researchers rearrange codes to reflect nonlinear storytelling rather than transcript order, participant input becomes especially important because the validity of the resulting temporal map depends not only on analytic coherence but also on whether the reconstructed temporal sequence is recognizable to those whose experiences are being represented.

With these linear and non-linear temporalities in mind, let us formally discuss the three modeling requirements of TIS: orthogonality, reducibility, and compatibility (Artale et al., 2007a).

Orthogonality

Orthogonality (Artale et al., 2007a) refers to the independent specification and retrieval of classes, relationships, and attributes. In GRATIS, classes are represented by actors and their qualitatively generated (or machine learning-identified) codes or labels. These classes are related across time, as shown in Eqs. (2 a–c) and (3) above, and each of these classes may be assigned attributes or characteristics (i.e., personal and nonpersonal attributes in the case of human actors and network-based attributes in the case of codes, as further explained in the Network Modeling section below). The property of capturing relationships across classes at different points in time allows GRATIS to meet the reducibility requirement of TIS.

Table 3.4 Example of nonlinear storytelling evolutions.

ID	Time	Information
Actor 1	X-n, 1	Label of X-n, 1
Actor 1	X-n, 2	Label of X-n, 2
Actor 1	X-n, 3	Label of X-n, 3
...	...	...
Actor 1	X, 1	Label of X, 1
Actor 1	X, 2	Label of X, 2
...	...	...
Actor 1	X + n, 1	Label of X-n, 1
...	...	...
Actor n	X-n	Label of X-n, 1

Reducibility

Reducibility (Artale et al., 2007a) specifies that we should be able to completely rebuild the temporal evolution of a database (i.e., reproduce, replay), at both the individual and group levels, with each of the snapshots captured in the relationships $actor_i$ and $code_j$ at time t, thus allowing us to go forward or backward in the evolution or devolution of information. It is this possibility of rebuilding the dataset that we are referring in GRATIS as *reproducibility*, rather than *reducibility*. This is so because although snapshots may be taken when analyzing the resulting time-evolving networks, users (which include researchers, research participants, and readers), are not constrained to observing only snapshots, instead they may access the entire evolution of the temporal database being analyzed. In this respect, as described in the software application, GRATIS offers play, forward, rewind functionalities () as well as speed control (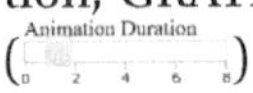

) to accomplish this re(pro)ducibility requirement of TIS.

GRATIS also allows the depiction of actors' activities with different "lifespans." This is naturally captured by the number of codes retrieved from each participant's discourses. For example, in Table 3.2, we can see that the essay written by Actor A resulted in two codes, whereas the essays written by Actors B and C resulted in four and three codes, respectively. The temporal representation of the evolution of these actor-to-code relationships shows that after the third global timestamp, only Actor B continued to provide information in the system, and in this case, this relationship was classified as "Research_&_Dissertation." This activity or contribution length is referred to as the *existence span* by Artale et al. (2007a).

Compatibility

The third requirement specified by Artale et al. (2007a) is compatibility, or the capability of accounting for, preserving, and retrieving "at each instance of time" nontemporal elements that may help contextualize the information being provided over time (p. 13). In GRATIS, this condition is met with the inclusion of personal, nonpersonal, and network analysis attributes computed as part of this integrative framework. Specifically, in our presentation of the GRATIS software application, we will describe the strategies that researchers may follow to include personal (e.g., age, ethnicity) and nonpersonal (e.g., employment status, college degree indicator) attributes of our participants that may help further contextualize their contributions—including the "tone" used when providing information, such as emotions, for example. We programmed GRATIS to allow these attributes to remain constant over their *existence span* or evolve over time. These attributes, an example of which is shown in Table 3.4, can be added to our resulting visualizations with a single click in the GRATIS user interface. In addition to these personal and nonpersonal attributes, we also provide an influence index that is automatically calculated by GRATIS—that is, researchers do not need to add this index as an attribute. Notably, this influence index is measured for both actors and codes and is retrieved via network analyses, as described in the next section. More to the point, note also that the size of the actors and codes will vary based on another measure of network influence (betweenness centrality) that, as discussed below, captures the connecting role that actors and codes have in the system (Lazega et al., 1995). This betweenness centrality is also automatically generated by GRATIS.

Another analytical feature related to compatibility is also depicted in Fig. 3.5, where we show an example of the first contribution of all participants listed in Table 3.2. In this case, by clicking on the links connecting each participant with their codes, we can observe the relevance of that code to their discourses. For example, Actor C mentioned "Skillset_Building" twice, and this code represented 66.7% of her/his contributions. As further described below, these pieces of information were derived from another social network analysis concept referred to as *degree centrality.*

In sum, GRATIS meets the three modeling requirements of TIS (Artale et al., 2007a), and these requirements are met by relying on the integration of qualitatively or computer-assisted coded data with the dynamic network analysis methods (Butts et al., 2020a) that we present next.

Network modeling

Network analysis and modeling are particularly useful ways to "deal with complex systems in the real world" (Mitchell, 2006, p. 1199), such as those we are modeling with TIS. In the context of GRATIS, network modeling is particularly useful in helping researchers make sense of multiple moving parts that evolve or devolve over time while meeting the three modeling requirements of TIS outlined above.

Operationally, network thinking emphasizes relationships among specific events, including connections among individuals and their actions or among broader categories, labels, or codes, as defined by researchers (Maroulis et al., 2010a; Maxwell, 2019). Formally, networks are a collection of *potentially interactive units* (E. D. Kolaczyk and G.Csárdi, 2014; M. Mitchell, 2006; Lazega et al., 1995). These units are typically referred to as *nodes* or *vertices* (e.g., actors, participants, or entities that participants may interact with or be ascribed to), and the connections resulting from their interactions are referred to as *edges* or *links* (Lazega et al., 1995). When the units configuring a given network are of the same *class* and hierarchy (e.g., students interacting with other students), they form a one-mode network. When the units configuring the network are different (e.g., professors ascribed to their employing universities or ascribed to a particular set of beliefs), they form *two-mode networks*.

GRATIS relies on two-mode networks, given our focus on the analysis of the evolution of *codes* (information, or first mode) provided by our research participants (the units of analysis, or second mode). Following Crossley (2010a), although network analyses can focus on individuals within the network, we can also observe groups of participants who clustered around certain topics or viewpoints during their interviews (referred to as "hot zones" above; see Alexander et al., 2019a).

Highlighting structure

GRATIS follows graph theory (Biggs et al., 1986a), dynamic temporal network analysis principles (Butts et al., 2020a), and global linear and non-linear timestamps (Artale et al., 2007a; Jensen et al., 1998b) to capture the time-based emergence of information simultaneously, even though each participant provided these pieces of information individually, at different times and places, and with different time or existence spans. The integration of these three methodological views enables the simultaneous depiction of all contributions taking place in the same global timestamp across *all* participants, regardless of temporal linear or non-linear decisions. Specifically, as depicted in Fig. 3.5, all first contributions of the participants are shown

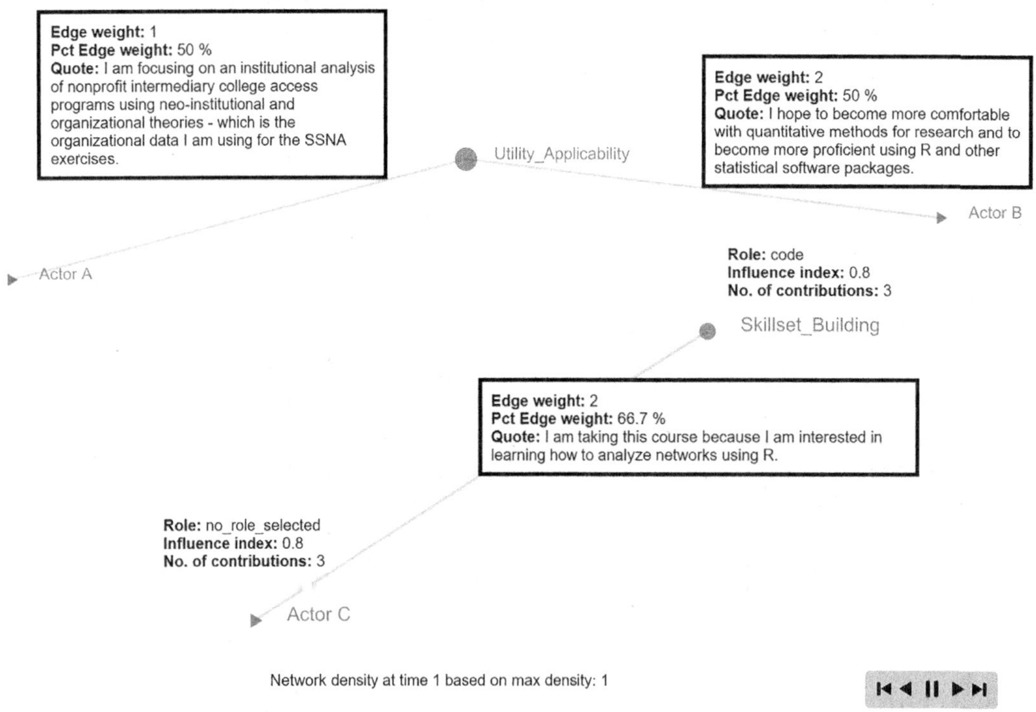

FIG. 3.5 GRATIS Network Representation of First Contribution.

concurrently at *time 1* (i.e., the first global timestamp), regardless of their actual location in the dataset (i.e., rows 1, 3, and 7 for Actors A, B, and C, as illustrated in Table 3.2), the actual time and place when/where they wrote these essays, and their "storytelling timespans."

Contextualization

As briefly described above, GRATIS allows the addition of attributes at the node (i.e., actor or code) and edge (links) levels. At the node level, we can add personal and non-personal attributes that may remain constant over the entirety of the data collection time or that may have changed during this same timeframe. For example, when visualizing coded interviews, coders may add a "tone" attribute (e.g., to depict whether the actor was surprised, angry, or sad when providing us with specific information, more on this below) to further contextualize the message of the code over and above its textual content. In addition to "tones," we may also capture participants' personal and nonpersonal attributes, all of which were programmed in GRATIS to change or not depending on our data sources and analytic decisions. For example, let us assume that we embarked in a multiyear longitudinal study, wherein we may have observed that during some parts of their storytelling some actors were married, but at some points they changed their marital status to separated or divorced. Similarly, we could have also seen that a participant did not have a college degree (time $X-n$ in Table 3.4, for example), but that, at some point in time, such a participant mentioned to be enrolled in graduate school (time $X+n$ also in Table 3.4).

Procedurally, these personal and nonpersonal attributes can be captured with two extra columns in our actor-code dataset, as presented in Table 3.5 (further described in the Data Formats section below), and added to our GRATIS plots with a single click, as shown in Figs. 3.4 and 3.5. Note, however, that GRATIS does not require long timespans to capture changes in respondents' contextualizing characteristics, which is exemplified by the "tone" attribute in Table 3.5 that can be added to each response in a qualitative interview.

See the complete dataset used in the examples shown in Figs. 3.4, 3.5, and A1 here: https://cutt.ly/N846wVl

Finally, GRATIS was programmed to automatically compute both the actors' and the codes' attributes based on network analysis measures, as described in the "*Detecting Actors and Codes Relevance*" subsection below.

With respect to attributes of the connections, in addition to the "Edge weight," "Pct Edge weight," described below, and text content or "quote" described above, the color of the connection shows whether the code just emerged or was the same as the one observed in the previous time. For instance, in Fig. 3.5, the green lines indicate that these connections just emerged as the first contribution across all participants. However, when the visualization is advanced to the next event (see Fig. 3.6), these lines will turn blue, indicating they were present in the previous time, and the new connections will be represented in green. If a participant offered information that was coded under the same code shown at time 1, that participant will have only one connecting line, and it will be dark blue. But if that participant instead offered information that was coded with a different label, that participant will have two connections, with the new one shown in green and the previous one shown in blue. As depicted in our user interface section, by default GRATIS shows the two most recent contributions over time, but the interface can be

Table 3.5 Example of attributes.

Provider	Code	Quote	Personal Attributes	Non-Personal Attributes
Actor A	Code J	Text of code J	Married: No Sex: Woman Ethnicity: "LatinX"	Employed: Yes College Degree: Yes **Tone: Calmed**
...	...	...	...	...
Actor B	Code F	Text of code F	Married: Yes Sex: Woman Ethnicity: "Black"	Employed: Yes College Degree: Yes **Tone: Upset**
...	...	...	...	...
Actor C	Code K	Text of code K	Married: No Sex: Woman Ethnicity: "White"	Employed: Yes College Degree: No **Tone: Calmed**
...	...	...	...	...
Actor C	Code L	Text of code L	Married: Yes Sex: Woman Ethnicity: "White"	Employed: Yes College Degree: Yes **Tone: Sad**

Actor	Code	Text
A	X	...
B	Y	...
A	Z	...

adjusted to show up to the 10 most recent contributions (This span may also be adjusted to go beyond 10 if researchers needed to.) (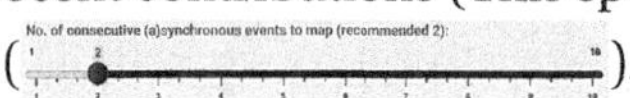). The recommended selection of two contributions is particularly useful with large systems or networks, where network visibility may be compromised given the abundance of actors and codes. Moreover, the observation of the two most recent connections enables users to understand what the previous most recent contribution of each actor was up to time t, thus enhancing the contextualization of the evolution of information.

Detecting actors and codes relevance

An important contribution of GRATIS in detecting structure in our coded/labeled dataset consists of its automatic detection of actors' and codes' relevance/influence based on social network analysis measures (Freeman, 1978). Specifically, GRATIS computes three measures of centrality that capture movement and structure in the network through actors' and codes' positions and paths or connections (Borgatti, 2006). These three centrality measures are degree, betweenness, and eigenvector, as we describe next.

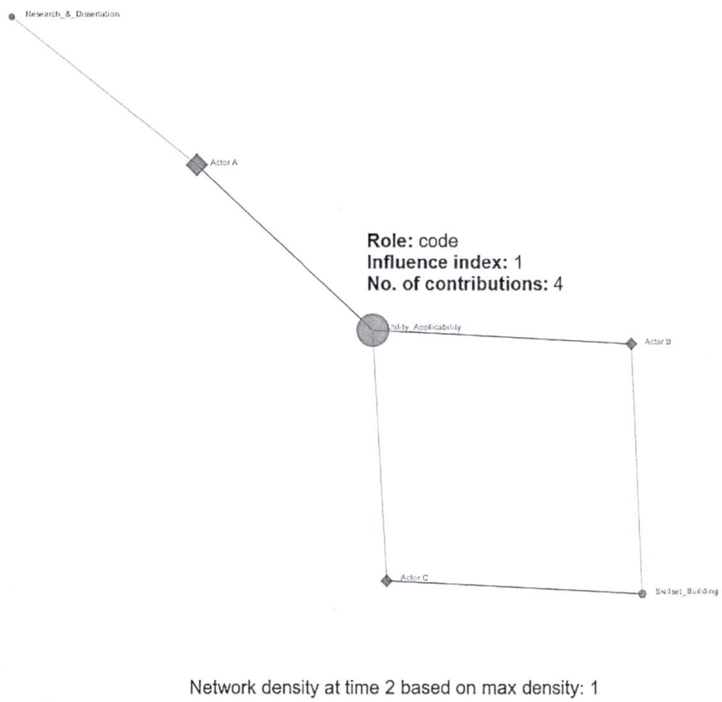

FIG. 3.6 Time 2 Events.

Degree centrality

This form of centrality consists of measuring the number and frequency of connections that a given actor has in the network (Borgatti, 2006). In terms of the *number* of connections, GRATIS computes degree centrality by counting the number of times an actor provided information that was eventually classified as a code—similar to the aggregated connections depicted in Eqs. (1a), (1b), and (1c) discussed in the background section. This means that the information "No. of contributions" displayed when clicking on an actor or code in the HTML output will show the total number of codes provided by that actor and the number of times a given code appeared in the network. Specifically, in going back to Fig. 3.5, the information displayed for Actor C indicates that this actor contributed a total of three codes, two of which were classified as "Skillset_Building." Similarly, when clicking on a code in this HTML output, we can see the number of times a given code was contributed across time by all actors. In the case of the code "Skillset_Building," the number of times textual information across all essays was labeled using this code was three, irrespective of who provided this information. This implies that degree centrality, as a counting device at the code level per se, does not reflect the relevance of that code to a given actor in particular. To address this limitation, and as noted above, we programmed GRATIS to also include code relevance or *frequency* per actor by clicking on the line(s) connecting each actor with their codes. In the case of Actor C in Fig. 3.5, the line connecting this actor with the code "Skillset_Building" reflects that this participant provided information labeled under this code with a frequency of two times (under "Edge weight") and that this code represented 66.7% of their total number of contributions (under "Pct Edge weight").

Before moving to the next centrality measure, let us note that although the actual content of the code (i.e., its textual information contained in the column called "Quote/Text Content" in Table 3.2) changes at each point in time in the GRATIS plot, these degree centrality measures do not—thus meeting the compatibility requirement of TIS (Artale et al., 2007a). Rather, these degree centrality measures are individual estimates of their aggregate values throughout each participant's contributions to the network. These estimates then allow us to further contextualize the relevance of each code for each actor when clicking on their connecting line, and the relevance of the actor or code for the network when clicking on that actor or code.

Betweenness centrality

This measure goes beyond frequency counts. Instead, it detects the extent to which a given actor or code falls in between other actors or codes (Lazega et al., 1995) thus effectively serving as a communication bridge in the system to ease the flow of information. In other words, betweenness centrality highlights *both* the participants who served as bridges of information *and* the codes that connected the individuals in this network. GRATIS uses betweenness centrality as an attribute to add size to each actor or code. To the extent that the size of a node (actor or code) grows at a given point in time, this indicates that such an actor or code is serving as a bridge of information. Accordingly, the software has standardized this measure to change as time progresses. Based on this dynamic change, researchers can observe which actors and codes were the most influential by serving as bridges of information at each point in time. Going back to Fig. 3.5, note that the biggest (in size) code at time 1 was "Utility_Applicability" for it connected two of the three actors in the system (Actors A and B). At the second event (see Fig. 3.6), this code connected all three participants. Fig. 3.6 also shows the different connections' colors described earlier.

Eigenvector centrality

This measure appears as the "influence index" in a dialog box when an actor or code is clicked in the GRATIS HTML output (see Figs. 3.4 and 3.5). Eigenvector centrality measures the overall influence of a given code or participant as a function of the relevance of its connections. Its influence index is standardized to range between 0 and 1, with 1 indicating the maximum level of influence, and it highlights individuals or codes that are central as a function of being connected with individuals and/or codes that are actively connected in the network (Borgatti, 2006; Freeman, 1978). Hence, an actor is central in eigenvector centrality if such an actor provided codes or established connections with other actors who were also active or central in the network (influence begets influence).

Given that, mathematically, the eigenvector centrality of a unit i is proportional to the sum of centralities of the units j to whom unit i is connected (Borgatti, 2006; Freeman, 1978), the level of influence of each actor and/or code may serve to further contextualize their contributions and strengthen our understandings. For example, note that Actor B was the most active contributor in the network with 4 codes in total—Table 3.2. Two of these codes were labeled as "Utility_Applicability." With these pieces of information, we can infer that actors connected with the code that was shared the most by the most active actor will increase their relevance or influence in the network. In this virtuous cycle, since Actor B was the most active actor and shared the most shared code, when clicking on this actor, we will note that she has the highest *influence index* (see Fig. 3.4). Similarly, code "Utility_Applicability" also shows this highest value (Fig. 3.4). In the case of Actor A, she had the lowest value in eigenvector centrality because she had the lowest number of contributions and only one of them was labeled as "Utility_Applicability." Following eigenvector centrality notions, if both contributions of Actor A were labeled under "Utility_Applicability," her influence index would have increased as well, even if her total number of contributions remained unchanged.

These TIS concepts and dynamic network analysis methods allow GRATIS to provide *free rides* and *inference support*, which, according to Shimojima (1996a) is the property of being over-specific in representing information, being self-consistent in information content, and generating inferences as a result of following notational rules (see also Green et al. (2006a)) such as those we presented in Eqs. (1a) to (3) above. From this view, in GRATIS, these free rides and inference support are achieved through the integration of TIS and dynamic network principles.

Procedural diagram, coding approaches, and data formats

Fig. 3.1 contains the procedural diagram guiding GRATIS. Overall, we can see that GRATIS follows an exploratory sequential analytic design (Fetters and Freshwater, 2015b) by moving from qualitative to quantitative phases and culminating with their integration, which we describe in the discussion section below. However, as part of these processes, also note that there are some analytic steps that although fundamental for GRATIS are not covered or implemented by our software application. These analytic steps include data collection (e.g., interviews, essays, ethnographic observations, social media posts), data preparation (e.g., transcriptions, cleaning, data formatting), data coding (via qualitative *or* machine-learning-based approaches, as discussed above), and a *well-grounded initialunderstanding* and *clear conceptual grasp* of the meanings of these codes. This latter point means that for the visualization to be meaningful, researchers *must first develop a clear understanding of the meaning of these codes*—even if as part of the GRATIS output they can easily retrieve the actual quote associated with each actor and code connection (see Fig. 3.5).

Because GRATIS visualizes the temporal evolution of coded information rather than raw text directly, the methodological integrity of the coding process is decisive. The dynamic maps do not generate meaning independently of coding; they formalize and render visible the temporal structure implied by the coding scheme. For that reason, GRATIS analyses are strongest when coding is conducted transparently, systematically, and with clear attention to the conceptual relationships among actors, codes, quotes, and temporal ordering. Whether the coding process is qualitative, machine-learning-assisted, or sequentially integrative, the analytic value of the resulting temporal visualizations will always depend on the rigor of the coding process that precedes them.

The second set of analytic steps that GRATIS covers is more quantitative in nature but fully relies on the results of the inherently qualitative (or mixed methods via machine learning classification) phase just described. This second set of quantitative analytic steps is based on the application of dynamic network analyses that are informed by TIS's tenets, as just described.

Edgelist data format

As discussed in the background section, GRATIS is flexible enough to handle qualitatively coded, machine-learning-coded, or computer-assisted coded textual data, provided that the resulting outputs preserve the actor-to-code structure required by the framework. The process of qualitatively coding data has received considerable attention in the literature (see Auerbach and Silverstein, 2003; Adu, 2019; V. Braun and V. Clarke, 2006; Saldaña, 2021, for example). However, although GRATIS requires these steps, it does not conduct this process. Nonetheless, in a similar way to how it can handle any form of computer-assisted text classification, it can also handle qualitative codes generated through any form of manual coding approach and collected via a myriad of qualitative methods, such as interviews, participant observations, and ethnographic observations, for example. In all cases (i.e., all forms of qualitative data collection and all forms of coding or labeling), the only requirement is to link actors to the codes they provided along with the content of those codes (please see Table 3.2 and Fig. 3.2). This can be achieved following qualitative coding procedures via software like Atlas.ti or NVivo, for example, or based on natural language processing via machine learning and supervised or unsupervised text classification using software like Leximancer, QDA Miner with the WordStat add-on, Machine Driven Classification of Open-ended Responses (MDCOR), or Latent Code Identification (LACOID) (González Canché, 2023a, 2023b), for example. This flexibility is based on the resulting coded outputs following an edgelist data format (Wasserman and Faust, 1994).

The data format presented in Table 3.2 and Fig. 3.2 is referred to as an *edgelist* or a collection of relationships (Csárdi and Nepusz, 2006a; Lazega et al., 1995). These relationships have two minimum components: one column in a data frame containing the *provider* of information, and another column representing the information being provided (i.e., code/label). In Table 3.2, these columns are "Provider" and "Code/Label," and in Fig. 3.2, they are called "text" and "code." GRATIS does not require that the dataset uploaded has these same column names or that this dataset follows the *provider and code* column order. Instead, researchers should conceptually know that these two columns are the minimal information required to conduct the GRATIS analyses. Once the data are uploaded to the GRATIS user interface, researchers will then *select these columns* from drop-down menus.

Given that coding software does not render the equivalent of column *t* in Table 3.2, GRATIS internally computes this column information with the row order uploaded to the user interface. Accordingly, researchers should be careful in correctly identifying the temporal evolution of codes by participants. In Fig. 3.2, this chronological order is observed with the numbers at the end of each actors' names, which reflect contribution number based on topic modeling (i.e., the "**2**" in Actor A.doc**2** identifies the second code provided by actor A.) (These actors' names are produced by machine learning procedures. Although we could preserve this linear order or make changes based on nonlinear evolution of events, these numbers at the end of actors' names should be removed before this database to GRATIS, otherwise ActorA.1 and ActorA.2 will be identified as two different actors. Accordingly, an important requirement to use GRATIS is that actors' and codes' names are consistent across the dataset.). With this in mind, GRATIS was programmed to follow the actor-code *row order* of appearance in a dataset as the chronological evolution of codes by actors, even if these contributions are not consecutive. For example

Although Actor A appears in the first and third rows, GRATIS will identify that their first contribution will be X and the second will be Z. In the case of Actor B, there will be only one contribution.

GRATIS aims to add as much context to the evolution of each participant's contribution as possible. With this goal in mind, GRATIS allows for the inclusion of attribute data at the actor level so that changes in the "tone" of delivery (i.e., emotion) are captured, for example. Another example involves depicting whether potentially important attributes may have changed. For instance, if a participant shared a story 10 years prior to the main events studied, and during that timeline (i.e., 10 years before) this participant was single, that marital status can be added to the analyses and updated in the present timeline if that status changed—though it can also remain the same across timelines. To demonstrate how to add this information, let us discuss Table 3.5 (see the complete dataset used in the examples shown in Figs. 3.4, 3.5, and 3.6 here: https://cutt.ly/N846wVl).

Table 3.5 is an expanded version of Table 3.2. In Table 3.5, we have the "Provider," "Code," and "Quote" columns (column t is not shown but is computed by GRATIS). In addition, we added two more columns: **"Personal Attributes" and "Non-Personal Attributes."** Notably, with the goal of adding as much context as possible and considering that the GRATIS outputs rely on HTML code, each cell may contain more than one attribute. Specifically, the "
symbol shown in Table 3.5, although optional, adds a line break to each attribute of interest. That is, to add gender and marital attributes in different lines, we may have the following: "Woman: Yes **
** Married: No," and when displayed, this would result in:

Women: Yes

Married: No

This output is shown in Fig. 3.4. Note that in the same figure, the attributes are shown in bold font, and this format may be added with the commands, with all text in between being bold in the resulting output. Once more, the use of <br> and these commands is optional but may enhance the clarity of the resulting visualizations. The full HTML-formatted input looks like this: **Woman:** Yes
 Married: No (an example of this data format can be accessed in González Canché, 2023c, which may be used to replicate Figs. 3.4 to 3.6).

The highlighted text in Table 3.5 represents changes in Actor C's personal and non-personal circumstances—as well as a change in her "tone." In her first recorded code, she was not married and had no college degree, whereas in her last one, she was married and had a college degree (similarly, her tone changed from calm to sad). In this example, we can assume that the data collection process spanned several years, or that the participant's storytelling jumped ahead in time by an unspecified number of years, both of which can be handled by GRATIS, as depicted throughout this paper.

To close this section, note that although the inclusion of multiple attributes (dynamic or time-invariant), along with the depiction of the content of the code, is highly recommended to add more context to the resulting visualizations, both attributes and quotes are optional (as indicated in the GRATIS user interface described next). Indeed, GRATIS requires only two columns (i.e., actor or provider and code/label) to render the output that visualizes the chronological evolution of code generation and decay.

GRATIs user interface

The execution of GRATIS requires the following steps, as shown in Fig. 3.7:

A. Load data or select a data example.

- **a.** By default, GRATIS asks you to load your data but also provides two data examples (i.e., the full set of coded short essays about the reasons for participating in a data science seminar and a subset of coded interviews on strategies employed by faculty members with short-term contracts to participate in savings and retirement plans).

B. Upload data

- **a.** To upload a database, search for a comma-separated values file stored locally. If you select a data example provided with GRATIS, load that database by selecting "Click to load [essay or interview] data example."
- **b.** Any uploaded database is then displayed in the GRATIS user interface.

C. Required information:

- **a.** Select the actor column (or document/essay ID).
- **b.** Select the code/label column.

D. Optional: (Attributes to aid in the contextualization of information evolution, displayed when clicking on an actor, code, or line/connection):

- **a.** Column denoting role (see example datasets: As shown in the example datasets, this is another optional column that can be added to the dataset. See the replication dataset here: https://cutt.ly/N846wVl). This column indicates specific **roles of the actors** (e.g., students, faculty, administrators) or documents (e.g., historical, legal). All codes will display the role "Code" in the visualization—see Figs. 3.4 and 3.5.
- **b.** Column denoting **personal attributes**. Actors with the same time-variant or invariant attributes will have the same color in the network, thus aiding in identifying groups. All codes will be orange in the visualization (colors can be changed in the "II Optional Adjustments" tab shown in Fig. 3.7).
- **c.** Column denoting **non personal attributes**. All actors with the same time-variant or invariant non-personal attributes will show the same shape in the visualization. All codes will be a circle in the visualization (shapes can be changed in the "II Optional Adjustments" tab).

- **d.** A column containing the **content of the code (quote)** that changes as time passes. This enables users **to retrieve, in real time, the information provided** by actors or collected from documents.
- **e.** You can modify the **number of consecutive events to map**. As stated above, the recommended number is two, especially if the groups are large. The current event will have a green line, and the immediately previous event will be blue.
- **f.** Finally, you can **select the transparency level** of the colors to be plotted, with 1 being solid and 0 being transparent. The recommended/default value is 0.8.

E. "Execute GRATIS" on the top panel to obtain two HTML documents.

- **a.** The first document relies on TIS and dynamic networks and depicts the complete evolution of information.
- **b.** The second document is an aggregated network depiction summarizing all the contributions made by each actor or retrieved from every document or essay.

These HTML files will be automatically launched in your computer's default internet browser and can be saved and distributed as standalone files. Their display does not need internet connection. Alternatively, they can be launched manually in the "III Execute GRATIS" tab. In addition, in this tab the local directory where those HTML files are stored may be retrieved to share in a webpage or distribute via email. (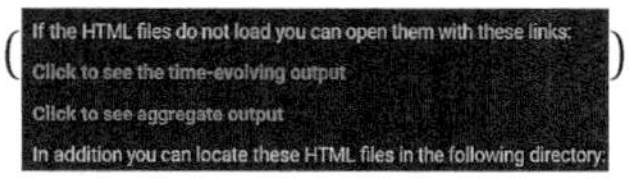

)

Software and source code access

As a product of this publication, we are offering full and unrestricted access to the GRATIS software with the goal of democratizing data science by removing all computer and programming requirements to implement the GRATIS methodology. A further advantage is that GRATIS does not require additional feature engineering beyond the coded actor-to-code structure itself, allowing researchers to move from coded data to temporal visualization with minimal extra

FIG. 7 GRATIS's User Interface.

preparation. Mac users can download GRATIS here: https://cutt.ly/cb5x1wW; PC users can download the software here: https://cutt.ly/Gb5xVBB. To execute the application: (1) extract the zip folder, (2) open the folder starting with "GRATIS-", and (3) double-click on the file called "GRATIS-Mac" or "GRATIS-win." The first time GRATIS is executed, you may be required to select the menu "View" and "Force Reload" to generate all paths to this newly extracted folder. If, after forcing reload, GRATIS does not display as shown in Fig. 3.7, close the window and execute GRATIS again; if needed, force reload once more. The interface will now load. Mac users are also recommended (and in some cases they do need) to place the GRATIS software (and all software products discussed in this book) in their applications folder for newer versions can only read and write files from this location. GRATIS is housed in a secure server location, and all executable files are virus-free but may require administrative permission to be deployed. Finally, each of these executable files for the Mac and Windows operating systems contains all the source code required to modify and/or expand the application's capabilities in the file called "app.R." This portability is methodologically significant. Because GRATIS outputs are generated locally, stored as standalone HTML files, and accompanied by accessible source code, the framework supports not only ease of use but also reproducibility, local control, and long-term inspectability. These same qualities make GRATIS especially compatible with the broader local analytic ecosystem advanced in this book, in which outputs can remain under the researcher's control while being revisited, compared, and interpreted in subsequent stages of analysis.

Discussion: Integration and contribution to mixed-methods research

GRATIS responds to the *integration challenge,* wherein the sum of the equally important qualitative and quantitative phases depicted in Fig. 3.1 is greater than their individual components (Fetters and Freshwater, 2015a, p. 116). As part of this integration, note that GRATIS does not quantify qualitative data; indeed, the qualitative meanings of these (manually or computer-aid-generated) codes are never altered, nor should they be lost or overlooked. Indeed, as we show in Fig. 3.1, as part of the qualitative phase and before moving to the quantitative one, we strongly recommend having a clear initial understanding of the meaning of these codes so that, when the GRATIS outputs are deployed (the aggregated and the time-evolving ones), researchers are prepared to continue with their meaning-building journey and comprehensively engage with the vast wealth of information contained in those interactive outputs.

Another integrative feature of GRATIS is its flexibility to model codes generated via "manual" or machine learning-based text classification procedures; hence, the qualitative phase depicted in Fig. 3.1 may itself be an integrative mixed-methods exercise as well (see González Canché, 2022c, 2022d, 2023a, 2023b).

Integrating GRATIS Outputs into Local Scholarly Brainstorming

One of the distinctive advantages of GRATIS is that its outputs are portable across analytic stages. The time-evolving HTML visualizations, aggregated network depictions, screenshots of key moments, and quote-linked temporal maps are useful not only for identifying processes and trajectories, but also for supporting subsequent interpretive work. In the broader ecosystem developed in this book, these outputs can be brought into ISARI, a local, fully offline, open-source, and multimodal brainstorming environment designed to support memoing, comparison, synthesis, and evidence-grounded writing.

This possibility is especially important because temporal visualizations often reveal patterns whose substantive meaning still requires careful interpretation. Researchers may wish to ask why certain actors or codes become central at particular moments, how a given discourse shifts across time, why some themes emerge, decay, or return, or how nonlinear storytelling should be translated into a coherent explanatory account. Because ISARI can work with textual notes, screenshots, figures, scanned pages, and other image-based analytic artifacts, researchers can upload GRATIS outputs in multiple forms—for example, screenshots of the interactive maps, exported figures, copied text from the HTML information panels, or notes derived from replaying the temporal visualizations—and use them as the basis for local scholarly brainstorming.

This integrative workflow should not be understood as a substitute for methodological rigor. On the contrary, it extends GRATIS's central logic. The outputs remain guides, not conclusions. ISARI can help researchers compare interpretations, generate analytic memos, identify possible explanatory contrasts, and begin drafting evidence-grounded prose, but responsibility for deciding what the temporal patterns mean remains with the researcher. In that sense, GRATIS and ISARI are methodologically aligned: one renders the evolution of information visible; the other helps the researcher think with that evolution under conditions of privacy, interpretive accountability, and local control.

Interactive reviewability as validity or representativeness checks

Based on data science, visualization, and text retrieval tools, the HTML time-evolving interactive outputs that GRATIS automatically renders, can be paused at each step as shown in Figs. 3.4, 3.5, and 3.6—and the speed of these visualizations may be controlled as well with a simple click (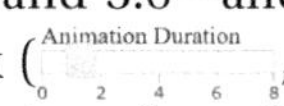

, wherein higher numbers imply slower evolution of connections). More to the point, when clicking on actors, codes, or their connections in these HTML visuals, their resulting attributes (see Table 3.5) are displayed to enhance both the story-telling attributes of our studies, as well as the interactive reviewability capabilities of these visuals.

As part of the orthogonality, reproducibility, and compatibility attributes of TIS, the textual content of each code (i.e., its quote), along with actors' attributes, can be displayed at each step of the process, and snapshots of relevant parts of the evolution of information may be taken for inclusion in final research papers. The selection of these snapshots (see Figs. 3.4, 3.5, and 3.6, for example) represents methodological decisions made by the research team. These decisions are constrained by the fact that, for publication purposes, no HTML outputs *can yet be* embedded in PDFs, which means that researchers should select specific points to highlight their findings and conclusions. *The selection of these snapshots poses a challenge associated with "imposing" our analytic choices on the final product* (i.e., *the final paper to be published).* However, for transparency and validity purposes, before making those decisions, we should seek our participants' input and try to validate whether our selection was optimal in capturing their experiences or whether there may be other snapshots that better represent their stories or perspectives.

Moving beyond our participants' feedback, we also encourage researchers to, *while protecting the identities of their participants*, upload the entire HTML GRATIS outputs to websites like GitHub (https://github.com/) so that readers of their final papers may interact with these outputs and possibly corroborate or even contest the conclusions presented by the research team. Because interactive temporal visualizations may reveal structural, contextual, or sequential clues that could weaken anonymity in small, distinctive, or otherwise sensitive samples, the public sharing of HTML outputs should always be treated as optional and governed by the same principles of confidentiality, data minimization, and participant protection that guided the original qualitative design. In many cases, local storage, restricted sharing, or the use of screenshots and selected snapshots may be methodologically and ethically preferable to full public release (see González Canché, 2025 for a formal discussion of ethics and trust in the age of qualitative data science).

This *interactive reviewability* process may not only strengthen the transparency and validity of our findings, conclusions, and recommendations but also, and as importantly or even more so, may spark other ideas and motivate curiosity among readers interacting with these HTML outputs. In this sense, GRATIS contributes to opening the analytic black box of qualitative research by making the temporal basis of interpretive claims more inspectable, revisitable, and, when ethically appropriate, open to participant and reader scrutiny.

An example of how these outputs may be displayed can be accessed at https://cutt.ly/vbjH1sY. This visual contains the complete evolution of about 40 hours of transcriptions resulting from interviews with 40 faculty members holding *unstable* faculty appointments (see González Canché et al., 2021). These transcripts document these faculty members' strategies for dealing with savings and retirement plans. From this perspective, while protecting the identity of our participants, GRATIS also enhances the transparency of the entire analytic process by allowing researchers and readers to interact with the main outputs and potentially create and recreate their own interpretations and analyses based on the evolution of information and continuity of events—i.e., *interactive reviewability* (Alexander et al., 2019a).

Limitations

Despite GRATIS's contributions to mixed-methods research, this analytic framework does not yet address communication exchanges among participants in focus or working groups, such as the VRM approach presented by Alexander et al. (2019a). Nonetheless, expansions of GRATIS's capabilities may be added to achieve this goal, which remains an area for improvement. In addition, González Canché (2022b) has shown that these analytic approaches are feasible to implement under the umbrella of mapping, organizing, and visualizing interdependent events (MOVIE).

Conclusion

GRATIS contributes to mixed-methods research by focusing on temporal data analysis and visualization that integrates qualitative and quantitative approaches to map the contextualized evolution of participants' contributions. The true value derived from GRATIS can only be attained with the full integration of high-quality qualitative and quantitative phases. Therefore, without quality assurance during data collection, preparation, and (qualitative or quantitative) coding, the tools provided by GRATIS will still fall short in accurately depicting the evolution of information that shapes our research findings and conclusions.

GRATIS integrates data science, visualization, mining, and retrieval tools into the analysis of qualitatively or computer-assisted coded textual data and does so without requiring any computer or statistical programming skills or programming knowledge from the research team. From this view, GRATIS aims to *democratize access to these data science tools*. This data science democratization goal is the main reason we also provide access to GRATIS's source code and documentation (see the subsection called "**Software and Source Code Access**" above) so that interested researchers may build upon GRATIS's current capabilities.

In terms of expansions, note that GRATIS was programmed to follow the order of information contained in the rows of the edgelist (Table 3.2 and Fig. 3.2). We purposefully programmed GRATIS to respect this order so that GRATIS may plot information beyond the linear or chronological code generation depicted in Table 3.2 and Fig. 3.2. As indicated above, researchers may instead purposefully seek to assemble a thematic timeline that does not follow start-to-finish or top-to-bottom code emergence in a document or transcript (as we showed in Table 3.1); rather, researchers may prefer to classify events based on themes (V. Braun and V.Clarke, 2006) regardless of their actual location in the analytic documents (e.g., interview transcript, essay, media posts) or based on their participants' storytelling or reconstruction of events across time, as shown in Table 3.4. To achieve a non-linear story timeline plot, researchers may carefully rearrange their edgelists' rows based on this new plotline. Procedurally, researchers may add a column to their coded datasets that captures the order each code may represent in their analyses. After all codes have been generated, researchers may simply sort or order this database (using Microsoft Excel, for example) based on this new column, going from 1 to N, where N is the last event being recorded given this non-linear or thematic coding scheme. After this sorting is completed and saved, researchers may upload this sorted database to GRATIS and execute GRATIS as described in this paper. The resulting HTML output will highlight this storytelling evolution of events, for, once more, GRATIS will identify this row ordering during the application of TIS.

Regarding other applications of GRATIS, researchers may pursue content analyses of media coverage by specific news outlets, for example. This analysis (as indicated by one of our reviewers) "may show the dynamics of their coverage over time and may help address questions such as: How long do crises echo over time? Does coverage really fade away as soon as there is nothing more scandalous to report on?" Our reviewer also mentioned, "There is certainly a myriad of other content-analysis topics where dynamics over time are of utmost relevance" and may be analyzed with GRATIS.

Finally, note that although the examples presented relied on short timespans, GRATIS can accommodate multiyear projects following the exact same rationale discussed herein. As suggested by one of our reviewers, our reliance on short-time depictions is strategic, as it may serve to combat "the common myth [or assumption] that longitudinal or time-based studies need to stretch over a long period of time." In this study, we demonstrated how GRATIS can accommodate textual data gathered in a variety of qualitative research settings. We also showed that even with short time frames, we can fully leverage the power of time in our sensemaking processes and produce a fully integrative research outcome.

GRATIS may also be used to observe how groups of individuals react to the same prompt. Here we could expose them to the same statement, say "January 6th did not happen" and ask them elaborate on this statement. Then, after coding their responses/narratives, we can apply GRATIS to map the evolution of their trend of thought, likely seeking for similarities and differences based on certain personal and non-personal attributes. This approach may also be employed in psycho- logical lab experiments. Within the broader infrastructure developed in this book, these temporally rich outputs may also serve as structured prompts for local scholarly reasoning in ISARI, where their evolving patterns can be compared, memoed, and translated into evidence-grounded narrative explanation.

Considering that deeper understandings may be achieved with the incorporation of time-based attributes that have shaped our participants' experiences and discourses, let us paraphrase (Alexander et al., 2019a) to state that we believe the time has come to start comprehensively leveraging the relevance of time in the analysis of qualitative evidence. In doing so, GRATIS strengthens transparency, enhances interactive reviewability, and helps open the analytic black box that has traditionally characterized qualitative (González Canché, 2024; González Canché, 2026) and mixed methods inquiry, particularly with respect to how researchers select, justify, and narrate the temporal basis of their findings and conclusions. GRATIS offers the possibility of strengthening interactive reviewability, transparency, clarity, and even accountability in qualitative and mixed methods research, and does so without posing any financial burden.

References

Adam, B. (1990). Time and Social Theory, Cambridge: Polity Press.

Adu, P., 2019. A Step-by-step Guide to Qualitative Data Coding, Routledge.

Alexander, E., Eppler, M.J., Bresciani, S., 2019a. Visual replay methodology: a mixed methods approach for group discussion analysis. J. Mix. Methods Res. 13 (1), 33–51.

Alexander, E., Eppler, M.J., Bresciani, S., 2019b. Visual replay methodology: a mixed methods approach for group discussion analysis. J. Mix. Methods Res. 15586001. 13 (1), 33–51. Available from: http://mmr.sagepub.com/. doi:10.1177/1558689816664479.

Artale, A., Parent, C., Spaccapietra, S., 2007a. Evolving objects in temporal information systems. Ann. Math. Artif. Intell. 50 (1), 5–38.

Artale, A., Parent, C., Spaccapietra, S., 2007b. Evolving objects in temporal information systems. Ann. Math. Artif. Intell. 1012-2443. 50 (1-2), 5–38. doi:10.1007/s10472-007-9068-z.

Auerbach, C., Silverstein, L.B., 2003. Qualitative Data: An Introduction to Coding and Analysis, Vol. 21. NYU press.

Biggs, N., Lloyd, E.K., Wilson, R.J., 1986a. Graph Theory, 1736-1936, Oxford University Press.

Biggs, N., Lloyd, E.K., Wilson, R.J., 1986b. Graph Theory. Oxford University Press, Oxford University Press, pp. 1736–1936.

Borgatti, S.P., 2006. Identifying sets of key players in a social network. Comput. Math. Organ. Theory 12 (1), 21–34.

Braun, V., Clarke, V., 2006. Using thematic analysis in psychology. Qual. Res. Psychol. 3 (2), 77–101.

Butts, C.T., Leslie-Cook, A., Krivitsky, P.N., Bender-deMoll, S., Almquist, Z., Hunter, D.R., Bender-deMoll, M.S. (2020a). Package 'networkDynamic.' https://cran.r-project.org/web/packages/networkDynamic/networkDynamic.pdf

Butts, C.T., Leslie-Cook, A., Krivitsky, P.N., Bender-deMoll, S., Almquist, Z., Hunter, D.R., Bender-deMoll, M.S., Package 'networkDynamic.' (2020b),

Canché, J.Y.B., Odle, T., Edgerton, A., Fahl, C., 2021. Employment Volatility in the Academic Workforce: Implications for Faculty Financial and Retirement Plans. TIAA Institute, Trends and Issues. doi:10.1177/15586898211051584.

Chang, T., DeJonckheere, M., Vydiswaran, V.G.V., Li, J., Buis, L.R., Guetterman, T.C., 2021b. Accelerating mixed methods research with natural language processing of big text data. J. Mixed Methods Res. 15586901. 15 (3), 398–412. Available from: http://mmr.sagepub.com/. doi:10.1177/15586898211021196.

Chang, T., DeJonckheere, M., Vydiswaran, V.V., Li, J., Buis, L.R., Guetterman, T.C., 2021a. Accelerating mixed methods research with natural language processing of big text data. J. Mix. Methods Res. 15586898211021196.

Charmaz, K., 2014a. Constructing Grounded Theory, sage.

Charmaz, K., Constructing grounded theory. (2014b),

Chomicki, J., Toman, D., 1998. Temporal Logic in Information Systems. Springer Science and Business Media LLC, pp. 31–70. doi:10.1007/978-1-4615-5643-5_3.

Chomicki, J., Toman, D., 1998. Temporal logic in information systems. In: Logics for Databases and Information Systems. Springer, Boston, MA, pp. 31–70.

Crossley, N., 2010a. The social world of the network: combining qualitative and quantitative elements in social network analysis. Sociologica 4 (1), 1–34.

Crossley, N., 2010b. The social world of the network: combining qualitative and quantitative elements in social network analysis. Sociologica 4 (1), 1–34.

Csárdi, G., Nepusz, T., 2006a. The igraph software package for complex network research. Complex Syst. 1695 (5), 1–10.

Csárdi, G., Nepusz, T., 2006b. The igraph software package for complex network research. InterJournal, Complex Systems 5.

Denzin, N.K., Strauss, A.L., 1988. Qualitative analysis for social scientists. Contemp. Sociol. 00943061. 17 (3), 430. doi:10.2307/2069712.

Elliott, V., 2018. Thinking about the coding process in qualitative data analysis. Qual. Rep. 23 (11), 2850–2861.

Fetters, M.D., Freshwater, D., 2015a. The 1+ 1= 3 Integration Challenge. J. Mix. Methods Res. 9 (2), 115–117.

Fetters, M.D., Freshwater, D., 2015b. Publishing a methodological mixed methods research article. J. Mix. Methods Res. 9 (3), 203–213.

Freeman, L.C., 1978. Centrality in social networks conceptual clarification. Soc. Netw. 1 (3), 215–239.

González Canché, M.S., 2022a. Network analysis of qualitative data: an integrative software application to visualize and assess similarities in participants' qualitative contributions. J. Mix. Methods Res. 16 (3), 373–377. doi:10.1177/15586898211051584.

González Canché, M.S., 2022b. Mapping, Organizing, and Visualizing Interdependent Events (MOVIE): a rigorous analytic framework and cost-free software application designed to model temporal and dynamic complex realist structures in social research settings. Methodol. Innov. 15 (3), 263–288. doi:10.1177/20597991221119012.

González Canché, M.S. (2022c). Latent code identification [LACOID]: a machine learning-based integrative Framework [and Open-Source Software] to classify big textual data, rebuild contextualized/unaltered meanings, and avoid aggregation Bias [Source Code]. 10.24433/CO.6576994.v1

González Canché, M.S., 2022d. Machine driven classification of open-ended responses (MDCOR): an analytic framework and no-code, free software application to classify longitudinal and cross-sectional text responses in survey and social media research. Code Ocean: A Computational Research Platform. doi:10.24433/CO.5785029.v3.

González Canché, M.S., 2023a. Latent Code Identification (LACOID): a machine learning-based integrative framework [and Open-Source Software] to classify big textual data, rebuild contextualized/unaltered meanings, and avoid aggregation bias. Int. J. Qual. Methods 22. doi:10.1177/16094069221144940.

González Canché, M.S., 2023b. Machine driven classification of open-ended responses (MDCOR): an analytic framework and no-code, free software application to classify longitudinal and cross-sectional text responses in survey and social media research. Expert Syst. Appl. 215. doi:10.1016/j.eswa.2022.119265.

González Canché, M.S., 2023c. Coded Data for Replication of Dynamic Analyses of Qualitative Evidence via GRATIS. Mendeley Data V1. 10.17632/c4jhckhhps.1

González Canché, M.S., 2023. Latent code identification (LACOID): a machine learning-based integrative Framework [and Open-Source Software] to classify big textual data, rebuild contextualized/unaltered meanings, and avoid aggregation bias. Int. J. Qualit. Meth. 16094069. 22. Available from: https://journals.sagepub.com/home/IJQ. doi:10.1177/16094069221144940.

González Canché, M.S., 2024. Graphical retrieval and analysis of temporal information systems (GRATIS): An integrative mixed methodology and open-access software to analyze the (non-)linear chronological evolution of information embedded in textual/qualitative data. J. Mix. Methods Res. 18 (1), 71–107.10.1177/15586898231166968.

González Canché, M.S., 2025. Ethics and trust in social network analysis: how data science and interactive visualizations impact anonymity, confidentiality, consent, and IRB procedures?. In Handbook of Social Network Analysis and Education (pp. 114–131). Edward Elgar Publishing. doi:10.4337/9781035312191.00013.

González Canché, M.S., 2026. Network analysis of qualitative data (NAQD): An analytic framework and software for qualitative data science visualization and hypothesis testing. Int. J. Qual. Methods. 25. doi:10.1177/16094069261439942.

González Canché, M.S., Bae, J.Y., Odle, T., Edgerton, A., Fahl, C., 2021. Employment volatility in the academic workforce: implications for faculty financial and retirement plans. TIAA Institute, Trends and Issues. doi:10.1177/15586898211051584.

Green, T.R., Blandford, A.E., Church, L., Roast, C.R., Clarke, S., 2006a. Cognitive dimensions: achievements, new directions, and open questions. J. Vis. Lang. Comput. 17 (4), 328–365.

Green, T.R.G., Blandford, A.E., Church, L., Roast, C.R., Clarke, S., 2006b. Cognitive dimensions: achievements, new directions, and open questions. J. Vis. Lang. Comput. 1045926X. 17 (4), 328–365. doi:10.1016/j.jvlc.2006.04.004.

Griffiths, T.L., Steyvers, M., 2004. Finding scientific topics. Proc. Natl. Acad. Sci. U. S. A. 101 (suppl 1), 5228–5235. https://www.pnas.org/content/101/suppl%5C_1/5228.

Ho, P., Chen, K., Shao, A., Bao, L., Ai, A., Tarfa, A., Brossard, D., Brown, L., Brauer, M., 2021a. A mixed methods study of public perception of social distancing: integrating qualitative and computational analyses for text data. J. Mix. Methods Res. 15586898211020862.

Ho, P., Chen, K., Shao, A., Bao, L., Angela, A., Tarfa, A., Brossard, D., Brown, L., Brauer, M., 2021b. A mixed methods study of public perception of social distancing: integrating qualitative and computational analyses for text data. J. Mix. Methods Res. 1558-6898. 15 (3), 374–397. doi:10.1177/15586898211020862.

Jensen, C.S., Dyreson, C.E., Bohlen, M., Clifford, J., Elmasri, R., Gadia, S.K., Grandi, F., Hayes, P., Jajodia, S., Käfer, W., Kline, N., Lorentzos, N., Mitsopoulos, Y., Montanari, A., Nonen, D., Peressi, E., Pernici, B., Roddick, J.F., Sarda, N.L., Scalas, M.R., Segev, A., Snodgrass, R.T., Soo, M.D., Tansel, A., Tiberio, P., Wiederhold, G., 1998b. The consensus glossary of temporal database concepts - February 1998 version. Lecture Notes in Computer Science (Including Subseries Lecture Notes in Artificial Intelligence and Lecture Notes in Bioinformatics). 1611-3349. 1399, 367–405. Available from: https://www.springer.com/series/558. doi:10.1007/bfb0053710.

Jensen, C.S., Dyreson, C.E., Böhlen, M., Clifford, J., Elmasri, R., Gadia, S.K., Wiederhold, G., 1998a. The consensus glossary of temporal database concepts—February 1998 version. In: Temporal Databases: Research and Practice. Springer, Berlin, Heidelberg, pp. 367–405.

Kolaczyk, E.D., Csárdi, G., 2014. Statistical Analysis of Network Data with R, Vol. 65. Springer.

Lazega, E., Wasserman, S., Faust, K., 1995. Social Network Analysis: methods and Applications. Revue Française de Sociologie. 00352969. 36 (4), 781. doi:10.2307/3322457.

Lynam, T., Damayanti, R., Rialine Titaley, C., Suharno, N., Bradley, M., Krentel, A., 2020. Reframing integration for mixed methods research. J. Mix. Methods Res. 14 (3), 336–357.

Maroulis, S., Guimerà, R., Petry, H., Stringer, M.J., Gomez, L.M., Amaral, L.A.N., Wilensky, U., 2010b. Complex Systems View of Educational Policy Research. Science. 0036-8075. 330 (6000), 38–39. doi:10.1126/science.1195153.

Maroulis, S., Guimera, R., Petry, H., Stringer, M.J., Gomez, L.M., Amaral, L.A., Wilensky, U., 2010a. Complex systems view of educational policy research. Science 330 (6000), 38–39.

Maxwell, J.A., 2019. Qualitative Research Design SAGE Research Methods Foundations, Sage Publications, Sage Publications.

Maxwell, J.A., 2019. Qualitative research design. In: Atkinson, P., Delamont, S., Cernat, A., Sakshaug, J.W., Williams, R.A. (Eds.), SAGE Research Methods Foundations. Sage Publications.

Miles, M.B., Huberman, A.M., 1994a. Qualitative Data Analysis: An Expanded Sourcebook, Sage.

Miles, M.B., Huberman, A.M., 1994b. Qualitative Data Analysis: An Expanded Sourcebook, Sage.

Mitchell, M., 2006. Complex systems: network thinking. Artif. Intell. 170 (18), 1194–1212.

Neale, B., 2015a. Time and the lifecourse: perspectives from qualitative longitudinal research. Researching the Lifecourse: Critical Reflections from the Social Sciences, 25–42.

Neale, B., Time and the lifecourse: perspectives from qualitative longitudinal research Researching the Lifecourse: critical Reflections from the Social Sciences. Policy Press, United Kingdom Policy Press, United Kingdom (2015b), 25–41, Available from: http://policypress.co.uk/Researching-the-lifecourse.

Neale, B., 2019b. Qualitative Longitudinal Research. SAGE Research Methods Foundations, Sage Publications, Sage Publications.

Neale, B., 2019a. The craft of qualitative longitudinal research. In: Atkinson, P., Delamont, S., Cernat, A., Sakshaug, J.W., Williams, R.A. (Eds.), SAGE Research Methods Foundations. London, UK: Sage Publications. 9781473995444. doi:10.4135/9781036212834.

Poth, C.N., Bulut, O., Aquilina, A.M., Otto, S.J., 2021a. Using Data mining for rapid complex case study descriptions: example of public health briefings during the onset of the COVID-19 pandemic. J. Mix. Methods Res. 15586898211013925.

Poth, C.N., Bulut, O., Aquilina, A.M., Otto, S.J.G., 2021b. Using data mining for rapid complex case study descriptions: example of public health briefings during the onset of the COVID-19 pandemic. Journal of Mixed Methods Research. 15586901. 15 (3), 348–373. Available from: http://mmr.sagepub.com/. doi:10.1177/15586898211013925.

Rubin, P.G., Canché, M.S.G., 2019. Test-flexible admissions policies and student enrollment demographics: examining a public research university. Rev. High. Educ. 10907009. 42 (4), 1337–1371. Available from: http://muse.jhu.edu/article/729346/pdf. doi:10.1353/rhe.2019.0068.

Rubin, P.G., González Canché, M.S., 2019. Test-flexible admissions policies and student enrollment demographics: examining a public research university. Rev. High. Educ. 42 (4), 1337–1371. doi:10.1353/rhe.2019.0068.

Saldaña, J. (2013a). The coding manual for qualitative researchers. sage.

Saldaña, J., The coding manual for qualitative researchers. (2013b),

Shimojima, A. (1996a). On the efficacy of representation. Indiana University. Retrieved from: https://philpapers.org/rec/SHIOTE-2

Shimojima, A., 1996b. On the Efficacy of Representation, Indiana University.

Strauss, A.L., 1987. Qualitative Analysis for Social Scientists, Cambridge university press.

Theodoulidis, C., Loucopoulos, P., Wangler, B., 1991a. A conceptual modelling formalism for temporal database applications. Inf. Syst. 16 (4), 401–416.

Theodoulidis, C., Loucopoulos, P., Wangler, B., 1991b. A conceptual modelling formalism for temporal database applications. Inf. Syst. 03064379. 16 (4), 401–416. doi:10.1016/0306-4379(91)90031-4.

Wasserman, S., Faust, K., 1994. Social Network Analysis: Methods and Applications, Vol. 8. Cambridge University Press.

4

Visual Evolution, Replay, and Integration of Temporal Analytic Systems (VERITAS)

Introduction

Scientific research involves the methodical identification and development of questions that require the systematic collection, analysis, and assessment of (primary or secondary) data to fully, or at least partially, address the questions posed and/or to develop new ones based on the understandings and insights resulting from this scientific endeavor (Çaparlar and Dönmez, 2016; Erol, 2017). Although this definition may apply to studying both natural and social issues, there are important differences in terms of the inherent complexity associated with building common understandings and knowledge. According to Williams and Dyer (2017), with regard to research in the natural sciences, physical laws and known properties of the natural world may help reduce (or at least more or less precisely control) the number of physical or environmental contingencies that may impact research findings and conclusions. On the other hand, when studying the social world, there are multiple, multidimensional, and endlessly dynamic competing systems evolving over time, with fewer opportunities to exert external control, that make social issues, and their corresponding research methodologies, complex (Byrne and Uprichard, 2012; Potter, 1996; Williams and Dyer, 2017).

This inherent complexity of the social world (Byrne and Uprichard, 2012), however, does not mean that such a world is chaotic, random, or unstructured (Williams and Dyer, 2017). Rather, structural forms do emerge and coexist in social settings. It is the emergence, maintenance, and decay of these structures that allow researchers to detect temporary stability and breaking points, providing them with the means to make predictions about continuity and change in social dynamics (Williams and Dyer, 2017; Alexander et al., 2019). Arguably then, the main challenge in the study of the social world consists of developing robust and consistent strategies or tools capable of tracing, mapping, or retrieving these structural forms in order to ultimately model this social complexity. When it comes to group analyses, the mapping of the processes that shape conversations and inform decisions and outcomes, such as inflection points, enables us to capture and replay, from the start to the end of the discussion, how patterns displayed during group debates, meetings (online or face-to-face), workshops, and decision-making or negotiation sessions unfolded.

Data Science, Interactive Visualizations, and Generative AI Tools for the Analysis of Qualitative, Mixed-Methods, and Multimodal Evidence. https://doi.org/10.1016/B978-0-443-21961-0.00001-2

Accordingly, the *overarching purpose* of this study consists of addressing the analytic and methodological challenge of mapping social dynamics by proposing an analytic framework, along with its corresponding software application, designed to map and visualize all participants' contributions and communication exchanges in relation to their specific times of emergence, continuation, and decay, thus illuminating processes and consequences of change or continuity. This analysis then opens the possibility of examining the evolution of trends and dynamics in group interactions, making the resulting reports not only more transparent and accurate but also leading to deeper understandings. In this sense, VERITAS contributes not only to dynamic description, but also to interpretive transparency by making the temporal basis of interactional claims more inspectable, revisitable, and documentable. To achieve this goal of modeling temporal and dynamic structures in the social world, we rely on *complex realism, complex systems, dynamic temporal network analyses, and data science and visualization techniques.* Together, these frameworks (*complex realism* and *complex systems*), along with a novel methodological approach (*dynamic and temporal network modeling*), constitute the foundations of **V**isual **E**volution, **R**eplay, and **I**ntegration of **T**emporal **A**nalytic **S**ystems (VERITAS), a rigorous analytic framework and cost-free software application designed to model complex structures emerging in the simultaneous and dynamic evolution of knowledge generation and communication exchanges in social settings and group discussions.

The relevance of these outputs extends beyond visualization and replay alone. Once the temporal evolution of communication exchanges and coded contributions has been rendered, researchers still need to determine what those patterns mean, which inflection points deserve closer memoing, how hot zones should be interpreted, and how the resulting dynamic structure should be translated into qualitative or mixed-methods argument. In the broader architecture of this book, VERITAS outputs can therefore be understood not as endpoints, but as structured guides that may also be brought into ISARI, a local, fully offline, open-source, and multimodal brainstorming environment designed to support scholarly comparison, memoing, synthesis, and evidence-grounded writing. In this way, VERITAS and ISARI are methodologically aligned: VERITAS makes interactional structure visible through time; ISARI helps researchers think with that structure under conditions of privacy, interpretive accountability, and local control.

To demonstrate VERITAS's performance and rigor in capturing and recreating the dynamic complexity of micro-level interactions, the analyses presented in this study relied on a publicly available data source on conflict resolution built from information gathered in the VERITAS "12 Angry Men." The second dataset is an eight-hour working group meeting sponsored by the National Science Foundation (NSF), wherein participants discussed strategies to address the shortage of faculty of color teaching STEM in public two-year colleges in the United States of America. All data elements, tools, and software are provided with this study to ensure these analyses are fully transparent and reproducible, allowing researchers to replicate the outputs and analyses presented in this chapter. VERITAS can trace and recreate the temporal elements embedded in existing qualitative databases (e.g., those generated with NVivo, MAXQDA, or Atlas.ti), even if they were created without considering their dynamic, time-evolving features, whose meaning-building relevance may help inform policy planning and action.

VERITAS applications are not restricted to small group interactions; instead, they can be extended to meso- or macro-levels, also following synchronous or group-based processes, whose tracing may help inform policy planning and action.

Social research settings and VERITAS's synchronous mechanics

The social case study settings discussed in this study were strategically selected to showcase the analytic power of VERITAS in capturing the structures contained in synchronous events, that is, those events evolving in the same place where researchers can observe processes and communication exchanges in real time. More generally, these synchronous events occur in the same places, whether virtual or in person, and at the same times wherein direct communication among participants is realized. What is common about these events is that, since communication is taking place among participants, the possibility of influence is real and may be observed and mapped.

Although the use of VERITAS is not limited to *qualitative social science research*, we use this research setting as the main case study for VERITAS for two interrelated reasons. First, this type of research is poised with the goal of understanding processes and reasons from a myriad of multiple and multilevel perspectives evolving over time almost simultaneously. As depicted in the introductory section, this complexity is typical of complex realism (Byrne and Uprichard, 2012; Williams and Dyer, 2017), and these types of complex systems are what VERITAS was designed to model. Second, *qualitative social science research has typically not relied on sophisticated, innovative, and cutting-edge data science and visualization tools*, certainly not at a comparable level to quantitative social science research. From this perspective, although VERITAS principles may also be applied to quantitative research settings, its application to qualitative research not only would demonstrate its analytic power in modeling the inherent complexity of capturing structural forms emerging in processes over time, but also, and perhaps even more importantly, qualitative researchers may likely benefit the most by applying VERITAS given the enormous complexity and multiplicity of moving parts characteristic of this type of research. Relatedly, as briefly noted below, currently available software for qualitative research is not yet capable of mapping the dynamic synchronous evolution of processes and reasons. By addressing this gap, VERITAS is likely to have a meaningful impact on the modeling of dynamic, complex structures embedded in (qualitative) social research settings.

Qualitative social science research

Anyone who has conducted even small qualitative studies (e.g., involving a dozen or more participants) knows that they generate an astonishing amount of data. Regardless of the specific methods of data gathering employed (e.g., diaries, interviews, debates, focus groups), when research memos have been generated, data collection has concluded, and data preparation (e.g., transcription) has been completed, the process of fracturing the data (Denzin and Strauss, 1988) into *labeled* categories or *codes* to ease their comparison and sense-making (Miles and

Huberman, 1994) typically renders hundreds or even thousands of pieces of valuable, meaningful, and ultimately temporally contextualized information (Coffey and Atkinson, 1996; Neale, 2019) that must be analyzed.

These labeled or coded pieces of information may account for text passages (quotes), images, photographs, or audio and video recordings that capture key concepts provided by research participants. These categories or codes may then be used to discover patterns and relations based on participants' shared or even disputed understandings of the issue under study (Smith, 1978). The resulting datasets are neither small nor easy to analyze. In addition to the abundance of information that may lead to thick descriptions (Holloway, 1997; Lincoln et al., 1985), these datasets *inherently* contain *temporal elements* (Neale, 2019) that can be used to better understand the specific contexts during which participants provided the insights that eventually inform our research findings and conclusions.

To be clear, note that the temporal attributes of qualitative data are present in all resulting databases regardless of the prospective (continuous and ongoing) or retrospective (historical) data collection strategies employed (Neale, 2019). Yet, despite their prevalence and meaning-building relevance, these temporal elements have largely been ignored or remained peripheral in most qualitative analyses and visualizations, not only due to the complexity of these data structures but also due to the lack of easy-to-use analytic tools, as we further elaborate below.

With this analytic complexity in mind, as demonstrated in this study, we can use VERITAS to discover patterns and relationships based on the temporal evolution of these labeled categories, highlighting, for example, participants' *shared or even disputed understandings* of the issue under study (Smith, 1978) as well as to unveil or capture *processes or reasons* as they were emerging during data collection (Alexander et al., 2019).

Based on this brief discussion, the main contribution of VERITAS is that this analytic framework focuses on capturing the *temporal elements* contained in social research settings (Neale, 2019), which are, of course, also present in the resulting labeled or coded pieces of information. In other words, VERITAS leverages the analytic power of these temporal elements to strengthen our understanding of the specific contexts during which participants provided the insights that eventually informed the research findings and conclusions in social research settings—as opposed to the current prevalent approach that, to a great extent or even completely, ignores these temporal elements and therefore discards their analytic power in modeling the dynamic complexity of social interactions.

VERITAS's synchronous mechanics

To contextualize the discussion of the conceptual lenses and methodological approaches described below, this subsection elaborates on the mechanics behind VERITAS synchronous settings. VERITAS captures the dynamicity of the complex social world by modeling the natural emergence of valuable information (key concepts, categories, or codes) produced by participants, regardless of the specific methods of data collection employed. Specifically, VERITAS can be applied to labels, codes, or categories obtained from group settings such as focus groups, debates, Zoom meetings, conversation analyses, and participant observations. Nonetheless, this framework may also be applied to archival or historical sources or documents, including blogs, social media interactions, news developments, or other forms of content analysis involving rigorous qualitative (or quantitative, via topic modeling, for example [see González Canché, 2022a for examples on quantitative coding in Mixed Methods Research and Gonzalez Canche (2022b) for an application of network analyses and hypothesis testing using "static" qualitative data]) coding/categorization. VERITAS builds from these *chronologically* categorized pieces of information to recreate the contexts and the flow of events over time and processes (Neale, 2019) that lead to the emergence of shared understandings and even potential disagreements embedded in the creation and recreation of discourses and knowledge (Alexander et al., 2019). This analytical flexibility allows VERITAS to handle synchronous, dynamic, temporal events in the form of group interactions. In a nutshell, VERITAS provides the means to analyze the processes that shape conversations by integrating the accuracy of visually elicited quantitative counts of contributions (ideas, opinions) and the thoughtfulness of qualitative reflections.

Accordingly, VERITAS addresses the goal of capturing and replaying, from the start to the end of the discussion, how patterns displayed during group debates, meetings (online or face-to-face), workshops, and decision-making or negotiation sessions unfolded. All participants' contributions and communication exchanges can be traced to specific times of emergence, continuation, and decay, which illuminate processes and consequences of change or continuity and open the possibility of analyzing the evolution of trends and dynamics (Artale et al., 2007) in group interactions (Alexander et al., 2019).

Synchronous temporal events

In synchronous events, VERITAS can be used to capture and recreate group dynamic interactions and communication exchanges taking place in real time, during juxtaposed and continuous time slots, and in the same space (even if this space is virtual, such as in a video call or Zoom meeting). Synchronous settings involve information and message exchanges (i.e., events) among research participants in real time, such as in focus groups, video calls or video-conferences, and more generally in any type of observation involving more than one research participant, where interactions are observed and there is a possibility of directly influencing the group's dynamics (Alexander et al., 2019). That is, these synchronous events involve any form of working groups or group interactions. In these synchronous instances, participants' actual exchanges of potentially interdependent and interrelated ideas and messages can be observed and recreated in a time-lapse.

In these settings, we can observe the first time a consequential concept was provided. This information allows us to identify who introduced this concept and potentially analyze more deeply the reasons or factors that led to it becoming consequential. For example, it may be that the originator of this concept held a privileged position in the group from a hierarchical perspective. It may also be that hierarchies did not matter and that the idea was groundbreaking and relevant in its own right, as we further discuss in our applied examples.

In sum, synchronous settings imply that direct communication exchanges among participants take place, thereby maximizing the potential for directly influencing one another's contributions.

Hot zones

The simultaneous depiction of how the information provided by research participants evolved as a group or network throughout the data collection process enables the detection of *hot zones* (Alexander et al., 2019) that signal the natural emergence of clusters of information, perspectives, and beliefs among participants. In other words, in this asynchronous analysis, to the extent that researchers (or moderators) allow participants to express their stories and episodic memories with enough detail and freedom by not being too restrictive with the questions asked or how the groups are moderated, categories of information or codes will organically evolve, depicting similarities but also differences in the evolution of codes or categories. (This organic and dynamic emergence may take place even when using semistructured interview protocols.)

This resulting depiction of hot zones via interactive and dynamic visualizations highlights structures resulting from subgroup formations and decays over time, as well as the recreation of the dynamic evolution of knowledge generation in social research settings. Finally, by highlighting the synchronous evolution of information, VERITAS not only provides greater transparency to the research analysis but also makes this process more interactive. Interactivity is achieved by enabling readers to create and recreate their own understanding of the evolution of these events and the validity implications of such an understanding of this evolution (Alexander et al., 2019).

Premise, purpose, and significance

Premise

The main premise of this framework is that the dynamic, temporal attributes of *social research data* are present in all resulting *qualitatively labeled or coded* databases, regardless of the prospective (continuous and ongoing) or retrospective (historical) data collection strategies employed (Neale, 2019; Alexander et al., 2019). Despite the prevalence and meaning-building relevance of these temporal elements, they have largely been ignored or remained peripheral in most qualitative analyses and visualizations. An important reason why these temporal pieces of data have been underutilized is the current lack of analytic tools that enable researchers to straightforwardly map, organize, and visualize the emergence and decay of these *dynamic temporal elements*. Indeed, even the most recent advancements in tools for visualizing qualitative data, as depicted by the CAQDAS Networking Project (see https://www.surrey.ac.uk/computer-assisted-qualitative-data-analysis/resources/software-information), typically stop at presenting a *snapshot in time* of shared meanings, instead of mapping the evolution of information, connections, and knowledge construction taking place in social research settings.

From a methodological perspective, the analysis of temporal elements in social research is currently a daunting endeavor, given the multiplicity of pieces of information moving simultaneously within the complexity of human interaction, knowledge production, and exchange (Byrne and Uprichard, 2012; Potter, 1996; Williams and Dyer, 2017). Relatedly, even when analytic tools are available to achieve the goal of mapping the emergence and decay of structure and information, *computer programming requirements* pose an almost insurmountable hurdle for the mainstream use of these analytic tools by the vast majority of social science researchers, particularly those interested in mixed methods and qualitative research.

Purpose

Based on the discussion presented so far, the twofold purpose of this methodologically innovative chapter is to provide an analytic framework—VERITAS—that dynamically captures, in time-lapse, the meaning-building processes followed by research participants in social research settings in both prospective and/or retrospective contexts. In addition, this paper seeks to *democratize access to sophisticated data science, visualization, and text retrieval tools* by removing all computer programming requirements to implement this groundbreaking analytic framework at no financial cost. Indeed, the distribution of the software that implements VERITAS is being made completely open-access and cost-free. In a nutshell, informed by complex realism and complex systems theory, operationalized via temporal dynamic network modeling, VERITAS enables analysts to trace and study how elements and events of synchronous research (e.g., focus group discussions, video calls or virtual meetings, class observations), or any setting where communication exchanges may be observed, evolve with a clear emphasis on visualizing the temporal emergence, stabilization, and decline of information as provided by research participants in the social settings studied.

Significance

Because VERITAS was intentionally built to work with the output resulting from currently available software tools for qualitative analysis (e.g., ATLAS.ti, NVivo, or MAXQDA), most qualitatively labeled, coded, or categorized datasets can be analyzed with VERITAS with *minimal or no extra* data *cleaning* or *preparation*. The most significant aspect of VERITAS as an analytic framework is that, as detailed next, VERITAS can be applied to ATLAS.ti, NVivo, or MAXQDA. (Note that VERITAS is purposefully made available free of charge. In line with the goal of expanding access to data science methods and analytic tools completely cost-free, another product of this paper is the development and distribution of a Microsoft Word Macro that enables using Word as a coding or labeling device, similar to specialized and commercial software like ATLAS.ti, NVivo, or MAXQDA. This macro can be accessed here: https://github.com/democratizing-data-science/word_as_a_coding_tool/, with installation instructions described on that website. To showcase the entire analytic process followed by VERITAS, from coding to visualization deployment, this study demonstrates the use of this macro to create labels/codes using two publicly available videos described in the methods section along with their coded transcripts. Readers interested in creating their own analyses may decide to eliminate the codes provided and create their own to conduct their own entire analytic processes.) VERITAS outputs *can be utilized even if* researchers built their datasets or created their codes or labels *without considering* their inherent temporal components. This added flexibility is relevant because not only can VERITAS be applied to future data collections or projects, but it can also be applied to already analyzed qualitatively coded databases. If VERITAS is used to *reanalyze* previous databases, possibly from already published studies, researchers may be surprised to see that new insights will surely be gained with the capturing and visualization of the temporal elements embedded in these databases. Indeed, it is quite likely that the resulting outputs, which fully depict the evolution of connections and information in time-lapse, may allow authors of published studies to write new papers with those same data while reaching new understandings and potentially strengthened conclusions and takeaways.

Another reason for VERITAS's significance is its added transparency to the analytic process. This added transparency is relevant because qualitative methodologists have argued for greater disclosure of how qualitative researchers conduct their analyses and reach their conclusions (Attride-Stirling, 2001; Lee and Fielding, 1996). Given that VERITAS focuses on the temporal evolution of information, the resulting analyses and visualizations depicting temporal components may enable everyone interacting with the resulting outputs to observe firsthand the flow of events (Smith, 1978) as they emerged over time in their actual contexts (Alexander et al., 2019). Moreover, since the outputs resulting from the proposed framework are interactive and offer detailed information about the contributions of each participant (including the content of the message or text quote, as depicted in the findings sections), research participants can observe their individual "performance" or contributions as well as their contextualized contributions with respect to the entire group and offer feedback or new insights from this observation exercise. In both instances, as an added validity check, researchers may ask participants to react to this "performance" to see if they are satisfied or agree with the depictions generated by VERITAS, a strategy referred to as interactive reviewability (Alexander et al., 2019). This participant-facing review is especially important because the validity of dynamic replay depends not only on structural coherence, but also on whether the reconstructed sequence of exchanges is recognizable to those whose interactions are being represented. Finally, readers themselves can form their own assessment of the thought processes and conclusions reached by the researchers through the observation of the interactive and dynamic outputs. Note, however, that since researchers themselves have been involved in data collection, cleaning, and classification, their input and expertise remain the most important assets in the analytic process, but with the outputs generated by VERITAS, this process is much more transparent. For that reason, VERITAS outputs should be treated as guides to interpretation rather than as interpretations in themselves. The replay of structure can show how communication and knowledge evolved, but researchers must still decide what those trajectories mean, how influence should be understood, and how those patterns relate to the broader aims of the inquiry.

All VERITAS' outputs are also portable across analytic stages. The replayable HTML visualizations, aggregate and temporally disaggregated sociograms, screenshots of key moments, and quote-linked interaction maps are useful not only for identifying dynamic structure, but also for supporting subsequent interpretive work. In the broader ecosystem developed in this book, researchers may bring these outputs into ISARI by uploading screenshots of the interactive maps, copied text from information panels, exported figures, or notes derived from replaying the temporal visualizations. Because ISARI can operate locally and work with both textual and image-based artifacts, these VERITAS outputs can become the basis for scholarly brainstorming, memoing, comparison of alternative interpretations, and evidence-grounded writing without moving sensitive materials outside the researcher's control.

Conceptual lenses: Complex realism and complex systems

The main conceptual tenets of VERITAS are Complex Realism and Complex Systems. The following discussion introduces the main concepts of complex realism, and subsequently, we will discuss complex systems and the operationalization of both lenses through temporal and dynamic network modeling.

Complex realism

As depicted in the introduction, the study of the social world is a challenging endeavor given the multiple, multidimensional, and endlessly dynamic competing systems evolving over time (Byrne and Uprichard, 2012; Potter, 1996; Williams and Dyer, 2017). VERITAS aligns with the notion of *general complexity* as presented by (Byrne and Uprichard, 2012), which, by involving nonlinearity in networks (or groups) of nested systems, offers reliable and valid models that represent *true social complex systems* rather than *simulations of these systems*—as in the case of *restricted complexity* (Byrne and Uprichard, 2012). VERITAS then seeks to capture and produce sociologically useful and meaningful descriptions of social world events in what George and Bennett, 2005, cited in Byrne and Uprichard, 2012) referred to as "process tracing." In the original work of George and Bennett, process tracing, as a method, "attempts to identify the intervening causal process—the causal chain and causal mechanism—between an independent variable (or variables) and the outcome of the dependent variable" (p. 206). In VERITAS, process tracing is operationalized with the visual depiction of knowledge generation and communication exchanges as observed in our social research settings. Certainly, the data input for VERITAS can be purposefully collected to achieve the goal of process tracing as depicted by George and Bennett, but VERITAS's intention in this paper is not to depict causal chains among dependent and independent variables. Rather, it aims to offer a transparent, reliable, rigorous, and fully reproducible depiction of how the information exchange and knowledge generation process evolved in our social research settings. From this view, by aiming to carefully reconstruct the history of knowledge evolution, VERITAS aligns more closely with the depiction of complex realism presented by (Byrne and Uprichard, 2012).

In VERITAS, the concept of "system traces" proposed by (Byrne and Uprichard, 2012) takes form as a network. In synchronous systems, there are multiple single cases brought together as part of the same discussion or event (e.g., multiple individual perspectives that may potentially converge into agreements or disputes). Each of these cases may be operationalized as a complex system (resulting in as many systems as there are research participants), and the analytic power of VERITAS consists of depicting all the independently or codependently created traces of each of these systems together in a simultaneous description of the "traces of the systems' trajectories through time" (Byrne and Uprichard, 2012).

Put differently, in synchronous settings, each group observed consists of a complex system configured by more than one knowledge producer whose traces, attributes, relationships, and coinfluence may be captured and reconstructed through time, to obtain a systematic and organized depiction of the evolution of relationships, dependence and influence of knowledge production and conflict and resolutions over time (Byrne and Uprichard, 2012; Alexander et al., 2019). In synchronous cases, the emphasis remains on accurately capturing these traces in the system(s), and the resulting outputs display all relationships and knowledge configurations that took place in each system(s)—this configuration resembles a dynamic, temporal multidimensional contingency table (or rather multiple tables), similar to the time-aggregated *truth table* produced by qualitative comparative analysis. The multiplicity of truth tables produced by VERITAS aligns with the depiction of the system(s) as having emergent properties through time. Accordingly, each trace captured at time *t* renders a truth table (network matrix) or contingency table. The subsequent trace observation at time, therefore renders another truth table, wherein the traces observed at time *t* may be the same or may have changed. From this perspective, "emergent effects can be explored [through time] by examining attribute [trace] configurations together these depictions help [us] rethink what difference and change, and conversely similarity and continuity, might mean with respect to exploring patterns of change and continuity within our data" (Byrne and Uprichard, 2012, p. 8). Therefore, and aligned with complex realism, in VERITAS both patterns of continuity and patterns of change are highlighted or visualized with interactive outputs.

In VERITAS, social categories (or classifications) precede all comparisons of patterns of continuity and patterns of change. As (Byrne and Uprichard, 2012) stated, our classifications are an essential part of the analytic process, for these systematic comparisons would not be possible without some form of measurement or classification. From this perspective, then, to the extent that our classification approach is accurate in capturing social complexity, our resulting analyses and outputs obtained from VERITAS will also accurately capture this complexity. This means that VERITAS strengthens rather than relaxes methodological responsibility. The dynamic maps do not generate meaning independently of coding; they formalize and render visible the temporal structure implied by the coding scheme and the tracing of interactional exchanges. For that reason, VERITAS analyses are strongest when coding is conducted transparently, systematically, and with clear attention to the conceptual relationships among actors, messages, codes, temporal sequencing, and context. If those upstream analytic decisions are weak, inconsistent, or poorly documented, the resulting visualizations may still look compelling while misrepresenting the actual dynamics of the social setting being studied. Conversely, when coding is rigorous and theoretically grounded, VERITAS can strengthen transparency, replayability, and interpretive accountability by making visible the structure that researchers are claiming to have observed. However, if our classification approach is weak or unsystematic, then our resulting analyses will fall short in providing useful evidence of the social setting studied. This latter point refers to the inherent challenges of coding textual data into labels or classes that accurately reflect the meaning of our participants' comments and contributions. Here is where, in projects with massive amounts of data, the use of machine learning or artificial intelligence tools may be useful in avoiding human inconsistencies or mistakes when creating those codes or labels. This is similar to the discussion of using GRATIS with machine-learned classes or codes in a previous chapter of this book.

Complex systems

The complex, multifaceted, and temporal-based evidence of participants' realities may be mapped (by identifying its elements through classification methods), organized (sequentially, chronologically, or otherwise), and visualized as this evidence evolves (or is chronologically generated). From this perspective, VERITAS consistently and systematically reproduces the classified evidence recovered from our synchronous or asynchronous research settings. VERITAS, then, *is an analytic framework and tool designed to ease the understanding of individually produced or interactively generated events by tracing and uncovering the processes that may have affected participants' experiences, outcomes, and standpoints.* Aligned with a complex realist perspective, VERITAS is founded on the notion that behaviors, meanings, and processes emerging during social interactions, along with their temporally contextualized *connections,* can be captured as they emerge and can be visualized and reconstructed following time-lapse and network principles. It is this organic evolution of meanings, behaviors, and connections over time that makes the use of complex systems, as operationalized through *temporal and dynamic network theory and methods* (González Canché, 2019; Wolf et al., 2021), a natural complement to complex realism. This approach allows for tracing, in time-lapse, these complex and dynamic exchanges of information and processes of knowledge production and sense-making.

Complex systems and network theory and methods

The elements and properties of complex systems may be better understood by first describing the notion of a system (Byrne and Callaghan, 2014; Mitchell, 2006; Zeng et al., 2017). In its simplest form, a system consists of a set of elements and relations. As these elements and relations grow in number, complexity, and variety, local interactions will organically emerge and yield new levels of organization. The relationships among these elements are nonlinear, indicating that a change in the number of elements and/or connections does not necessarily result in a change in the output or outcome. For example, if during a working group observation we notice that some members had to leave before the meeting ended (see (González Canché, 2019), the observed dynamics may not necessarily be affected by these departures—if the party who left was rather quiet and nonparticipative. Rather, the roles played by the members who left (and those who stayed) are what contribute to changes in output or dynamics. In other words, nonlinearity indicates that the output is not proportional to input changes (Sanchez et al., 2019).

Another important attribute of complex systems is feedback (Zeng et al., 2017), wherein outputs generated as the process develops can be routed back as inputs to trigger more outputs. This can be observed in information exchanges where meanings can be cogenerated or, in some cases, driven mainly by one participant (or subgroup) who is influential enough to convince the entire group to adopt a particular viewpoint. However, the evolution of this process of change (i.e., convincing other participants) is complex. For instance, in one of our applied examples, we will observe the interactions taking place among 12 jurors (from the film "12 Angry Men"), wherein at the beginning of their journey, only one of these jurors was not convinced that the accused was guilty and eventually, all of them changed their vote from guilty to not guilty. Accordingly, complex systems also involve the properties of spontaneous order and the emergence of new levels of organization. These properties refer to the natural emergence of subgroups or subtopics absent from central control (Zeng et al., 2017). That is, even if the groups or organisms observed are subjected to some form of external control (i.e., group moderators or individuals holding some form of hierarchical status), subgroup formation based on the numerosity of relationships and potential changes in the number of elements (addition of new elements or departure of some participants) will happen, eventually leading to order, organizational fracturing, and reconfiguration.

As depicted above, these complex trajectories of patterns of continuity and patterns of change (Byrne and Uprichard, 2012) follow a similar structure, eventually reaching a climax in terms of information production, and then decay, based on the potential extenuation of episodic memories, information innovativeness, or even timeliness. With this information, VERITAS is capable of highlighting or capturing the emergence of spontaneous order or hot zones (Alexander et al., 2019) and subgroups or levels of organization. VERITAS, as an analytic strategy, can also highlight how certain participants took longer to extenuate their episodic memories, the usefulness of their contributions—and whether the production of more information actually led to a better understanding of the phenomenon under study. Certainly, then, more information does not necessarily mean more meaningful contributions if these longer pieces of information are redundant, repetitive, or even irrelevant.

Complex systems thinking, then, is useful for analyzing different types of relationships and communities interacting synchronously or asynchronously. Its use allows for the visualization of the structure of the system (group dynamics) or systems (multiple contributions) being studied. Because these systems are configured by elements and connections (Maroulis et al., 2010), complex systems can naturally be operationalized as dynamic and temporal *complex adaptive networks*. Indeed, although the examples so far have centered around social settings, more common applications of complex systems include the study of cellular processes, the immune system, and traffic and transportation systems (Maroulis et al., 2010; Mitchell, 2006; Zeng et al., 2017). In each of these applications, the emphasis is on how the parts configuring these systems are interconnected and can be represented as temporal dynamic networks. Relatedly, given that *network principles and methods* operationalize VERITAS—as well as complex realism and complex systems—the following section describes the main features of temporal and dynamic network modeling according to their application in this proposed framework.

Methods: temporal dynamic network thinking and modeling

Network thinking and network modeling are particularly useful to "deal with complex systems in the real world" (Mitchell, 2006, p. 1199). Temporal network thinking and modeling help researchers make sense of multiple, multidimensional, and dynamic competing classifications (Byrne and Uprichard, 2012; Potter, 1996; Williams and Dyer, 2017) evolving synchronously or asynchronously in time and space (González Canché, 2019). Operationally, dynamic network thinking emphasizes relations among specific events at a specific time (Butts et al., 2020), including connections among individuals and their actions or among broader social categories or classifications (Byrne and Uprichard, 2012) as defined or measured by researchers (Maroulis et al., 2010).

Formally, networks are a collection of potentially interactive units (Kolaczyk and Csárdi, 2014; Mitchell, 2006; Lazega et al., 1995). These units are typically referred to as *nodes* or *vertices* (e.g., actors, participants, or entities with which participants may interact or to which they may be ascribed), and the connections resulting from their interactions are referred to as *edges* or *links* (Lazega et al., 1995). When the units configuring a given network are of the same type and hierarchy (e.g., students interacting with other students), they form a *one-mode network*. When the units configuring the network are different (e.g., professors ascribed to their employing universities or ascribed to a particular set of beliefs), they form *two-mode networks*.

VERITAS relies on both one-mode and two-mode networks. When the analyses focus on the information provided by participants (i.e., codes or labels), the main form of relationships captured is among participants and the information (i.e., codes) they provide. This form of analysis makes it possible to depict participants across time who cluster around certain topics or viewpoints and those who do not (Alexander et al., 2019). These relationships can be expressed as $Actor_{i,t} \rightarrow Code_{j,t}$, where the subscript allows us to trace the time when each relationship was realized. Moreover, VERITAS may also focus on the synchronous analysis of group dynamics and discourse patterns (Alexander et al., 2019), with emphasis on participants' communication exchanges—rather than on the production of ideas, the analyses take the form $Actor_{i,t} \rightarrow Actor_{j,t}$ (or one-mode network). This latter specification or operationalization enables capturing the directed exchange of information among these actors, also at a particular time t.

Data format and transformations

The time dimension captured by VERITAS makes the storage of information in traditional matrix formats (or rectangular databases), where participants are in the rows and variables (attributes, social categories, or classifications) are in the columns, impractical. This is because, as described in the conceptual lenses section, each time component may be expressed as a *truth table*, leading to as many *truth tables* as time traces are captured in the system(s). To address this complexity, VERITAS instead initially stores all time traces of information in two main columns: sender and receiver connections, wherein each of these connections were observed at time t. Sender is the provider or source of information (or $Actor_i$ as noted above), and receiver may be another participant in the group discussion being observed (or $Actor_j$ as also noted above), to whom the sender directed the information captured at time t. However, in a two-mode network, where the emphasis is placed on tracing the evolution of information provided by participants, the receiver column accounts for the labels or social categories (or classifications) that enable us to trace the evolution of events at time. This information was depicted as $Code_j$ above. Unless otherwise specified by users, the temporal component is retrieved from the chronological appearance of these connections in a database—as detailed in the examples below.

These two columns (sender and receiver) are referred to as a collection of links or edges, also known as *edgelists* (Csárdi and Nepusz, 2006; Lazega et al., 1995). However, recall that traces of these events over time t require us to rely on contingency tables captured at each point in time. VERITAS was designed to capture this multiplicity of *truth tables* based on advanced feature engineering techniques in the form (see González Canché, 2019)

$$\text{edgelist} \rightarrow \text{graph} \rightarrow \text{matrix} \quad (1)$$

where, from an edgelist in the forms or $Actor_{i,t} \rightarrow Actor_{j,t}$, VERITAS retrieves the matrix version of these contingency tables (affectively a myriad of truth tables) at each point in time (see Fig. 4.1 explained below).

In addition to its methodological properties, another value added of (effectively temporal) edgelists is that this data format is the default output of all widely available qualitative coding software (e.g., ATLAS.ti, NVivo, MAXQDA, or R for Qualitative Data Analysis). Accordingly, when applied to qualitative data sources, no extra cleaning is required to use VERITAS. Nevertheless, note that as part of this study, we are providing a Microsoft Word macro that mirrors the coding process followed by the ATLAS.ti, NVivo, MAXQDA, and RQDA software applications. To exemplify this process, assume we have two participants, each debating about what the policy of the United States in Afghanistan should be—more on this case selection below. Access to the entire *coded/labeled* transcript of this analysis, along with the data source, is provided in the application section below; however, for now, let us simply focus on the first three relationships per expert (i.e., human actors) recorded in that transcript.

These relationships are shown in Fig. 4.1 To maximize the chance that no extra costs are associated with the replication of these procedures, we have also developed a Microsoft Word macro that works in both Mac and Windows operating systems. This macro can be accessed here https://github.com/democratizing-data-science/word_as_a_coding_tool/ with installations instructions for Mac and PC described in that site. In short, this macro was designed to use Word as a coding device. This macro transforms all comments (added by selecting a text, Review menu, New Comment ) in a Word document (observation or interview transcript, for example) into a table that is automatically exported to Microsoft Excel (To load this table to VERITAS, please save it first as a comma separated value file, which is the data format read by VERITAS.). The execution of this macro and the example of this table is shown also in Fig. 4.1 and contains the following columns:

- Page and line: the page number and line number where the quote (selected text in the transcript) associated with a comment (i.e., code) can be found, to ease access to the original information.
- Quote: content of the code or label
- Author_or_coder_name: information about the coder (retrieved from Microsoft's username)
- Coding_date: date when the code or label was created
- Sender: The provider of information, which we have referred to as $Actor_i$
- Receiver: The label or code, referred to as $Code_j$ above, or the target of information, referred to as $Actor_j$ (As depicted in, the result of exporting these codes to Excel will result

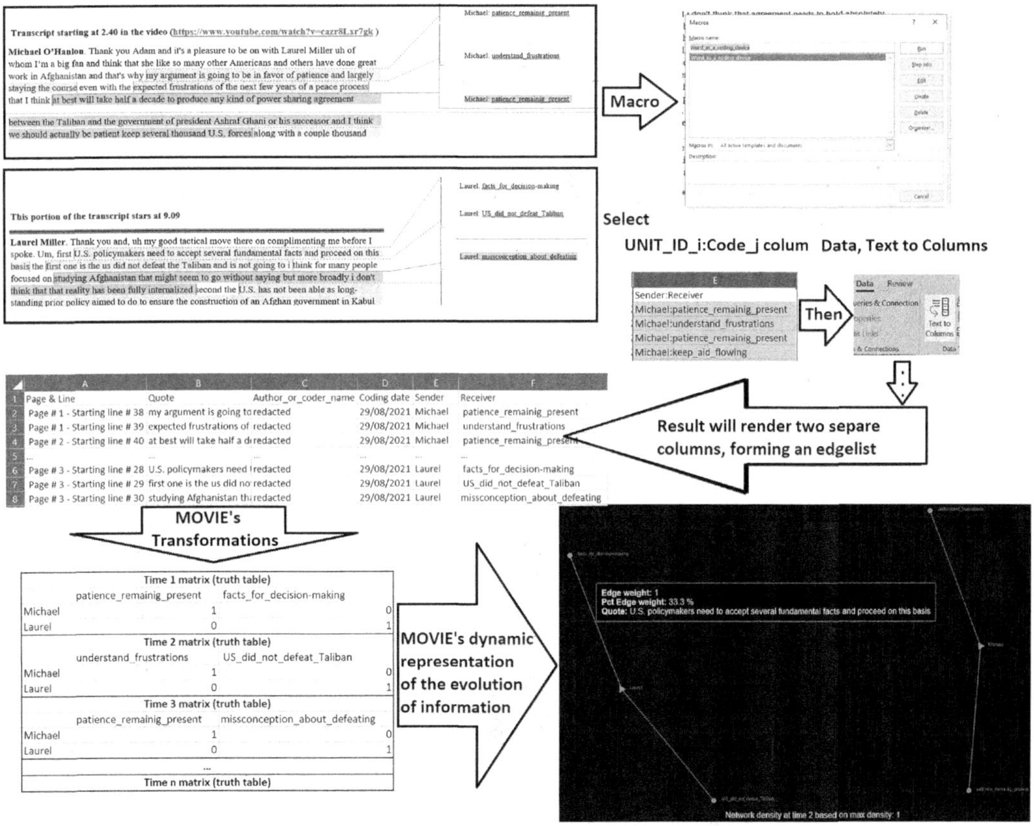

FIG. 4.1 Conceptual representation of data formats and the flow of the analytic process.

in a column in the form: "Sender:Receiver." To separate the content of this column, select the entire column, then select Data, Text to Columns.)

Accordingly, the edge list or list of connections in this table that would allow us to trace the evolution of information is contained in the columns "Sender" and "Receiver."

Procedurally, each comment starts with the name of the actor (Laurel and Michael, in this case) followed by a colon (or ":"). The colon serves as a text delimiter to separate the columns Sender and Receiver, which by default are rendered as one column by our macro. To separate the content of this column, select the entire column, then choose Data, Text to Columns, and select ":" (without quotes) as the delimiter—see Fig. 4.1. Note that the indicator of time is automatically computed by VERITAS and is inferred from the row number of the Sender and Receivercolumns, capturing the chronological emergence of the codes or labels in the transcript. For example, row number 2 in Fig. 4.1 indicates that Michael's first recorded contribution (as shown in his transcript) was "patience_remaining_present." Similarly, for Laurel, her first contribution was "facts_for_decision_making," stored in row 6 in this example. VERITAS was designed to detect the minimum row number per actor's contribution. Specifically, had Laurel's first contribution been in row 20, this value would not have affected the chronological mapping of her contributions. Relatedly, the three dots after the third contribution of Michael (row 5) were added simply to symbolize that Michael had more than three contributions in the original transcript—these dots are not added in the database loaded to VERITAS. The actual database loaded to VERITAS that renders Fig. 4.2 can be accessed here: https://cutt.ly/QWgtf3u.

Finally, note that our macro allows and encourages users to be explicit about the "tone" or "stance" of participants when creating the labels. For example, if a given text and the associated audio or video evidence indicate that the participant who provided this text was "angry" or "agitated," we could add this information when creating the code or label as follows. Let us assume "actor 1" provided "text 1" and that she was "happy" when sharing this information. In our macro, the resulting tone or stance may be captured by adding a comment using the following structure: "actor 1:code j: happy". The period outside the quotation marks indicates that we did not need to add that period in that comment. The resulting cell decomposition employed by our macro results in three columns. In the first, we have "Sender"; in the second, we have "Receiver"; and in the third, we have "Tone or Stance." If we did not include the latter information, the column "Tone or Stance" will be empty, but if we added information, this will populate this column. In our explanation below, we will show how the addition of "tone" or "stance" may be useful in adding more context to our analyses.

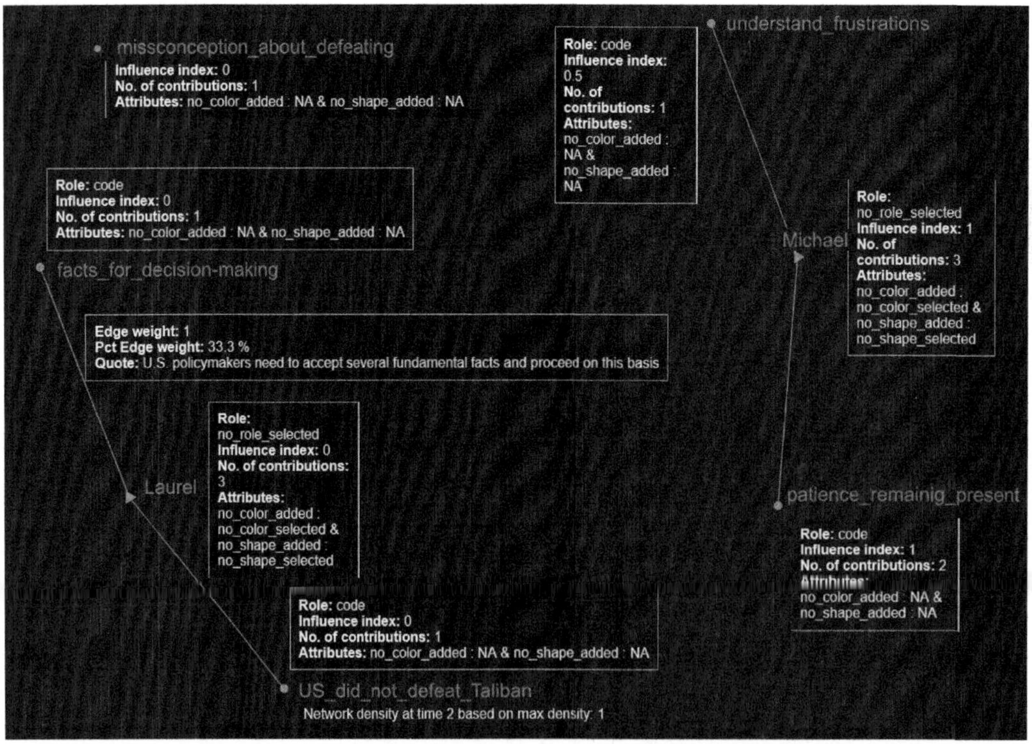

FIG. 4.2 Example of VERITAS dynamic HTML output. Access to interactive output here https://cutt.ly/xtLA04Qm.

The transformations automatically conducted by VERITAS are reflected in the matrices (or truth tables), which render one contingency table of actor by code/label per time observed. Note that as the number of actors increases, the number of rows also increases and, if each actor has a unique contribution at a given time, the number of columns equals the number of rows (as in our example). However, if there is repetition in the codes or labels, wherein two actors mention the same code at the same time, the number of rows and columns will differ. Finally, as further detailed below, VERITAS's user interface (UI) renders interactive, dynamic HTML outputs containing all contributions over time, in time-aggregated (Fig. 4.3) and temporally disaggregated forms (Fig. 4.4), with this latter output also enabling the retrieval of the specific quote traced at a given time (original text from the transcript contained in the column called "Quote"). More on this information is provided below.

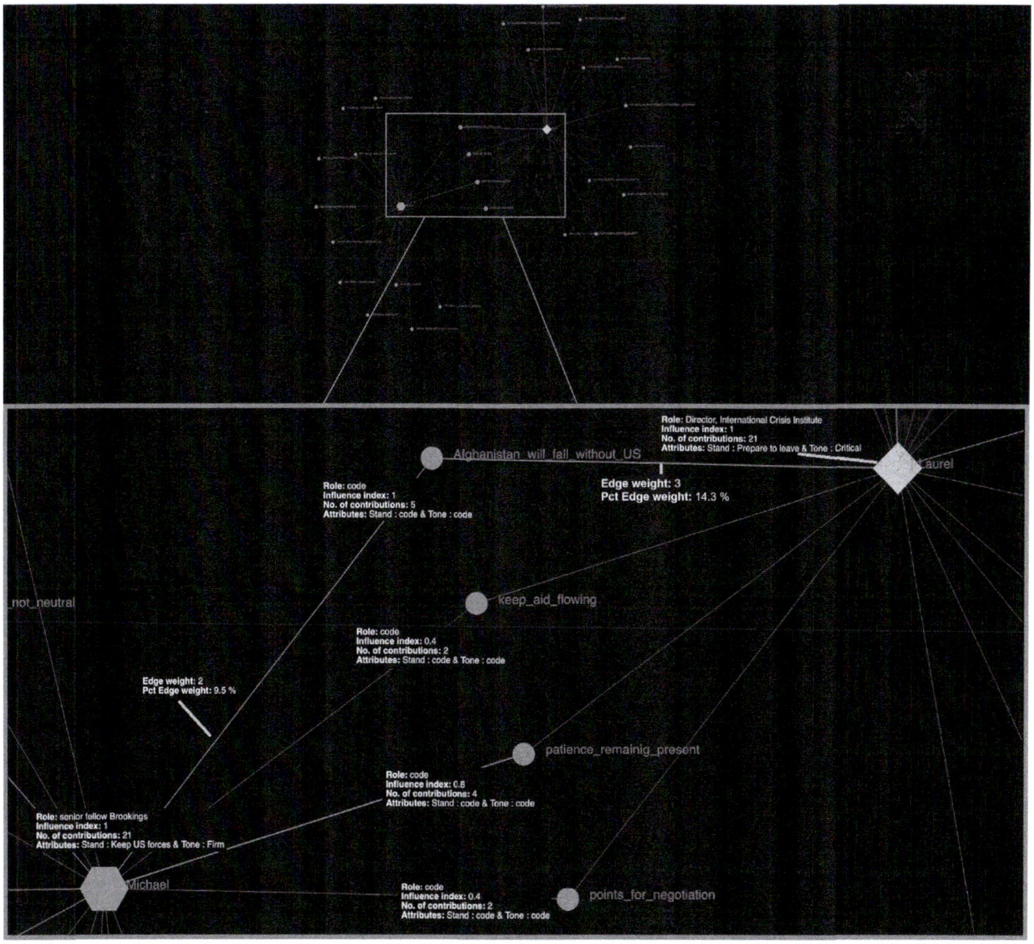

FIG. 4.3 Time-aggregate asynchronous debate output. Access to dynamic output here https://cutt.ly/QtLA9smv.

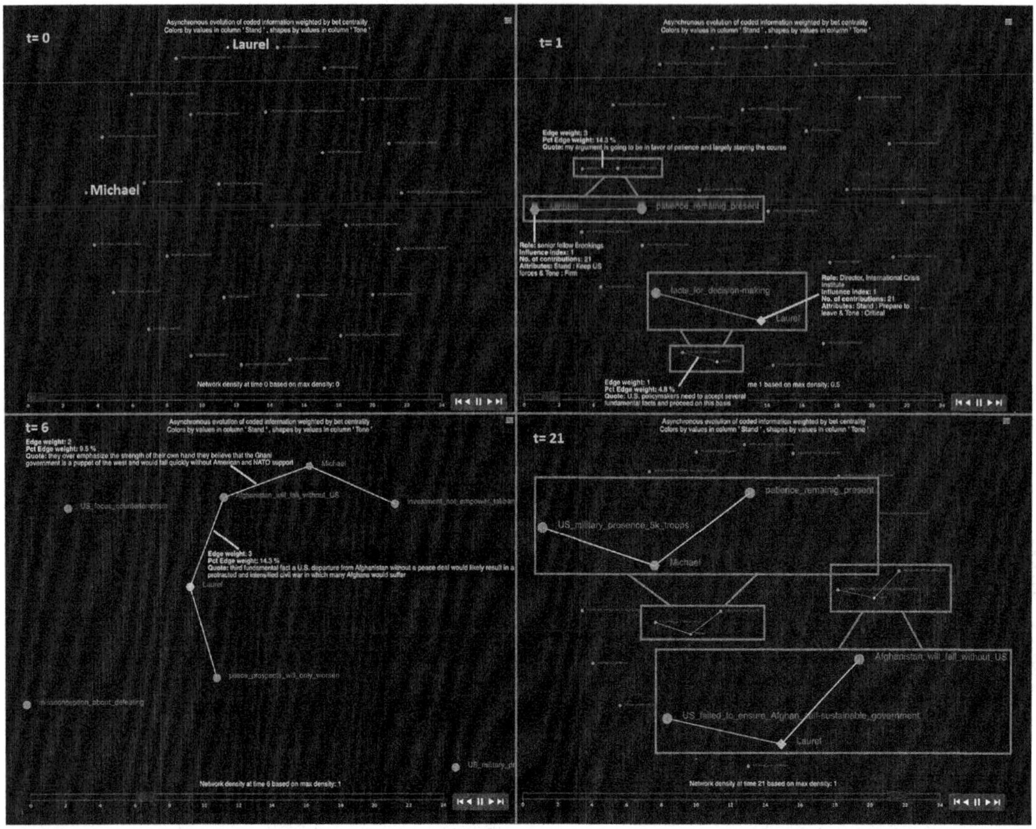

FIG. 4.4 Dynamic temporal asynchronous debate output. Access to dynamic output here https://cutt.ly/xtLA04Qm.

Highlighting structure through social network measures

In its visual representation of data, VERITAS follows graph theory principles (Biggs et al., 1986) and indices of relevance in the system, also known as centrality (Freeman, 1978). Centrality is one of the most important concepts in network analysis because it enables an understanding of the overall network structure through each actor's individual relevance or influence in the network (Freeman, 1978). Centrality measures (also sometimes referred to as *influence*) identify movement and structure in the network through actors' positions and paths or connections (Borgatti, 2006). VERITAS computes three centrality measures to highlight structure in the system by identifying sets of key units in the network who play different roles, as highlighted by these centrality measures: degree, betweenness, and eigenvector. The following discussion is a summary of each of these measures.

Degree centrality reflects the number of connections that a given actor has in the network (Borgatti, 2006). In this project, degree centrality is computed by counting the number of times that an actor provided information that was eventually classified as a code or by the number of messages exchanged with another actor in the group discussion. In our small example, we purposefully limited this number to the first three contributions. Accordingly, both Michael and Laurel have a degree centrality of three. The interactive HTML outputs (presented below) allows users to click on actors and codes to show these centrality measures. With respect to degree centrality, the information labeled "No. of contributions," displayed when selecting an actor, indicates the total number of codes (contributions) provided by that actor. For example, in Fig. 4.2, in the dialogue box near actor Laurel, "No. of contributions: 3" indicates that during this subset of her reactions to the prompt "role of the US in Afghanistan policy" (formally described below), Laurel provided 3 pieces of information that were eventually qualitatively labeled or coded. Similarly, clicking on each code generates a dialog box showing the number of times *all* participants provided information classified under that code. Take "facts_for_decision_making" for example, in this case the "No. of contributions: 1" indicates that this code was only mentioned once among all participants (even though there are only two in this brief example). To be as comprehensive as possible, note that the connecting line of an actor to a code (or edge) also conveys information derived from these degree centrality measures in VERITAS. In following with this example, note that actor Laurel and code "facts_for_decision_making" have an "Edge weight" value of 1 (Note that this value will be reflected in the thickness of the connecting line as depicted in our applied examples.). This indicates that, of the 3 codes provided by Laurel, 1 corresponded to the code "facts_for_decision_making," representing 33.3% of her total number of contributions (or $1/3=.333$). This means that the information "Pct Edge weight" reflects a derived degree centrality measure of the relevance of this particular code for a given actor.

The next centrality measure included in these frameworks is *eigenvector centrality*. This measure accounts for the overall influence of a given code or participant. As depicted in Fig. 4.2, this measure also appears in a dialog box (as "Influence Index") when an actor or code is selected. Eigenvector highlights individuals or codes that are central as a function of being connected with individuals and/or codes that were active in the network (Borgatti, 2006; Freeman, 1978). Hence, an actor is central in eigenvector if such an actor provided codes or established connections with other actors who were also active or central in the network (influence begets influence). Mathematically, the eigenvector centrality of a unit i is proportional to the sum of centralities of the units j to whom unit i is connected (Borgatti, 2006; Freeman, 1978). Eigenvector is standardized to range from 0 to 1, with 1 indicating the maximum level of influence. The level of influence of each actor and/or code may serve to further contextualize their contribution to our understanding and strengthen our qualitative evidence. However, these numbers *must not replace* the nuances of researchers' understandings gained before and during data collection and analyses. As depicted in Fig. 4.2, Michael has the highest eigenvalue centrality among human actors. In the case of codes or labels, the code "patience_remaining_present" has the highest possible value of 1, which reflects that this code was the most relevant among all other codes. Note that eigenvector goes beyond counts, as in the case of degree centrality. That is, although both Laurel and Michael were constrained to only showing their first three codes (hence both have a degree of three), Laurel had an eigenvector of 0 because the codes she mentioned were not shared by the other member of the group—or only Michael in this case. Nonetheless, note that in the case of Michael, his influence index value of 1 is explained by him having shared one code ("patience_remaining_present") twice during this short example. In this case, both Michael's centrality and the centrality of this code shared twice were the most influential actors from an eigenvector centrality perspective, because Michael provided a code that was shared more than once and because the code was connected to the actor who provided the code shared the most. Hence, this particular dynamic may be classified as self-selection or privilege reproduction. Finally, to further highlight the notion of "influence begets influence," note that although the code "understand_frustrations" was only mentioned once (as in the case of all three codes provided by Laurel), the influence index of this code was 0.5, whereas all the codes provided by Laurel had a value of 0. This 0.5 value is explained by this code having been provided by the actor with the highest eigenvector centrality in this system. This example serves to highlight three important aspects: (1) these measures of influence are more powerful when contextualized and understood, (2) their interpretation may not simply be reduced to numeric values, and (3) perhaps more importantly, this self-selection issue is most concerning with limited data (three codes and two actors in this case). As the system grows, the resulting complexity will lead to influence measures that are not affected by outliers or atypical instances. This latter point is illustrated in our examples section, wherein when considering all codes provided by both Laurel and Michael, both of them were equally influential.

Finally, VERITAS also calculates *betweenness centrality*. This measure captures the extent to which a given actor or unit falls between other units (Wasserman and Faust, 1994), thus serving as a bridge between actors or units to ease the flow of communication (Borgatti, 2006; Freeman, 1978). Accordingly, betweenness centrality highlights *both* the participants who serve as bridges of information *and* the codes that connect individuals in a network. The dynamic HTML outputs use betweenness centrality to adjust the size of each actor or code. From this perspective, larger actors or codes indicate that such actors or codes are serving as bridges of information. VERITAS has standardized this measure to evolve as time progresses. Based on this dynamic change, researchers can observe which actors and codes were the most influential from a centrality perspective at a given point in time. Figure 4.2 shows that both Laurel and Michael have a larger size (and shape) than the codes or labels (more on this below) because they fall between their respective codes. Betweenness centrality size changes will be more evident in systems that have more participants and codes.

Adding time-evolving connections and actor attributes

The connecting lines in VERITAS (which may connect actors to codes or actors to actors) allow researchers to add the actual content of the information traced at time *t*. In the case of $Actor_{i,t} \rightarrow Code_{j,t}$, the line includes the quote or content of the code at time *t*, in the case of $Actor_{i,t} \rightarrow Actor_{j,t}$ the connecting line contains the content of the message or information exchanged among these actors also at time *t*. In both cases, these contents change/evolve or are traced over time. For example, in Fig. 4.2, clicking on the line connecting Laurel and the code "facts_for_decision_making" displays the actual content of that code at time 2 (stored in the column "Quote" as depicted in Fig. 4.1). As time progresses, the content of these connections evolves to ease the information retrieval and meaning-building process. For example, if you observe the content of the first quote provided by Michael under "patience_remaining_present" (Fig. 4.1), this content is different in his third contribution, which is also classified under this same label. Accordingly, the respective line content at times 1 and 3 are different in the dynamic HTLM output, even if the connection is the same (Michael and this code), reflecting that it was traced at different times.

In addition to the "Edge weight," "Pct Edge weight," and text content or "quote" described above, the color of the connection shows whether the code just emerged or was present in the previous time. Green lines (or orange lines in high-contrast settings, as shown in Fig. 4.2) correspond to codes that just emerged at time *t*, and dark blue lines (or violet lines, also in high-contrast settings in Fig. 4.2) indicate that the connection emerged during time *t-1*. As depicted in the UI section, by default, VERITAS shows the two most recent contributions over time, but the UI can be adjusted to show up to the 10 most recent contributions with a simple click—more on this below. The selection of two contributions is particularly useful with large systems or networks, where network visibility may be compromised given the abundance of actors. Moreover, the observation of the two most recent connections enables users to contextualize the most recent contribution of each actor at time *t*, given their *t-1* contribution.

In addition to network-based attributes calculated by VERITAS and the line attributes just described, VERITAS accommodates *time-evolving* attributes at the node (i.e., actor or code) levels. These attributes, while optional, may be added to highlight the roles and influence levels represented in these relationships. At the node level, users can add participants' time-invariant and dynamic attributes. Time-invariant attributes include characteristics that did not change during the data collection process, such as age (if the project lasted less than a year), ethnicity, gender, or any particular role played in the group that remained constant during the study. Dynamic attributes capture changes, and these changes may include differences in standings on a given issue or even the tone used when sending a message or providing information that was subsequently categorized in a code. In the conflict resolution example (i.e., the "12 Angry Men" example), we added a "Tone" attribute to depict whether the message was sent with an aggressive, nonaggressive, or frustrated tone (tone can also be considered a theme or code family). We could also have added "*stand*" during this coding process (or added it after the fact to include both tone and stand) to capture whether, at a particular time, the stand of the participant (juror) was guilty or not guilty. Finally, in our conflict resolution example, we added a role column indicating whether the participant was a juror or the foreman—this latter attribute did not change in this case.

All attributes that change over time will evolve in the VERITAS dynamic outputs. Nonetheless, in addition to time-evolving visualizations, the outputs also include a summary IR aggregated network depiction. This depiction contains the first attribute per actor captured in the database—see the interactive outputs section below. When researchers select columns with these "sender" attributes, the resulting network visualization will represent them via color or shape variation in the HTML rendering. VERITAS currently allows up to 11 colors.

These colors may be conceptualized as categorical attributes. That is, if researchers are interested in mapping ethnicity and this indicator has up to eleven categories, VERITAS is programmed to automatically map these colors to their corresponding categories. Note that if no color attribute is selected, VERITAS automatically sets different colors for actors and codes. All codes are blue by default (in the high-contrast setting depicted in Fig. 4.2, or orange in a nonhigh-contrast setting) and green or pink (in high-contrast and nonhigh-contrast settings, respectively) for actors. These colors may be changed in the "optional adjustments panel" in the VERITAS UI. The "Role" information displayed when selecting a code will always be "code" because all changing attributes are added for actors or senders of information only. That is, if a role attribute is added as a different column, similar to the example depicted with seniority and sector, one of these role attributes will be displayed when clicking on an actor but will remain as "code" when selecting a code. In our example in Fig. 4.2, we did not add any color or shape attributes, but we will demonstrate the addition of these attributes in the complete analysis. For now, it may suffice to mention that when no shape attributes are added, codes will be circles and actors (senders) will be triangles. Shapes may also be changed in the "optional adjustments panel" in the VERITAS UI.

Application examples

To demonstrate the capabilities of VERITAS, this section presents two examples of analyses. Two main data sources are used. The first is a publicly available YouTube video that was fully transcribed and coded using the Microsoft Word macro, which allows employing Word as a coding device. The second data source corresponds to a working group sponsored by the NSF, as illustrated below. The selection of these sources was made to provide as much clarity and transparency to the analytic processes as possible. All readers will have access to all the tools and elements required (transcripts included) to replicate these analyses or create their own, using their own coding or labeling schemes.

Before describing these datasets in more detail, let us first elaborate on the types of analyses that VERITAS handles. Essentially, for each group observation we have two types of analyses: one focused on information exchanges amongst actors in the form $Actor_i \rightarrow Actor_j$. The second focuses on actors' contributions, regardless of who was the target. This takes the form $Actor_i \rightarrow Code_j$. To elaborate more on these foci, we present Table 4.1. Here we show an extract of the NSF sponsored project we will describe below. Note that these codes are coming from the same transcripts, but the process of building the edgelists changes. In the relationships showing ID, Code, and Text/Quote we can link each actor to the codes they provided and the content of that code (i.e., text or quote). This will enable the mapping of the evolution of participants' contributions throughout the focus group or working group event. Nonetheless, by observing the videos or listening to the recordings we may be able to also observe whether a given actor was sending that message to another actor in the working group. In this case, the relationship involves the actors and the message in the form of From, To, and the text or quote also in Table 4.1. In VERITAS the connecting line in either form, may include the content or

Table 4.1 Data structure for dynamic network analyses.

Time	Id	Code	Text/Quote	From	To
1	**Actor1**	Team effort	Text team effort …	Actor2	Actor1
2	**Actor1**	Future faculty	Text future fac …	Actor1	Actor2
3	Actor2	Why future?	Text why future …	Actor2	Actor1
4	**Actor1**	Problem identification	Text problem …	Actor1	Actor2
5	Actor2	Why future?	Text why future …	Actor2	Actor1
6	**Actor1**	Evidence-based	Text evidence …	Actor1	Actor2
7	**Actor1**	Future faculty	Text future fac …	Actor2	Actor1
8	Actor2	Why undergraduate?	Text why underg …	Actor1	Actor3
9	**Actor1**	Future faculty	Text future fac …	Actor3	Actor1
10	Actor2	Graduates too?	Text graduates …	Actor1	Actor3
11	**Actor1**	Graduates too?	Text graduates …	Actor3	Actor1
12	Actor3	Why undergraduate?	Text why underg …	Actor1	Actor3
13	Actor3	Diversification	Text diversificat …	Actor3	Actor4
14	**Actor1**	Advising	Text advising …	Actor4	Actor2
15	Actor3	Science identity	Text science iden …	Actor2	Actor3
16	Actor3	Advising	Text advising …	Actor3	Actor2
17	Actor1	Shared understanding	Text shared und …	Actor2	Actor10
18	Actor1	Shared agenda	Text shared agen …	Actor10	Actor2
19	Actor1	Shared metrics	Text shared metr …	Actor2	Actor3
20	Actor4	Diversification	Text diversificat …	Actor3	Actor2

Notes: Columns Time, Id, and Code are one independent edgelist in the form Columns Time, From and To are another independent edgelist in the form.

quote of the connection as captured by the content of column "Text/Quote," in addition to other attributes as we have discussed above. We recommend extra caution when creating these actor to actor communication exchanges for they involve us being certain that we were accurately capturing such exchanges.

Data sources

The first example is a scene of the movie "12 Angry Men" which depicts a complex and dynamic generation of information and exchange of ideas and rebuttals among a group of 12 jurors. The scene analyzed can be accessed here https://www.youtube.com/watch?v=HK7CfffM1OQ.

The transcript with codes for the synchronous $Actor_{i,t} \rightarrow Code_{j,t}$ setting can be accessed here https://cutt.ly/ZWlGlBy (the replication database included in VERITAS can be accessed here https://cutt.ly/1RJQKKt) and the transcript with codes for the synchronous $Actor_{i,t} \rightarrow Actor_{j,t}$ setting can be downloaded here https://cutt.ly/2WlGbNt (to access the replication database click here https://cutt.ly/5RJQCUS).

The second case study uses data retrieved for a NSF-sponsored project. This project addressed the issue of the lack of representation of faculty of color in STEM fields at community colleges. The data analyzed is based on observations and transcripts from an eight-hour focus group of highly influential stake-holders and education policy-makers, most of whom are located in Washington, D.C., at their headquarters. In this academic setting (as opposed to a movie scene like in our first database), we have moderators and research participants. Accordingly, in this case, we can observe interactions among participants and between participants and moderators. Basic training in designing and conducting focus groups emphasizes that an indicator of successful implementation of a focus group occurs when participants, not interviewers or moderators, are the ones providing researchers with relevant information (González Canché, 2019). If moderators or interviewers are the most active participants, then the content generated may be highly influenced by the moderators and may not necessarily capture the perspectives of the focus group participants. From this point of view, and given the dynamic nature of these data-generating processes, one can rely on dynamic network analysis to assess the extent to which meaningful interactions in focus groups took place mostly among participants or whether moderators led such interactions.

From the previous description, when analyzing research data, we can also ask the following questions:

- Is there evidence to suggest that relevant and meaningful exchanges of information took place among participants or between participants and moderators/interviewers? (one-mode transformed version)
- When and by whom was knowledge created and further discussed? (Two-mode network: participants and codes.)
- Is there evidence to suggest that moderators influenced the type of content generated by actively engaging in the knowledge production? Or were the participants the main generators and discussants of this knowledge? (two-mode network: participants and codes)

Findings: VERITAS HTML interactive, dynamic outputs

VERITAS renders two main HTML outputs. The first is the time-aggregated analyses of all $Actor_i \rightarrow Code_j$ or $Actor_i \rightarrow Actor_j$. These time-aggregated depictions offer a summary of all connections observed, and clicking on each code, actor, and link (line connecting actor and codes/labels) renders dialogue boxes as those depicted in Fig. 4.2, with the exception of the quote content of each line reflected in Figs. 4.1 and 4.2 precisely due to the time aggregation (all dynamic temporal analyses render time varying quote visualizations). The second output is the dynamic temporal depiction of these elements. VERITAS's software architecture integrates all quantitative and qualitative data elements along with temporal/dynamic network visualization to provide fully interactive and integrated outputs. These outputs provide controls that pause or rewind the transitions. The farthest right and left buttons included at the bottom of the HTML output (see  controls in this HTML rendering https://democratizing-data-science.github.io/MOVIE/Debate_temporal_dynamic_analyses.html, for example, further described below) will advance the animation forward or backward *one transition point* at a time. The buttons to the right and left of the pause button show the evolution of interactions from zero to N. The backward or rewind button, located to the left of the pause button, initiates the *devolution* of these interactions until time zero. The pause button stops the animation at any point in between zero to N, and researchers can zoom in on any *participant* or *hot zone* and display message content or participants' attributes to contextualize these interactions.

Synchronous $Actor_{i,t} \rightarrow Code_{j,t}$ Analyses (Conflict Resolution, 12 Angry Men)

Time aggregated output

This output is shown in Fig. 4.5. The HTML version can be accessed at https://cutt.ly/BtLA4I0C. The resulting time-aggregated depiction indicates that this system (inside the yellow square) is configured by three subsystems, wherein two participants were disconnected from the major system (a network containing 10 of the remaining jurors) as a result of having provided at least one contribution that was not shared by other jurors. This indicates that none of the contributions shared by Actors 1 and 7 were included in the network, hence the peripheral or disconnected status of these two actors. In terms of betweenness centrality, note that Actors 8 and 5 were the most central human actors, meaning that they served as bridges connecting the flow of information or discussion in this social setting. Nonetheless, neither of these two actors had the highest eigenvector centrality value. The actor with the highest eigenvector centrality value was Actor 4, which is explained by his direct connection with the single most influential code (with a value of 1 in the influence index) and with the highest connecting power (betweenness centrality, as indicated by its size) in this movie scene: "did_old_man_run?" This latter code was mentioned 12 times in this system, and Actor 4 himself mentioned this code four times out of his nine contributions (or 44% of his contributions).

A particularly powerful code that, during the analysis, kept drawing our attention was provided by Actor 8. At some point, he stood up in anger and desperation, mentioning, "what is this `love-your underprivileged-brother-week?'" As avid readers of qualitative research, we have always wondered how researchers use a quote or code to title their studies. In the absence of VERITAS as an analytic tool, we would have likely selected this code to describe the social circumstances of this scene, given its sociological strength. However, if the goal is to help the reader observe the most crucial aspect of the issue under study, the title of this study, at least pertaining to the video analyzed, should be related to "did_old_man_run?" The debate and

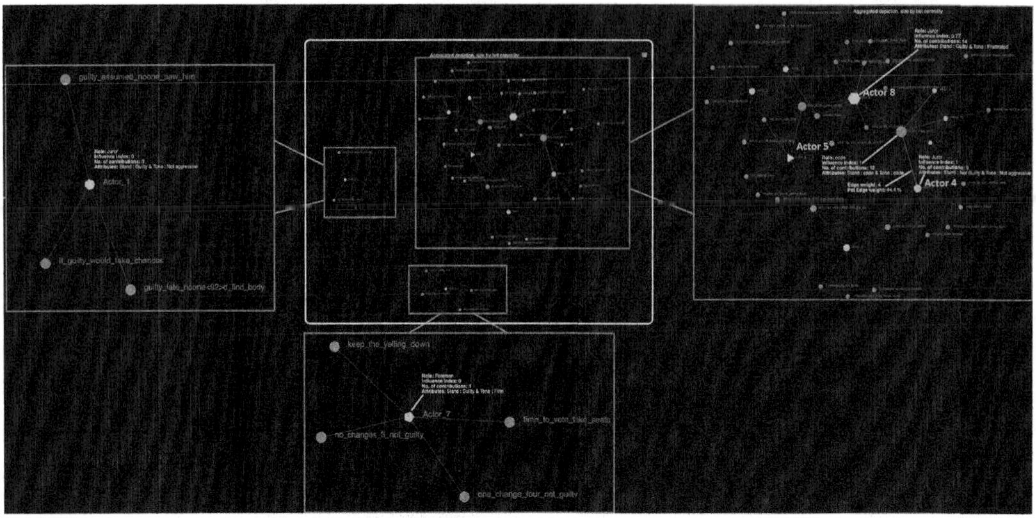

FIG. 4.5 Time-aggregated synchronous debate output (actor to code). Access to interactive version here https://cutt.ly/BtLA4I0C.

pivotal trend of thought that this code captures was the single most important factor that made the angriest and perhaps "informally" the most influential juror realize that he, and everyone else, should not trust the testimony of the "old man."

Note that, although the time-aggregated visualization that VERITAS rendered is useful in providing a summary of the state of the system, this level of understanding and analysis loses or forgoes more nuanced details about the evolution of information that may help strengthen our grasp of processes and reasons.

Temporal dynamic output

Fig. 4.6 presents the temporal dynamic analysis of the evolution of these contributions. Different from the asynchronous analyses, wherein all first contributions are plotted simultaneously, in the synchronous analysis, the contributions are presented one at the time following their actual temporal evolution, as observed in the group interactions. Given that the system analyzed is not configured by many actors, instead of plotting only the two most recent contributions, as in the case of the asynchronous example, in this depiction we selected to plot instead the ten most recent contributions. This is achieved by shifting the slider provided in the VERITAS's UI by clicking in the number desired or grabbing the handle and moving it to from 2 (default value) to 10 (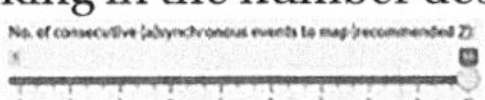). Once more, we invite readers to interact with this HTML output here https://cutt.ly/4tLSesjS.

Despite the richness associated with analyzing each time-based change, research papers typically face space limitations. Based on this, in this example, we demonstrate a strategy that can be used to condense these dynamic structures while still making a compelling case to showcase the analytic power of VERITAS. The analysis of Fig. 4.6 focuses on depicting the most active players in 40 of the 65 recorded contributions, as shown in the transcript (see https://cutt.ly/ZWlGlBy). During the first 10 events, there were four contributing jurors: Actors 1, 2, 3, and 4.

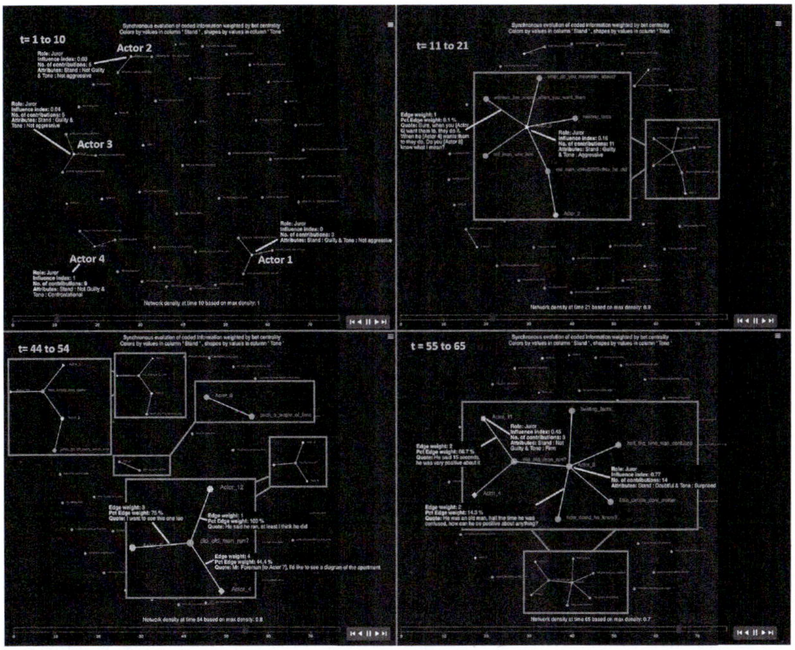

FIG. 4.6 Dynamic temporal synchronous debate output (actor to code). Access to dynamic output here https://cutt.ly/4tLSesjS.

As indicated in the aggregated analyses, for Actor 1, these were his only contributions, and no other juror provided information coded under these labels. Actor 4, on the other hand, became the most influential player in this social setting. The subsequent 10 interactions (times 11 to 21) were dominated by Actor 5, and it was during this time period that the moderator of the jurors (Actor 7) had to ask Actor 5 to "keep the yelling down."

It was during the 44 to 54 time points that a breaking point could be observed. This point refers to the raising of a question that eventually changed the course of action at the system level: "did the old man run?" Once more, when interacting with the HTML file, we can retrieve the actual content of this code. For Actor 6, the first content of this code was the realization that it was not quite feasible for the old person to actually run. Eventually, Actor 6 had three contributions coded under this label, representing 75% of his total inputs. For Actor 12, his only contribution was coded under this category and consisted of sharing that, "He (the old man) said he ran, at least I think he did." Both contributions prompted Actor 4 to request more information about the apartment layout to assess the feasibility of having run (in 15 seconds).

In the closing part of this depiction (time 55 to 65), we can see that Actor 8, an enthusiastic proponent of finding the accused guilty of murder, dominated the discussion. His interventions were focused on convincing the other jurors that "little details do not matter" and inadvertently, Actor 8 ended up realizing that the main witness was "confused half the time," which invalidated his attempts to convince the jurors and even himself about condemning the accused for having killed his own father.

Time aggregated output

The time-aggregated depiction of actor-to-actor interactions is shown in Fig. 4.7. Recall that this analysis is relevant when researchers are interested in understanding the evolution of communication exchanges in the system. Once more, the HTML version is available here: https://democratizing-data-science.github.io/MOVIE/time-aggregated_12-angry-actor-actor.html. The shape of the actors is based on their tone attribute described above. The color indicates their stance, and the size is based on their betweenness centrality. As in both previous examples (see Fig. 4.5), these attributes are based on the first instances observed in the system. Specifically, Fig. 4.7 shows that Actor 8 had a nonaggressive tone in his first intervention, but the dynamic analysis presented next will reflect that all but one of his remaining interventions had an aggressive tone. This summary also shows that Actor 1 is the only actor who did not send any messages (his contributions shown in Fig. 4.5 were not directed at anyone in particular,

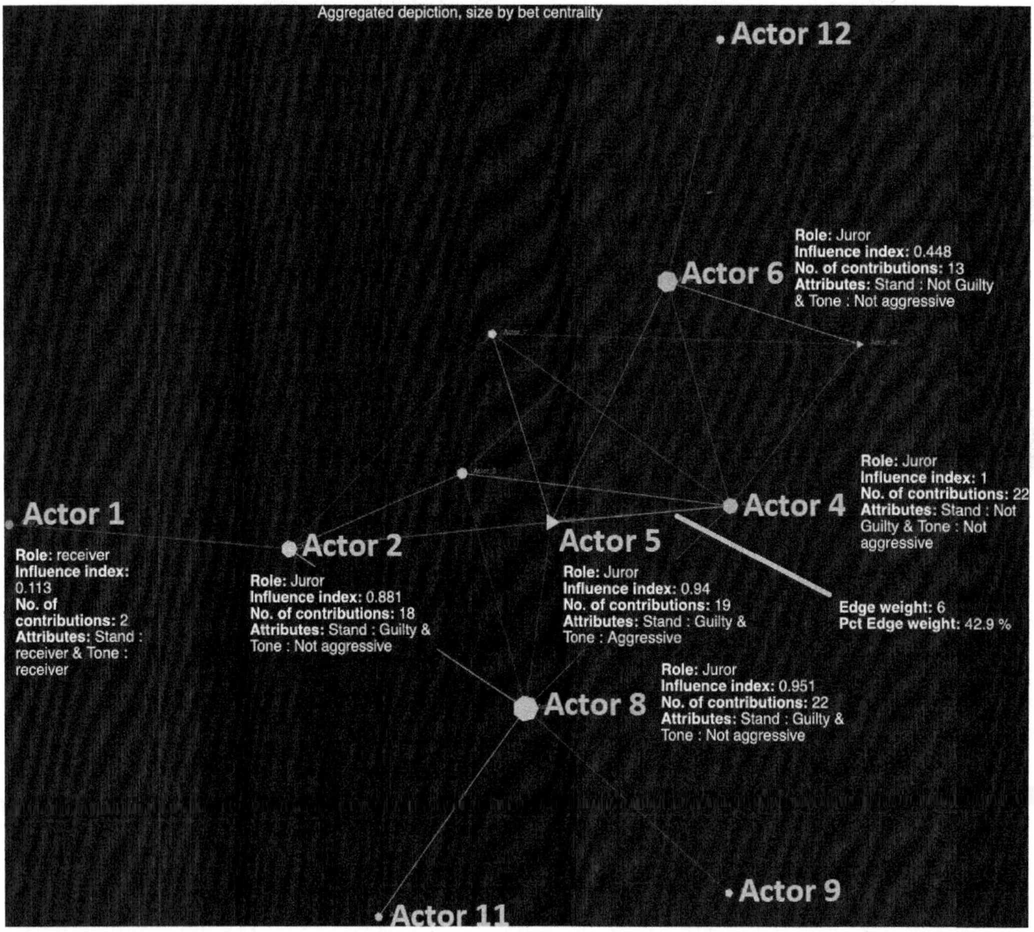

FIG. 4.7 Time-aggregated synchronous debate output (actor to actor). Access to dynamic output here https://cutt.ly/MtLSrIOL.

as shown in the video clip: https://youtu.be/HK7CfffM1OQ?t=2), but merely served as the receiver of two messages, from Actor 2 in this case. This explains Actor 1's stance, role, and tone as a receiver in this analysis.

As described above and shown in Table 4.1, note that, although these connections come from the same transcript of the movie scene analyzed above, the emphasis in the detection of interactions and communication exchange is placed on who was the target or sender of a message in the system. This new emphasis means that the coded pieces of information presented here may not necessarily match those discussed in the $Actor_{i,t} \rightarrow Code_{j,t}$ analyses for these communication exchange framework was designed to highlight other structural forms in the system. Nonetheless since the origin of information is the same, there must be resemblances in the findings and conclusions. Specifically, note that Actor 4 remained the most influential juror from an eigenvector centrality perspective. This influence is explained by his direct relationships with Actors 5 and 8, who themselves were highly influential. Indeed, Actor 8 was the main bridge in the system. Thanks to his presence Actors 9 and 11 were not disconnected from the conversation. Actor 4 was also in direct contact to Actors 2 and 6, both of whom, like Actor 8 helped not quite active actors (1 and 12) remain connected to the main system. Actor 8 had the second highest influence index with a value of 0.951. Had this actor been directly connected with Actor 6, he would likely have reached the maximum eigenvalue of 1.

Temporal dynamic output

Before presenting the temporal dynamic output of the evolution of these communication exchanges in Fig. 4.9, let us discuss an important analytical decision with clear visualization implications. This decision may be observed when studying the transcript and coding schemes provided in this Word file: https://cutt.ly/2WlGbNt. Note that the video analyzed clearly allows us to observe that when a juror directed a message, this message, in some instances, was directed specifically to one juror and, in other instances, to more than one juror—one at the beginning of the dialog and another at some point during the same uninterrupted dialog. To illustrate this point, see Example A in Fig. 4.8. The analytical decision here was that, instead of taking the entire dialog as a single message sent to Actor 2 by Actor 3, this message was decomposed into the three main components that we, as coders and analysts, may consider relevant to analyze. The visualization implication is that, for this dialog there were three messages sent from Actor 3 to Actor 2 instead of one aggregated message. As usual, each link will contain the specific message subset. Given that the quote would be displayed in the resulting output, coders who prefer to take this entire dialog as a single whole message being sent would not lose too much information, as readers may still see that the dialog contained more than one point or message—more on this below.

Example B in Fig. 4.8 is similar in that the dialog opens with Actor 5 rebutting assertions made by Actor 4. The first two messages in this dialog are directed to Actor 4 (which, once more, could have been aggregated as one), whereas the last message is directed to Actor 2. Accordingly, in this dialog this disaggregation was necessary to convey the change of recipient in this communication exchange. In sum, Example B in Fig. 4.8 could have been coded as two links instead of three. There is no rule of thumb regarding this decision, but our preference

Example A

Actor 3. Two things [to Actor 2]. One, in his state of panic he may not have heard the scream, perhaps it wasn't very loud. Two, if he did hear it, he may not have connected it with his own acts. Remember, he lived in a neighborhood where screams were fairly commonplace.

Actor 3 to Actor 2
Actor 3 to Actor 2
Actor 3 to Actor 2

Example B

Actor 5. What do you mean Dell [Actor 4]? What are you talking about? Didn't the old man see him running out of the house? he's [Actor 4] twisting the facts I'm telling you. Tito (Actor 2), didn't the old man see the kid running out of the house at 12.10? Well, did or didn't he?

Actor 4 to Actor 4
Actor 5 to Actor 4
Actor 5 to Actor 2

FIG. 4.8 Depiction of the level of aggregation when labeling, coding, or social categorizing.

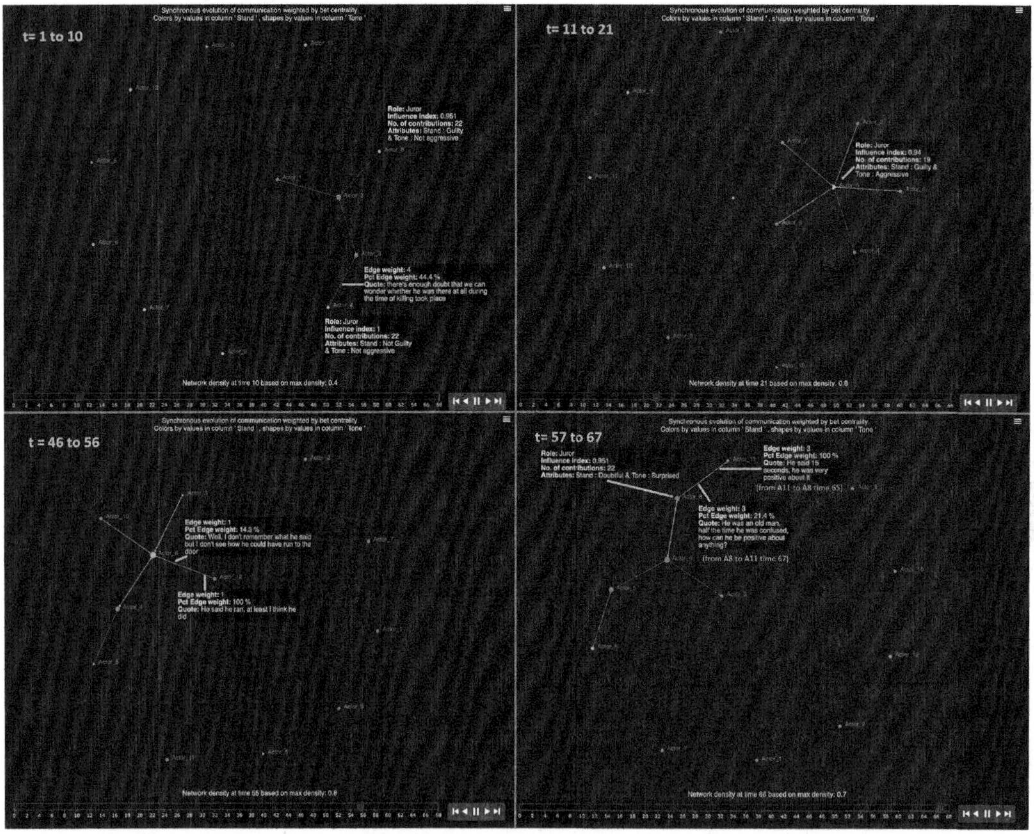

FIG. 4.9 Dynamic temporal synchronous debate output (actor to actor). Access to dynamic output here https://cutt.ly/EtLSyT33.

is to have more disaggregated analyses to enhance transparency and clarity in the content of communication exchanges. The more disaggregated the message exchanges are, the more opportunities our centrality measures will have to trace patterns in these complex systems. That is, although Actor 3 clearly delivered three important pieces of information to Actor 2 in

Example A, we would not have captured these complexities with an aggregated depiction of this communication exchange, and the resulting level of influence captured by centrality measures would have been different.

The dynamic temporal output of these communication exchanges is represented in Fig. 4.9. The HTML version can be accessed here: https://cutt.ly/EtLSyT33. Similar to our previous analysis presented in Fig. 4.6, we present the 10 most recent communication exchanges per snapshot. Our first depiction accounts for times 1 to 10 and reflects that during this time period, Actors 2 and 3 were the highest connectors or bridges of information. During the time periods 11 to 21, Actor 5 clearly became the most important broker of information exchanges. This is the time period when he was asked to stop yelling since he got very agitated, and his tone was aggressive (see triangle shape and information box). The next snapshot accounts for time points 46 to 56. In this case, Actor 6 became the most important bridge, communicating and facilitating the flow of information. This is the point in the exchange of ideas where the single most important question emerged ("did_old_man_run?"). Note that VERITAS allows the retrieval of multiple communication exchanges represented in a single link. For example, Actor 12 told Actor 6, "He said he ran, at least I think he did," to which Actor 6 replied, "Well, I don't remember what he said, but I don't see how he could have run to the door." To retrieve the message sent by an actor when there are actual exchanges, just click on the link closest to the actor of interest. That is, for Actor 6's message, the click should be closer to this actor than to Actor 12 in Fig. 4.6.

Finally, we present time points 57 to 67. In this depiction, Actor 4 has the highest betweenness centrality (brokerage) role, followed by Actor 8. The dynamic representation of VERITAS, which allows the dynamic retrieval of betweenness centrality, offers a more nuanced understanding of this measure. That is, recall that in Fig. 4.7, Actor 4 did not reach the highest value in betweenness as captured by the time-aggregated analyses; however, throughout the evolution of message exchanges, and, as a matter of fact, during the most crucial moment from a verdict-change standpoint, Actor 4 held the most impactful broker role—as highlighted by its size in the network. It is in the last interaction when Actor 8 realized he could not trust the testimony of the "old man," as reflected by his message: "He was an old man, half the time he was confused, how can he be positive about anything?" This is reflected in his color and shape (doubtful and surprised, respectively).

Focus or working group data

The final set of analyses used focus group data that are dynamic, given the possibility of observing the evolution of interactions among participants and the qualitative codes generated. Here, we address the question regarding when, who, and how knowledge was generated, which serves to assess whether the codes were primarily generated by participants (i.e., experts) or the moderators (i.e., researchers).

As mentioned above, the structure of the data analyzed also follows an edgelist form, with actors being linked to their codes as shown in columns two and three of Table 4.1, and actors being linked to one another given their meaningful interactions, as depicted in columns four and five, also shown in Table 4.1. The fundamental difference between the data structure required to conduct the analyses presented in the previous subsection and the data structure required for the analysis of dynamic networks is the inclusion of moderators in this process. For clarity, we are also presenting Table 4.2, wherein we identify participants' roles based on being experts or moderators. Whenever possible, note that video recording would facilitate the process of identifying who said what and when, but often audio recording would suffice to accomplish this goal as well.

In Table 4.2, we show attributes at the actor level. Roles indicate that there were three moderators in this focus group (Actors 1, 8, and 11, which are represented in light gray in the dynamic sociograms). The Counts columns show how active the actors were in providing the network with codes. The column AC shows that Actor1, a moderator, provided a total of 14 codes, which was the highest code contribution among moderators in this network. The other two moderators contributed a combined total of six codes. This information demonstrates that Actor1, a moderator, was more influential in setting the "tone" of the conversations than the other two moderators. This static information, however, does not address the questions posed regarding whether moderators dictated the information being discussed. Finally, the counts under AA_from and AA_to reflect the number of interactions in which a given participant sent information to discuss a code and the number of interactions where a given participant was the target of an exchange based on the code that such participant provided to the network. In this respect, it is worth noting that Actor 6 was both the participant who provided the highest number of codes (22 in total) and the person who interacted the most from both the sender and receiver perspectives. In this case, note that although a given actor can react to a given code, the information such an actor provides may or may not warrant inclusion into an existing code, yet it could be captured in the dynamic exchange.

Table 4.2 Roles of focus group participants.

		Counts	
Roles	**AC**	**AA_from**	**AA_to**
Actor1:Moderator	14	15	15
Actor2:Participant	18	18	17
Actor3:Participant	20	17	17
Actor4:Participant	7	9	9
Actor5:Participant	8	13	13
Actor6:Participant	22	21	22
Actor7:Participant	5	8	8
Actor8:Moderator	1	1	1
Actor9:Participant	6	4	4
Actor10:Participant	10	8	8
Actor11:Moderator	5	7	7

Synchronous $Actor_{i,t} \rightarrow Actor_{j,t}$ Analyses (Working Group Data)

Fig. 4.10 contains the information required to address the research question, concerning whether there is evidence to suggest that the meaningful exchanges of information took place among participants or among participants and moderators. This figure is comprised of eight snapshots of the dynamic network that captured the exchanges that took place in the focus group. The light grey circles represent moderators, and the dark grey ones represent participants. The size of the circle is weighted by participants' betweenness centrality and this centrality changes over time given the level of activity that a given node has in the network serving as a bridge between other actors.

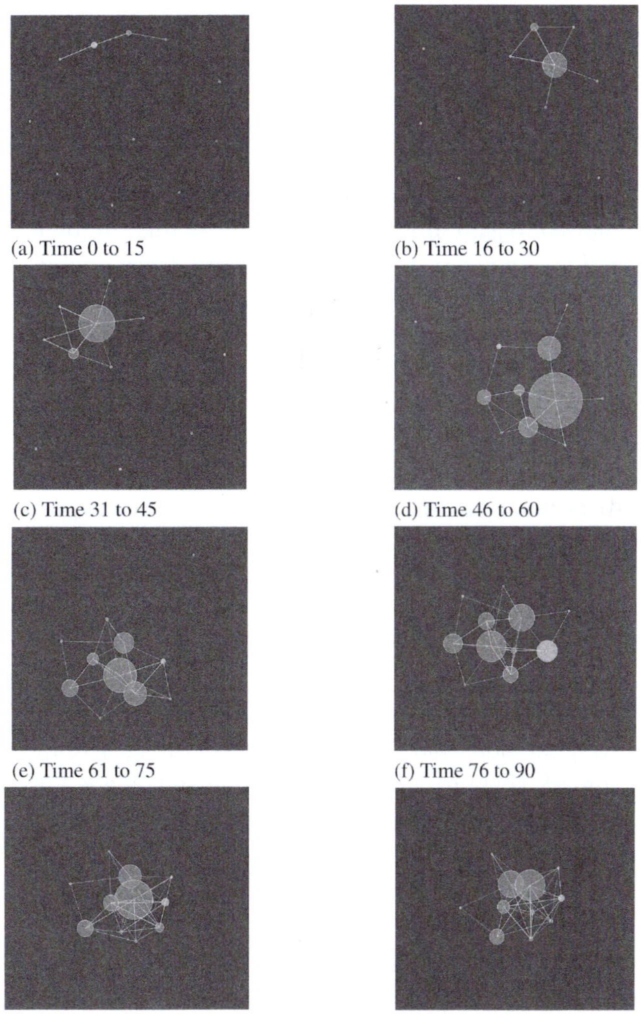

(a) Time 0 to 15
(b) Time 16 to 30
(c) Time 31 to 45
(d) Time 46 to 60
(e) Time 61 to 75
(f) Time 76 to 90
(g) Time 91 to 105
(h) Time 106 to 125

FIG. 4.10 Representation over time of the evolution of interactions (actor by actor).

Fig. 4.10A shows that during the first 15 interactions the most central player in terms of serving as a connector was a moderator. This is also reflected in Table 4.1 under the column "Id," highlighted in magenta, with eight of the first 15 code contributions to this network. Note also that, at this point, most of the actors remained disconnected. The reason for this initial influence of the moderator is based on their role in explaining the overarching goals of the focus group and requesting initial reactions. This indicates that, so far, it is too early to conclude that moderators were not taking advantage of participants' wealth of knowledge. Moving forward up to time 30, note that the role of the moderator was relegated to becoming an observer rather than an active contributor. At this point, there was one participant who was actively contributing and serving as the main receiver and sender of information among participants. These behaviors remained consistent up to time 45. Note also that the other two moderators had not had meaningful interactions yet at that point. Up to time 60, a second moderator joined the conversation, but overall, one can see that the discussion up to this point was mostly participant-driven with all participants having higher levels of influence and interaction than the two active moderators. At this point, note that the third moderator was the only actor in the network who remained disconnected. The snapshot corresponding to times 61 to 70 shows no major changes; nonetheless, the subsequent snapshot shows that one moderator started to gain momentum in terms of their influence. The transcripts of these focus groups and field notes show that, at this point, the moderator asked participants to break into groups to write a report about the aspects of the discussion that were the most relevant for them and their organizations. Note also that, because of this activity, the previously isolated moderator became integrated into the network, as they were in charge of one of the small groups formed. The two last snapshots show no drastic change, with the exception that the most active moderator did not have a betweenness centrality level as high as at the beginning of the focus group session.

Based on this discussion, one can then state that there is enough evidence to conclude that even though the conversations were influenced by one moderator in particular, this level of influence is not considered to have been an important detractor from participants expressing their opinions and being active contributors. Indeed, the last graph clearly shows that the exchanges were heavily influenced by participants. Although this analysis is relevant in terms of highlighting dynamic exchanges, it still does not help to address the question of who generated knowledge and when it was generated. This goal is achieved with the next figure. However, before moving on, note that the online version of Fig. 4.10 includes a video of the entire exchanges recorded. These exchanges can also be downloaded from this link (https://cutt.ly/le9m62kr). Users can generate an HTML file using VERITAS with the data we are providing here: https://cutt.ly/We9Qt3xb. Note also that VERITAS has this database preloaded under the option "Group Analyses of actor-actor interactions" in the UI, as we describe in our closing section.

Before starting, and to be as specific and clear as possible, let us consider Table 4.1, which contains a subset of these relationships under the "Actor Code Relations". The column time, under row 1, indicates that Actor 1 was the first actor in the network who provided the information that was eventually coded under "Team effort." The second row indicates that Actor 1 was also the first actor in the network who provided the code "Future faculty." From this information, it must be noted that the numbers 1 and 2 do not refer to exact timing; rather, they refer to a time-ordered set of contributions or global timestamps, as described in our GRATIS chapter.

Fig. 4.11 contains the actor by code relationships registered in this focus group. In this dynamic network a given actor who provides a code or who mentions information that was eventually classified as a code was connected to such a code. Here moderators are represented

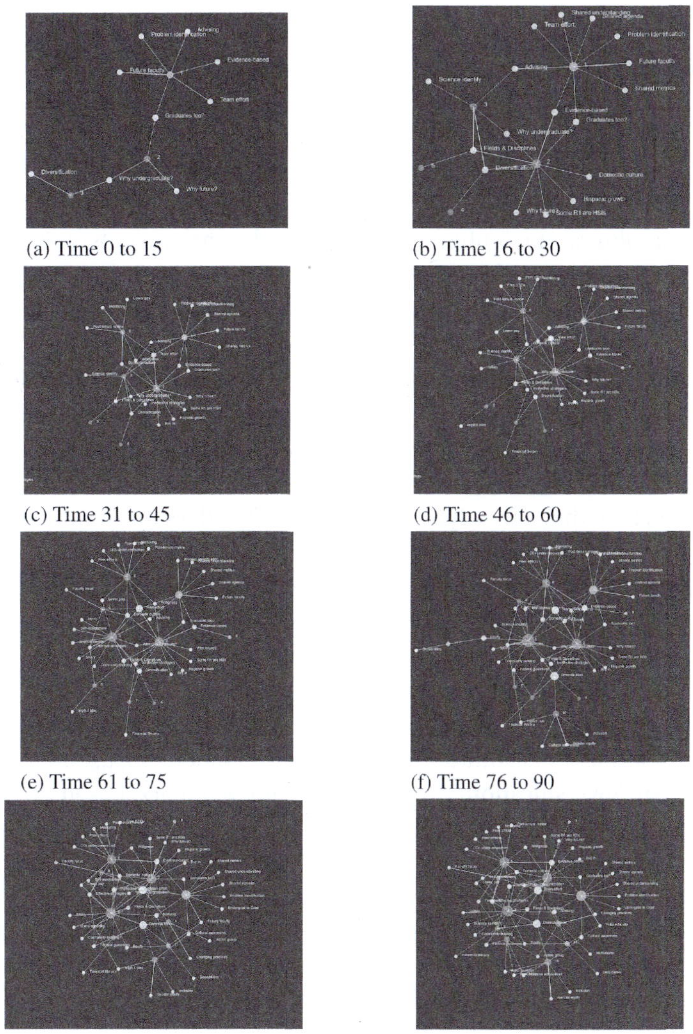

(a) Time 0 to 15
(b) Time 16 to 30
(c) Time 31 to 45
(d) Time 46 to 60
(e) Time 61 to 75
(f) Time 76 to 90
(g) Time 91 to 105
(h) Time 106 to 117

FIG. 4.11 Representation over time of the evolution of interactions (actor by code).

with numbers 1, 8, and 11, participants are remaining numbers, and codes have names. The first snapshot indicates that five of the nine codes that emerged as of time 15 were provided by a moderator. This depiction is consistent with Fig. 4.10 and Table 4.1. At this point there were only two other participants, one of which provided codes that were much more related to questions and these questions basically challenged the focus group's charge. Participant 3 in this graph began the conversation about diversification. The next snapshot, containing interactions up to time 30, shows that the moderator added new codes (shared understanding and shared agenda) to which no one in the network replied or commented about so far. These codes may have been provided in response to the multiple questions asked by participant 2, who similarly to the moderator is being highly influential in providing codes and enabling exchanges. By time 45 the network is more complex and at this point participant 2 has the highest betweenness centrality measure, a pattern that remains until time 60. At this point a second moderator joined the conversation.

With respect to the codes, "whiteness," "domestic culture," "advising," and "team effort" are at the center of the network. This position indicates that these codes were highly discussed by participants. By time 75, participant 3 is the most influential actor and is referring to "science identity," "domestic culture," and "culture building" as relevant concepts/codes. In this respect, it is worth noting that the content of these codes reflected a level of concern or discomfort related to training international students in STEM fields, as these students are likely to "outwork domestic students and are happy to do so while working for much less salary and benefits." This participant also mentioned that these international students cannot relate to undergraduates since they do not know the American ways, music, or TV shows. By time 90, participant 3 was again the most influential actor, as this person was constantly the target of other participants' comments, who reacted to the examples just provided. At this point, the code "diversification" is the most discussed. However, note that it is only at this time that "gender equity" was brought up but was not commented on further by anyone else in the network. This finding is considered a point for concern. Interestingly, the participant who mentioned "gender equity" also provided codes that, while important and pertinent to the discussion and the charge of the focus groups, remained at the periphery. These codes referred to "cultural awareness" and "inclusion." This peripheral position indicates that no other participants considered these points worthy of further elaboration or discussion. Finally, by time 105, the network is complete, as no new code emerged and all participants and moderators contributed to the conversation at least once. The final state of the graph shows that "diversification," "team effort," and "evidence-based" were the codes that were more discussed by participants. Participants 3, 6, 2, and one moderator were the most influential in this knowledge exchange. As with our previous example, these exchanges can also be downloaded from this link: https://cutt.ly/2e9Qp8v3. In addition, users can recreate this depiction in HTML format using VERITAS by uploading this file: https://cutt.ly/Wc9Qt9xb to the UI. Like in the previous case, note that VERITAS has this dataset preloaded under the option "Group Analyses of Actor-Code Evolution" in the UI, as we describe in our closing section.

The previous discussion aimed to address questions regarding who created and discussed knowledge in this dynamic network. Overall, the evidence suggests that only one of the moderators generated a variety of codes. However, five of these codes were not further discussed by the participants, which indicates that most of the codes or knowledge generated came from the participants themselves rather than being set by the moderators.

In closing, as briefly elucidated above with the comments made on participant 3, these sets of analyses are informative but only become truly relevant when analyzed based on the actual content of the code. From this perspective, field notes and transcripts, as well as the knowledge and experience of researchers who partook in the focus group and/or were in the field, will undeniably contribute to a better and more robust understanding of these dynamic exchanges. Once more, let us close this section by restating that VERITAS (and Graphical Retrieval and Analysis of Temporal Information Systems (GRATIS)) is not meant to replace the researcher's expertise gained prior to or during data collection. Theoretical lenses, previous knowledge and literature, and, if necessary, additional data collection are fundamental aspects to better understand a given phenomenon.

Discussion, closing remarks, and software access

The decision to select a publicly available video as a data source, along with a working group in a typical research setting to be analyzed in this chapter, was made to provide readers with different perspectives and settings wherein VERITAS can be used. This decision also serves to demonstrate the tools and elements needed to critically assess actors' contributions and exchanges rigorously using VERITAS, which functions as an innovative methodology and software to model complex processes in social settings. Providing transcripts, codes, and Microsoft Word tools to fully replicate the analyses—and hopefully even conduct new ones—was a fundamental goal of this chapter. That is, while it is likely that even with the same transcripts researchers may find slightly different results, it is unlikely, if not impossible, that they would find completely different or contradictory results to the ones depicted here.

As described in our conceptual framework, VERITAS aligns with the notion of *general complexity* as presented by (Byrne and Uprichard, 2012) by providing *reliable* and *valid* models that *represent true social complex systems* (Byrne and Uprichard, 2012). VERITAS aims to capture and produce *sociologically useful and meaningful descriptions* of social world events via "process tracing." VERITAS, then, makes the pursuit of causality—or, at the very least, objectivity or full reproducibility of our shared understandings—more feasible to attain, and with this, the possibility of truly influencing policy planning and actions using *rigorous qualitative evidence.*

With regard to objectively capturing the complexity of a system, to the extent that readers who were able to watch the YouTube video referenced in our example found our analyses "redundant" when compared to the original video, this would be *a clear win* for VERITAS. Recall that in actual research settings and papers, it is unlikely that videos would be shared, likely due to participants' confidentiality protections. Accordingly, insofar as our depictions achieved with VERITAS were so accurate that the tracing of the evolution of these interactions or contributions truly resembled or mirrored the actual events taking place in those videos, the more powerful VERITAS's impact may become in bringing clarity, transparency, and rigor to social science research, particularly in settings that require an understanding of processes and reasons, such as in qualitative research and case studies.

Before closing, it is worth mentioning some of the overarching questions that can be asked in synchronous settings. Note that these questions are just general examples, and researchers can employ this analytic framework to address their own substantive questions in whatever way they see fit for their projects.

Examples of questions that can be addressed with *VERITAS in synchronous* analyses include:

1. When considering the processes of information exchanges among participants, especially when there are group moderators present in the focus of working groups,
   a. Were meaningful exchanges of information driven primarily by experts or by moderators/interviewers? [if the latter were driving these exchanges this may be evidence of bias or overt influence of experts' standings]
   b. How did this process evolve, and when did it start to decay?
   c. What can the observance of this process mean in our meaning-building endeavor?
   d. What might be the most transcendent or impactful moments or exchanges?
2. When considering the evolution of codes or categorized pieces of information,
   a. When and by whom were these codes created and further discussed? [Moderators or experts again.]
   b. How did this process evolve, and when did it decay?
   c. Was there redundancy or evidence of a few actors influencing the group's perspective?
3. In both instances, we can also ask who the most influential actors were in general (aggregated depiction) and over time.

Note that the outputs generated from VERITAS are tools or maps in the analytic process and do not replace the expertise of the qualitative researcher, who remains the most important instrument in social research settings (Brodsky, 2008). We hope that, with the use of VERITAS, new questions will emerge that may prompt researchers to revisit the collected data or potentially even return to the research sites in order to further enrich their understanding. Within the broader infrastructure developed in this book, these same replayable and quote-linked outputs may also serve as structured prompts for local scholarly reasoning in ISARI, where dynamic patterns, interaction sequences, and key inflection points can be compared, memoed, and translated into evidence-grounded qualitative and mixed-methods explanation.

##Integrating VERITAS Outputs into Local Scholarly Brainstorming

One of the strengths of VERITAS is that its outputs are portable across analytic stages. The replayable HTML visualizations, aggregate and temporally disaggregated sociograms, screenshots of key moments, and quote-linked interaction maps are useful not only for identifying the emergence, maintenance, and decay of interactional structure, but also for supporting subsequent interpretive work. In the broader methodological ecosystem developed throughout this book, these outputs can be brought into ISARI, a local, fully offline, open-source, and multimodal brainstorming environment designed to support scholarly memoing, comparison, synthesis, and evidence-grounded writing.

This possibility matters because the replay of group dynamics, however rigorous, does not determine its own meaning. Researchers still need to ask why certain actors became brokers at specific moments, how hot zones should be interpreted, which exchanges became turning points, how moderator influence should be evaluated, and how the resulting temporal structure should be translated into qualitative or mixed-methods explanation. Because ISARI can work with textual notes, screenshots, exported figures, scanned pages, and other image-based or document-based artifacts, researchers can upload VERITAS outputs in multiple forms—for example, screenshots of interaction maps, copied text from quote panels, exported frames of key moments, or notes derived from replaying the dynamic visualizations—and use them as the basis for local scholarly reasoning.

This workflow should not be mistaken for automation of interpretation. On the contrary, it extends VERITAS's methodological purpose. The outputs remain guides, not conclusions. ISARI can help researchers compare interpretations of dynamic patterns, generate analytic memos, test explanatory contrasts, and begin drafting evidence-grounded prose, but the responsibility for deciding what these interactional trajectories mean remains with the research team. In that sense, VERITAS and ISARI are methodologically aligned: one renders the evolution of knowledge generation and communication visible through time, while the other helps researchers think with those outputs under conditions of privacy, interpretive accountability, and local control.

VERITAS user interface execution and software access

VERITAS's UI is shown in Fig. 4.12. In that figure, we can see that all the interactive outputs (both aggregate and dynamic, as discussed above) can be achieved with a few clicks. In the following paragraphs, we describe each of these steps.

The execution of VERITAS requires the following steps.

1. Select a method (actor-to-actor or actor-to-code).

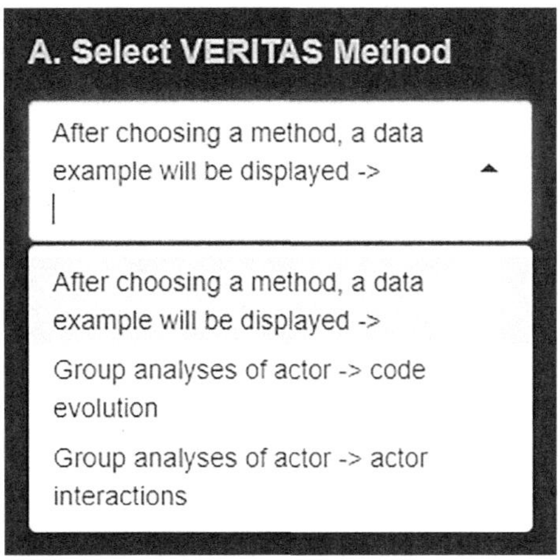

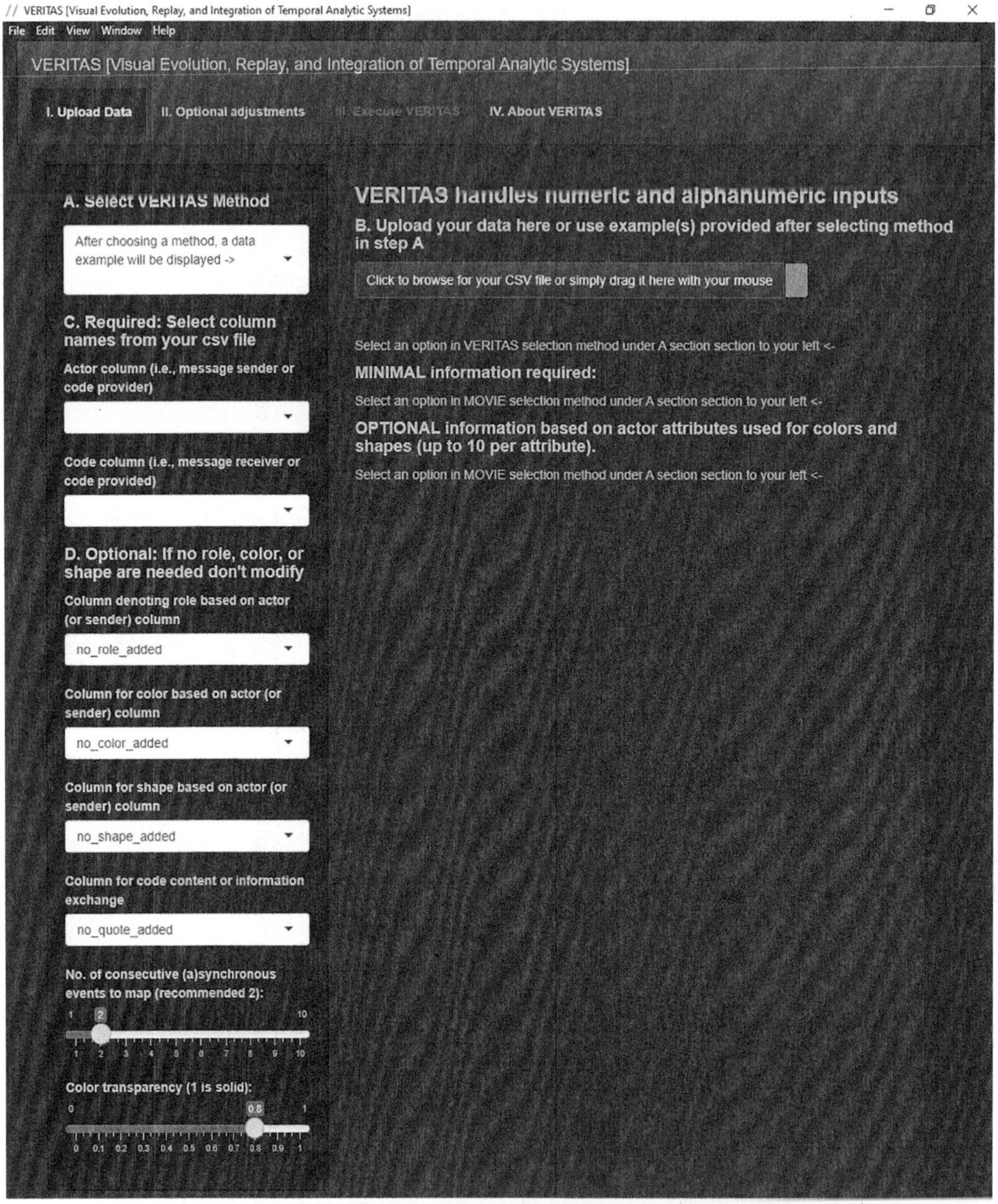

FIG. 4.12 VERITAS user interface.

2. Load the data displayed as indicated in the icon, or upload your own dataset.

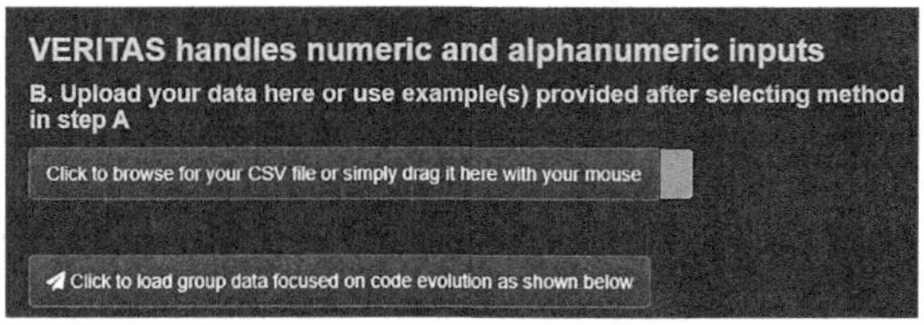

3. Select the sender or actor column (Sender if using our Microsoft Word macro), as shown in the image after point 4 below.
4. Select the receiver or code column (Receiver if using our Microsoft Word macro).

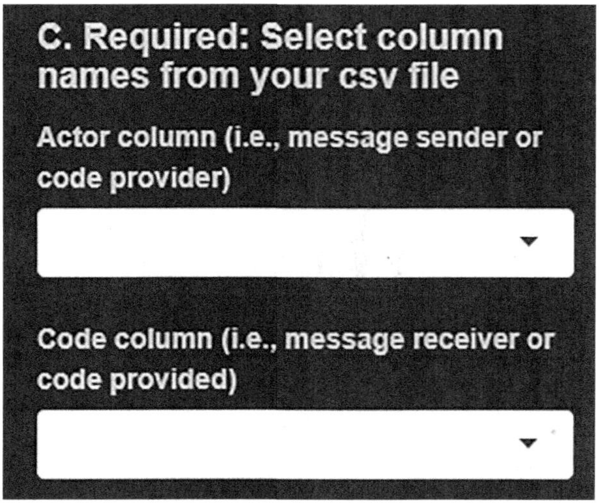

5. OPTIONAL:
   a. If your actor/sender column has attributes, select the corresponding columns in the role, shape, or color menus (see the figure after point "d" below).
   b. If the text of the message sent or the quote of code is available (column "Quote" in our Microsoft Word macro), users can select the corresponding column in the menu for code content (see the figure after point "d" below).
   c. Researchers can map the number of consecutive events to trace. We recommend two, especially if the network has many units (see the figure after point "d" below).
   d. Finally, researchers can select the transparency level of the colors to be plotted.

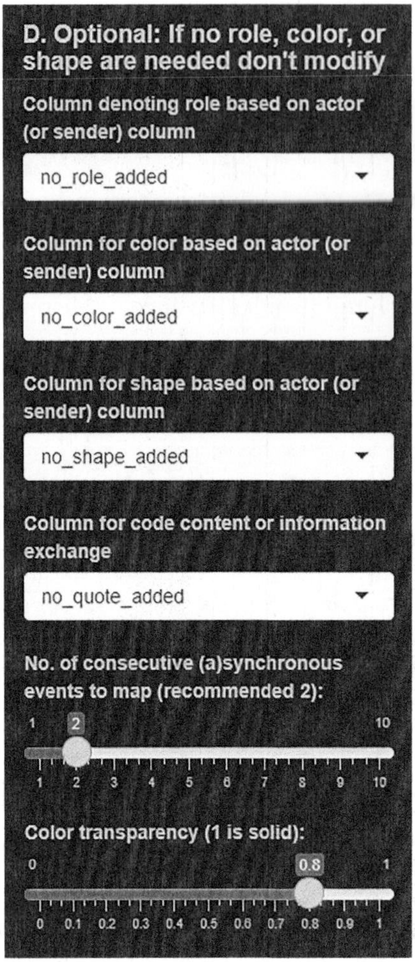

6. Click "Execute the VERITAS" on the top right panel

7. Two HTML files will be launched on your computer's default internet browser. These files can be saved and shared via email. They can also be uploaded to GitHub or your personal website. Because VERITAS outputs may preserve original quotes, actor roles, temporal sequencing, and interaction patterns that could make individuals or small

groups more identifiable, the public sharing of HTML files should always be treated as optional and governed by the same principles of confidentiality, data minimization, participant protection, and contextual sensitivity that guided the original research design (see González Canché, 2025 for a deeper discussion on how data science applications to qualitative research may impact ethics and trust). In many cases, local storage, restricted sharing, or the use of selected screenshots rather than full public release may be methodologically and ethically preferable. Researchers can take snapshots of important events that may help open the "black-box" qualitative analytic process. Within the broader analytic architecture of this book, those same snapshots and replayable outputs may also function as prompts for local AI-supported scholarly reasoning in ISARI, where researchers may compare alternative readings of dynamic interaction patterns, memo explanatory contrasts, and begin translating those patterns into evidence-grounded prose. In addition, researchers can upload the HyperText Markup Language (HTML) files to their website repositories, as demonstrated in the examples presented in this study.

Software access

VERITAS is available on Mac and Windows operating systems. All processes are conducted locally, meaning that no data is uploaded to any server, which serves to protect the anonymity and confidentiality of these analyses. This local architecture is not only a technical feature; it is also an ethical one (see González Canché, 2025). Because VERITAS can preserve quote-level material, actor-level attributes, and temporally ordered patterns of interaction, keeping both the analytic and downstream interpretive workflow on the researcher's own machine is methodologically preferable for many qualitative and mixed-methods projects. That same logic extends to later stages of interpretation, including when VERITAS outputs are examined in ISARI as part of an offline scholarly workflow. In both cases, local execution helps preserve confidentiality, reduce unnecessary exposure of sensitive relational material, and maintain the research team's responsibility for governing how interactional evidence is stored, revisited, and interpreted. This is especially important in small-group, institutionally sensitive, or otherwise identifiable settings, where replayable dynamic outputs may reveal more about participants than a static quotation or summary ever could.

- Access to Windows application: https://cutt.ly/7npPods
- Access to Mac application: https://cutt.ly/Ybcdrhh
- Access to peer reviewed Coce Ocean: https://doi.org/10.24433/CO.6298987.v1

Software execution of Mac and Windows versions

1. Extract the ZIP folder.
2. Open the folder starting with "VERITAS-Application-[Win or Mac]."
3. Double-click on the file called "VERITAS-Application-Mac" or "VERITAS-Application-Win."
4. The first time VERITAS is executed, you may be required to select the menu "View" and "Force Reload" to generate all paths to this newly extracted folder.

- If, after forcing a reload, VERITAS does not display as shown in the online version, close the window and execute VERITAS again.
- If needed, force reload once more. The interface will now load.

5. This process is only needed the first time VERITAS is executed on a computer.
6. VERITAS is housed in a secure server location and requires no installation (just execution). All executable files are virus-free but may require administrative privileges to be deployed.

Mac users are also recommended (and in some instances they do need) to place the VERITAS software (and all software products discussed in this book) in their applications folder for newer versions can only read and write files from this location.

VERITAS is offered free of charge as part of Democratizing Data Science. This openness is methodologically significant. Because VERITAS runs locally, stores outputs in reusable HTML formats, and makes its source code publicly available, the framework supports reproducibility, inspectability, and long-term usability. These same qualities make VERITAS especially compatible with the broader local analytic ecosystem advanced in this book, in which outputs can remain under the researcher's control while being revisited, compared, and interpreted in subsequent stages of analysis. All source code available to expand this software application is located at https://github.com/democratizing-data-science, a novel analytic movement that seeks to remove computer programming prerequisites or requirements that may prevent the use and application of data science, machine learning, and data visualization tools in qualitative and mixed methods research.

References

Alexander, E., Eppler, M.J., Bresciani, S., 2019. Visual replay methodology: a mixed methods approach for group discussion analysis. J. Mixed Methods Res. 13 (1), 33–51.

Almila, Erol, 2017. How to Conduct Scientific Research? Noro Psikiyatri Arsivi. 1300-0667. 54 (2), 97–98. doi:10.5152/npa.2017.0120102.

Artale, A., Parent, C., and Spaccapietra, S., 2007. Evolving objects in temporal information systems. Ann. Math Artif. Intell. 50 (1-2), 5–38. doi:10.1007/s10472-007-9068-z.

Attride-Stirling, J., 2001. Thematic networks: an analytic tool for qualitative research. Qual. Res. 1468-7941. 1 (3), 385–405. doi:10.1177/146879410100100307.

Biggs, N., Lloyd, E.K., Wilson, R.J., 1986. Graph Theory, 1736-1936, Oxford University Press.

Borgatti, S.P., 2006. Identifying sets of key players in a social network. Comput. Math. Organ. Theory. 1572-9346. 12 (1), 21–34. doi:10.1007/s10588-006-7084-x.

Brodsky, A., 2008. Researcher as Instrument the SAGE Encyclopedia of Qualitative Research Methods. Sage, pp. 766 767.

Butts, C.T., Leslie-Cook, A., Krivitsky, P.N., Bender-deMoll, S., networkDynamic: dynamic extensions for network objects. (2020).

Byrne, D., Callaghan, G., 2014. Complexity Theory and the Social Sciences: The State of the Art, Routledge.

Byrne, D., Uprichard, E., 2012. Useful Complex Causality. Oxford University Press. doi:10.1093/oxfordhb/9780195392753.013.0006.

Çaparlar, C.Ö., Dönmez, A., 2016. What is scientific research and how can it be done? Turk Anesteziyoloji Ve Reanimasyon Dernegi Dergisi. 1305-614X. 44 (4), 212–218. Available from: http://www.journalagent.com/z4/download_fulltext.asp?pdir=tard&ppdf=2&plng=eng&un=TARD-34711. doi:10.5152/TJAR.2016.34711.

Coffey, A., Atkinson, P., 1996. Making Sense of Qualitative Data: Complementary Research Strategies, Sage Publications, Inc.

Csárdi, G., Nepusz, T., 2006. The igraph software package for complex network research. InterJournal Complex Syst. 1695 (5), 1–9.

Denzin, N.K., Strauss, A.L., 1988. Qualitative analysis for social scientists. Contemp. Sociol. 0094-3061. 17 (3), 430. doi:10.2307/2069712.

Freeman, L.C., 1978. Centrality in social networks conceptual clarification. Soc. Netw. 0378-8733. 1 (3), 215–239. doi:10.1016/0378-8733(78)90021-7.

George, A.L., Bennett, A., 2005. Case studies and theory development in the social sciences. MIT Press. Available from https://mitpress.mit.edu/9780262072571/case-studies-and-theory-development-in-the-social-sciences/.

González Canché, M.S., 2019. Geographical, Statistical, and Qualitative Network Analysis: A Multifaceted Method-Bridging Tool to Reveal and Model Meaningful Structures in Education Research. Springer Science and Business Media LLC, pp. 535–634. doi:10.1007/978-3-030-03457-3_12.

González Canché, M.S., 2022a. Latent Code Identification (LACOID): a Machine Learning-Based Integrative Framework [and Software] to Efficiently and Reliably Classify Big Textual Data and Rebuild Contextualized, Unaltered Meanings. Int. J. Qual. Methods. 22. https://journals.sagepub.com/doi/10.1177/16094069221144940. doi:10.1177/16094069221144940.

González Canché, M.S., 2022b. Network analysis of qualitative data: an integrative software application to visualize and assess similarities in participants' qualitative contributions. J. Mixed Methods Res. 16 (3), 373–377. doi:10.1177/15586898211051584.

González Canché, M.S., 2025. Ethics and trust in social network analysis: how data science and interactive visualizations impact anonymity, confidentiality, consent, and IRB procedures?. In: Handbook of Social Network Analysis and Education, (pp. 114–131). Edward Elgar Publishing. doi:10.4337/9781035312191.00013.

González Canché, M.S., 2019. Geographical, Statistical, and Qualitative Network Analysis: A Multifaceted Method-Bridging Tool to Reveal and Model Meaningful Structures in Education Research. In: Paulsen, M.B., Perna, L.W. (Eds), Higher Education: Handbook of Theory and Research. Higher Education: Handbook of Theory and Research, vol 34. Springer, Cham. doi:10.1007/978-3-030-03457-3_12.

Holloway, I., 1997. Basic Concepts for Qualitative Research, Blackwell Science.

Kolaczyk, E.D., Csárdi, G., 2014. Statistical Analysis of Network Data with R, Use R!, Vol. 65. Springer, New York. doi:10.1007/978-1-4939-0983-4.

Lazega, E., Wasserman, S., Faust, K., 1995. Social network analysis: methods and applications. Revue Française de Sociologie. 0035-2969. 36 (4), 781. doi:10.2307/3322457.

Lee, R.M., Fielding, N., 1996. 'Qualitative data analysis: Representations of a technology: A comment on coffey, holbrook and atkinson'. Sociol. Res. Online. 1 (4), 15–20. doi:10.5153/sro.1326.

Lincoln, Y.S., Guba, E.G., Pilotta, J.J., 1985. Naturalistic inquiry. Int. J. Intercult. Relat. 0147-1767. 9 (4), 438–439. doi:10.1016/0147-1767(85)90062-8.

Maroulis, S., Guimerà, R., Petry, H., Stringer, M.J., Gomez, L.M., Amaral, L.A.N., Wilensky, U., 2010. Complex systems view of educational policy research. Science. 0036-8075. 330 (6000), 38–39. doi:10.1126/science.1195153.

Miles, M.B., Huberman, A.M., 1994. Qualitative Data Analysis: An Expanded Sourcebook, Sage.

Mitchell, M., 2006. Complex systems: network thinking. Artif. Intell. 0004-3702. 170 (18), 1194–1212. doi:10.1016/j.artint.2006.10.002.

Neale, B., 2019. Qualitative Longitudinal Research SAGE Research Methods Foundations, Sage Publications.

Potter, J., 1996. Representing Reality: Discourse, Rhetoric and Social Construction. SAGE Publications Ltd. doi:10.4135/9781446222119.

Sanchez, M.A., Aguilar, L., Castañón-Puga, M., Rodríguez, A. (Eds.), 2019. Applied Decision-making: Applications in Computer Sciences and Engineering, Vol. 209. Springer.

Smith, L.M., 1978. An evolving logic of participant observation, educational ethnography, and other case studies. Rev. Res. Educ. 0091-732X. 6 (1), 316–377. doi:10.3102/0091732X006001316.

Wasserman, S., Faust, K., 1994. Social network analysis: Methods and applications (Vol. 8). Cambridge University Press. doi:10.1017/CBO9780511815478.

Williams, M., Dyer, W., 2017. Complex realism in social research. Methodol. Innov. 2059-7991. 10 (2), doi:10.1177/2059799116683564.

Wolf, S., González Canché, M.S., Coe, K., 2021. A complex systems network approach to quantifying peer effects: evidence from Ghanaian preprimary classrooms. Child Dev. 92 (6), e1242–e1259. doi:10.1111/cdev.13608.

Zeng, A., Shen, Z., Zhou, J., Wu, J., Fan, Y., Wang, Y., Stanley, H.E., 2017. The science of science: from the perspective of complex systems. China Phys. Rep. 0370-1573. 714-715, 1–73. Available from:. http://www.elsevier.com/locate/physrep. doi:10.1016/j.physrep.2017.10.001.

PART III

Topic Discovery and Language Intelligence Frameworks

5

Latent Code Identification (LACOID)

Introduction

After qualitative evidence has been collected and all textual data from interviews, focus groups, site visits, social media posts, and other sources have been compiled into textual databases, researchers need to *label, code, or classify* the information to ease comparison and draw shared meanings (Coffey and Atkinson, 1996; Miles and Huberman, 1994). Nonetheless, labeling or classifying textual data and qualitative evidence is an expensive and consequential challenge (Coffey and Atkinson, 1996; Miles and Huberman, 1994); put simply, the rigor and consistency behind the construction of these labels may ultimately shape research findings and conclusions (Belotto, 2018).

A multifaceted methodological conundrum when addressing this challenge is *the need for human reasoning for classification* that leads to *deeper and more nuanced understandings*; however, this same *manual human classification* comes with the well-documented increase in classification *inconsistencies and errors*, particularly when dealing with vast amounts of documents and teams of coders (Chang et al., 2021; Poth et al., 2021). An alternative to human coding consists of machine learning assisted techniques based on data science, visualization, and natural language processing (NLP). These techniques offer quantitative tools for data labeling and classification that are cost-effective and consistent but are prone to losing *participants' meanings or voices* for two main reasons: (1) these classifications typically aggregate all text inputs into a single topic, code, or tag; and (2) the words that configure texts are analyzed outside of their original contexts (Eickhoff and Wieneke, 2018).

From this view, the analytic conundrum and challenge in text classification of big data entails both taking advantage of the consistency in the classification process that is gained with quantitative tools, but doing so *without* altering participants' inputs, contributions, and experiences, and *without losing* contextual elements of those experiences. Moreover, this efficiency and reliability should not happen at the *expense of losing our human reasoning and input,* which are fundamental to recovering or rebuilding meanings and to strengthening our understandings.

To address this challenge and analytical conundrum, we present an analytical framework and free software tool designed to leverage the power of machine learning, text mining, and advanced text visualization to classify textual data or qualitative evidence while fully utilizing the benefits of human reasoning in both this classification and during our sense-making processes. From this perspective, the goal of this study is to address *the following question*:

How can researchers classify vast amounts of qualitative evidence effectively and efficiently without losing context or the original voices of research participants, while leveraging the nuances that human reasoning brings to qualitative and mixed methods analysis?

Data Science, Interactive Visualizations, and Generative AI Tools for the Analysis of Qualitative, Mixed-Methods, and Multimodal Evidence. https://doi.org/10.1016/B978-0-443-21961-0.00003-6

Our proposed analytic framework was designed to mirror, as closely as possible, the *line-by-line coding employed in human/manual code identification,* relying instead on latent Dirichlet allocation, text mining, Markov chain Monte Carlo (MCMC) simulations, Gibbs sampling, and advanced data retrieval and visualization. More importantly, these analytic procedures are conducted via a no-code, cost-free software application (see González Canché, 2022a) that (1) offers a comprehensive set of outputs providing complete transparency of the classification process and (2) aids in recreating the *contextualized, original, and unaltered meanings* embedded in the input documents, as provided by participants.

Meaning and context preservation

Our emphasis on "original and unaltered meanings" is particularly important for our proposed analytic framework. As briefly mentioned, currently available classification approaches using text mining (see Feinerer, 2020) and machine learning require the analysis of words outside their original context (Eickhoff and Wieneke, 2018). During standard and traditional machine learning text classification processes, these decontextualized words are also normalized (i.e., altered) to facilitate their mathematical handling and statistical classifications. Notably, these classified outputs do not attempt to retrieve or preserve original texts; instead, the potential loss of meaning is treated as a price to be paid for the gain in time efficiency. This implies that text analysts must form their understanding by relying only on these modified and decontextualized resulting classified texts. LACOID was developed precisely to resist that reduction: its outputs are designed not only to classify, but also to preserve the conditions under which interpretation can remain accountable to participants' original language, textual location, and contextual meaning.

Our proposed analytic framework differs from the traditional—and still frequently used—classification approach in that original texts containing our participants' voices and thoughts are never altered. This preservation is not only a technical feature but an epistemic commitment, because it keeps latent code identification accountable to the actual language through which participants expressed their experiences, concerns, and reflections. Instead of altering input texts, these original narratives are preserved and displayed as part of the standard outputs containing classified information. More to the point, the actual original location of each classified piece of information, as situated in each original participant's discourse (i.e., interview transcript, essay, social media post), is also provided in our standard outputs, so that qualitative and mixed-methods researchers in charge of (re)building knowledge may easily retrieve not only unaltered meanings but also the specific context where those meanings were originally offered by our participants (see Table 5.1 and Fig. 5.1).

Table 5.1 Example of original and modified texts.

ID	Code	Text mined text	Original text
doc1.doc.1	V1	*keep average sentence length document around 20 25 words*	*Keep the average sentence length of your document between 20 and 25 wordsto ensure readability and maintain a balanced flow of information without overwhelming the reader.*

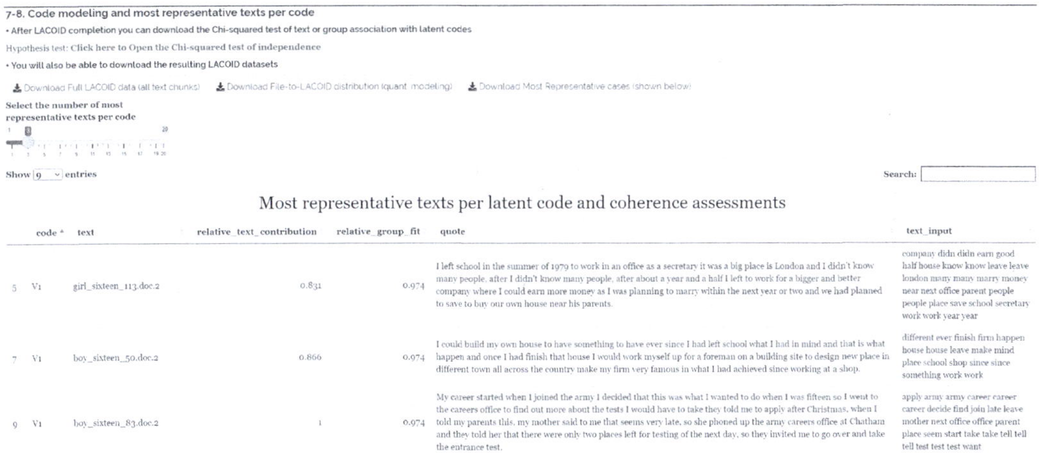

FIG. 5.1 Main LACOID output showing most representative texts, relative Text contribution, relative group fit, original/unaltered Text, and text-mined Text.

Types of textual data sources handled

Our analytic framework, along with its corresponding free software application, supports the classification of unstructured data or qualitative evidence in the form of interview transcripts, focus group transcripts, essays, policy briefs, open-ended responses in survey research, and even social media posts. To illustrate the software's capabilities and its suite of outputs, we provide access to a publicly available textual database consisting of 153 essays, as provided by Pahl (2012), which can be accessed through González Canché (2022b). As part of our analytic framework, instead of offering a single classification for each document uploaded, we first decompose each input file (i.e., essay) into dozens, hundreds, or even thousands of texts, and these resulting texts are then ready to be classified. This document-to-text decomposition is important because it makes it possible to retrieve a multiplicity of meanings per input document (i.e., essay) instead of automatically classifying each document into a single topic, which is how traditional machine learning text classifications currently work. (This *one code per document* is a procedural limitation of most machine learning text classification approaches, which we describe as latent code aggregation bias (LCAB) below and which our framework and software tool aim to overcome.)

As part of this demonstration, we illustrate how this classification process can be completed in minutes rather than weeks or months. The relevance of these outputs extends beyond classification efficiency alone. Once latent codes have been identified, their most representative texts have been retrieved, and file-to-code distributions have been estimated, researchers still need to determine what those patterns mean, how competing interpretations should be compared, which cases deserve closer memoing, and how the resulting structure should be translated into qualitative argument. In the broader architecture of this book, LACOID outputs can therefore be understood not as endpoints, but as structurally rich guides that may also be brought into ISARI, a local, fully offline, open-source, and multimodal brainstorming environment designed to support scholarly comparison, memoing, synthesis, and evidence-grounded writing. In this way, LACOID and ISARI are methodologically aligned: LACOID classifies and organizes large-scale textual evidence while preserving original voices; ISARI helps researchers think with those outputs under conditions of privacy, interpretive accountability, and local control. However, we also note that the use of generative or reasoning AI tools are completely optional, for researchers can use all LACOID outputs as guidelines or maps to form their own understandings. In short, our goal is to illustrate that the resulting classification is only the first step of our analytic journey. Once these classes or codes are retrieved, the expertise of the research team is fundamental to bringing these classes or codes "alive" by rebuilding their meaning based on the unaltered and contextualized original voices of research participants. And this process can be aided with AI as a brainstorming parner but the final analyses and conclusions may also be achieved without AI support.

Purpose of the study

The purpose of this paper is to showcase all the elements required to address the methodological conundrum of *how to rely on machine learning and text classification while minimizing the loss of context and leveraging the nuances that human reasoning brings to the analysis of qualitative evidence.* Accordingly, in this study, we illustrate all the practical analytic steps guiding our analytic framework and provide a hands-on example, concentrating primarily on the outputs of the analysis and its interpretation, to formally introduce this framework and software tool as Latent Code Identification (LACOID).

We offer access to the textual database required (González Canché, 2022b) along with the software tool (González Canché, 2022a, Mac https://cutt.ly/jc7n3OT, and Windows https://cutt.ly/wc7nNKF) needed to replicate the analyses presented, so that qualitative and mixed-methods researchers can experience firsthand the entire analytic process. We hope that this opportunity to become familiar with the analytic framework and free software tool results in its widespread application.

Structure of the manuscript

In the following section, we present a summary of human code identification (HUCOID), understood as the process in which human coders are trained to identify codes or classes in qualitative data. We also describe computer-assisted techniques based on data science, machine learning, and NLP, which we refer to as *machine learning code identification* (MACOID). In addition, we explain how we conceptualize the integration of HUCOID and MACOID and how such an integration benefits the entire analytic process. In this respect, given that our proposed framework builds on quantitative and qualitative methods that are integrative rather than competing, we also offer a brief discussion of the epistemological lenses driving this integration, where both approaches are classified as holding a fully mixed equal-status design (Alexander et al., 2019; Fetters et al., 2013; Jason and Glenwick, 2015; Teddlie and Tashakkori, 2009).

An important contribution of the proposed analytic framework is our goal to reduce or avoid code aggregation bias, which consists of classifying large amounts of text under a single code when such text may be configured by a myriad of classes or codes. Accordingly, in a subsequent section, we present our analytic strategy specifically designed to minimize or avoid this issue. In this study, we refrain from overly technical details regarding NLP and topic modeling, which can be found elsewhere (see González Canché, 2022c). Instead, we focus on the practical applications of the framework and software, as well as the relevance of each of its outputs for our meaning-building process.

Human and machine learning code identifications

In this section, we discuss the two approaches that we, as qualitative and mixed-methods researchers, have typically employed for classifying text data and qualitative evidence. Note that instead of being specific about strategies or methodologies for qualitative or quantitative coding, we focus on the overall processes in both the manual (human-driven) and machine (computer-driven) code-identification processes.

Human code identification

The predominant approach to addressing this labeling or classification challenge consists of qualitatively, or manually, coding textual data. Although this manual code identification process greatly benefits from humans' nuanced and contextualized understandings (Poth et al., 2021), it remains a time-consuming and expensive process that is prone to human errors and inconsistencies, particularly when dealing with large-scale qualitative projects (Chang et al., 2021; Poth et al., 2021) and in situations that require the rapid retrieval of understandings of participants' experiences and actions to make decisions or implement strategies (Ho et al., 2021) (Recent examples of these latter situations are times of crises like the COVID-19 pandemic (Chang et al., 2021; Ho et al., 2021; Poth et al., 2021).

Note that although computer-assisted qualitative data analysis (CAQDAS) software such as ATLAS.ti, NVivo, or MAXQDA may ease this human or manual labeling or coding challenge (Humble, 2019; Tesch, 2013), the coding process remains time-consuming and labor-intensive because both the process and the final product fully depend on researchers' individual inputs and coding decisions. That is, CAQDAS software is not yet capable of being trained to *learn and identify text pieces that may be good candidates for specific codes or labels.* From this view, CAQDAS software remains completely dependent on human inputs, which are subject to coding inconsistencies and errors, especially when dealing with vast amounts of documents and teams of coders (Chang et al., 2021; Poth et al., 2021).

A common strategy for minimizing subjectivity and room for interpretation in human code assignment is through intercoder agreement (Belotto, 2018; Eickhoff and Wieneke, 2018; Miles and Huberman, 1994). In this validation process, "team members separately code (a sample of) the given data set [and then] compare and discuss the assigned codes and applied coding rules and recode the data according to the agreed solution... [to] correct for discrepancies in individual judgment, and for joint mistakes that become apparent during analysis" (Eickhoff and Wieneke, 2018, p. 904). From this view, the intercoder agreement test/training may increase coding reliability, but once more, depending on the scale of the project, the costs in both time and resources may become unaffordable, particularly when dealing with time-sensitive information. Relatedly, despite this training effort, the final product is, once more, still subject to inconsistencies, particularly in this era of big data (González Canché, 2022c).

MACOID

An alternative, yet seldom-used, analytic approach to tackling this labeling challenge relies on machine- and data science–driven tools involving NLP, text and data mining, machine learning, and topic modeling to *identify patterns in unstructured* data (Chang et al., 2021; Grün and Hornik, 2011; Ho et al., 2021; Lynam et al., 2020; Poth et al., 2021). These computer- and statistically-powered analytic tools are designed to deal with large-scale textual or written data sources and are capable of classifying vast amounts of textual data or unstructured information faster, better, and more consistently than humans (Chang et al., 2021; Eickhoff and Wieneke, 2018; Janasik et al., 2009; Lynam et al., 2020; Poth et al., 2021) by mathematically learning word-to-text distributions and offering the probabilistically most suitable classification solution based on thousands of learning trials.

In terms of the performance of these quantitative techniques when handling time-sensitive data, a recent special issue of the *Journal of Mixed Methods Research* (see Fetters and Molina-Azorin, 2021) demonstrated how these computer-assisted techniques helped researchers deal with the deluge of COVID-19 data that needed to be rapidly and rigorously analyzed to strengthen policy development.

Limitations of MACOID

Despite these important contributions of MACOID, this analytical strategy has two main limitations. One is procedural, and the other is conceptual and analytical, as we describe next.

Conceptual and analytical limitation

The conceptual limitation relates to the *analytic conundrum* we described in the opening section. This analytic conundrum, associated with the use of these techniques assisted by machine learning and data science, is that even though these tools are *excellent at applying rules to detect patterns* better or more consistently than humans (Chang et al., 2021; Ho et al., 2021; Poth et al., 2021), their main limitation is the "absence of human reasoning around classification" (Poth et al., 2021, p. 351), which may lead to overlooking nuances and context in the classified texts. From this perspective, MACOID's outputs would not convey any social meaning and would remain abstract and nonsensical without human intervention.

To address this limitation, we offer a balanced analytic framework that, after textual data or qualitative evidence is collected, relies first on MACOID as a strategy to consistently apply rules across thousands of *decomposed text inputs* and then, second, on human input to create or recreate the social meanings embedded in these outputs. From this perspective, rather than treating MACOID and HUCOID as competitive frameworks, our analytic approach integrates them. More specifically, we can express this integration as a purposeful reliance on human reasoning via *guided qualitative analyses* of the *quantitative outputs* rendered by MACOID to retrieve the *meaningfulness* of the original and contextualized preprocessed information.

Procedural limitation

Despite MACOID's analytic power and the possibility of applying these computer-assisted tools using open-source, cost-free software (like the R Project for Statistical Computing), it poses a procedural limitation; namely, its use and application remain conditioned on statistical and computer programming expertise. For example, Chang et al., (2021) mentioned that "deep linguistic analysis of text-based research data is still reserved for investigators or research teams with a computer science background" (p. 400). This computer-programming knowledge represents an important hurdle that has prevented qualitative and mixed-methods researchers with no training in NLP and computer or statistical programming from leveraging the benefits of machine learning–driven tools to classify textual data.

This study addresses a procedural limitation by completely eliminating computational, technical, and financial barriers associated with implementing and using MACOID (NLP, machine learning, and text classification and visualization tools) to conduct fully integrative analyses of large-scale textual and qualitative data. To this end, we offer a completely *cost-free* and *computer programming-free* rigorous software application (Software access, local execution, and reproducibility details are provided in the chapter's software access note and accompanying source-code materials.) that fully implements the *mixed methods integrative* framework guided by the following epistemological lenses.

Epistemological lenses

The main premise of our analytic framework is that, despite the added accuracy and efficiency of using computer-assisted NLP techniques to identify *latent codes* (latent refers to unobserved factors that make observed indicators—traditionally columns in databases—correlated, as in factor analysis), in textual databases, the input data are unstructured, which requires text mining and normalization procedures (presented in the appendix) to create document-term matrices. The columns in these document-term matrices are populated by the "cleaned" vocabulary in our texts (which are sentences or paragraphs), and the goal of topic modeling via Latent Dirichlet Allocation is to identify the probability of a given sentence or paragraph (represented in the rows of this document-term matrix) being part of a latent topic or code. As in factor analyses, these latent codes have no inherent meaning; this meaning needs to be retrieved from the sentences or paragraphs that have been statistically identified as belonging to such a latent code. However, those techniques might fall short in yielding relevant findings and valuable conclusions without qualitative and nuanced understandings of the contextualized meanings of those latent codes based on the original and unaltered textual data. Considering this premise, we find it important to discuss the epistemological lenses driving the MACOID and HUCOID integrations, a framework we henceforth refer to as LACOID.

Understanding epistemology as the nature, origin, and limits of human knowledge (Stroll and Martinich, 2022), LACOID is guided by the epistemological belief that knowledge can be built from machine learning and text classification analyses of qualitative evidence, but *only if* this output is fully integrated with the qualitative analysis of this *classified, unaltered, and contextualized* qualitative information. This implies that the original voices, as expressed by research participants, must be preserved and contextualized in their original document inputs, including the location of these texts and words within their original and unaltered discourses.

From this perspective, the resulting qualitative analysis may be enriched and contextualized because LACOID was built to never lose the original voices and contexts that give qualitative evidence its meaning and nuances. Empirically, with this context preservation as our priority, participants' unmodified voices must retain their original position in each document uploaded to LACOID, thus strengthening the reconstruction of contextualized meanings embedded in those machine-learned latent codes.

Accordingly, LACOID's epistemology represents a fully mixed, equal-status design, wherein rigorous, valid, and useful knowledge is achieved only by integrating the quantitative analysis of qualitative evidence with the qualitative analysis of the resulting classified codes. From this perspective, machine classification does not eliminate the need for qualitative interpretation; it intensifies the importance of it. The latent codes identified by LACOID only become analytically useful when researchers return to the original texts, compare the most representative passages, assess overlaps and distinctions across codes, and reconstruct meaning in relation to the purposes of the study. This same epistemological commitment also clarifies the role of local AI in the broader ecosystem of the book: systems such as ISARI are not meant to decide what latent codes mean, but to support researchers as they compare, memo, synthesize, and draft interpretations grounded in the original voices that LACOID preserves. Overall, this integrative and holistic analytic strategy implies that if one of these approaches is conducted poorly, the entirety of the analysis may fall short in reconstructing the meaning and knowledge embedded in our qualitative evidence.

Moreover, note that although LACOID will always consistently identify latent codes without altering the original meanings or texts from input documents, this does not mean that the resulting knowledge will always be useful or informative. If the data-gathering process for these input documents lacks methodological rigor, or if data are collected without internal consistency or validity, LACOID will not be able to fix or solve this data-input quality problem. More to the point, although LACOID's algorithms do not suffer from bias toward certain groups due to its unsupervised classification nature (which means that the learning process starts without any preconceived notions), if the qualitative data do contain these biases (i.e., if our participants' discourses are themselves biased toward certain groups), the classified results will still reflect these issues, precisely because LACOID does not alter meanings or original texts. That is, bringing these two points together, LACOID, by itself, would not know whether the input information meets a certain quality standard or whether this input information is subject to biases. Accordingly, quality assurance and equity issues require the expert and qualitative analysis of the research team because LACOID is not programmed to detect these highly consequential issues. This, once more, reflects the relevance of a fully mixed, equal-status design (Alexander et al., 2019; Fetters et al., 2013; Jason and Glenwick, 2015; Teddlie and Tashakkori, 2009), where both qualitative and quantitative perspectives are truly important in the knowledge-building process. Fig. 5.2.

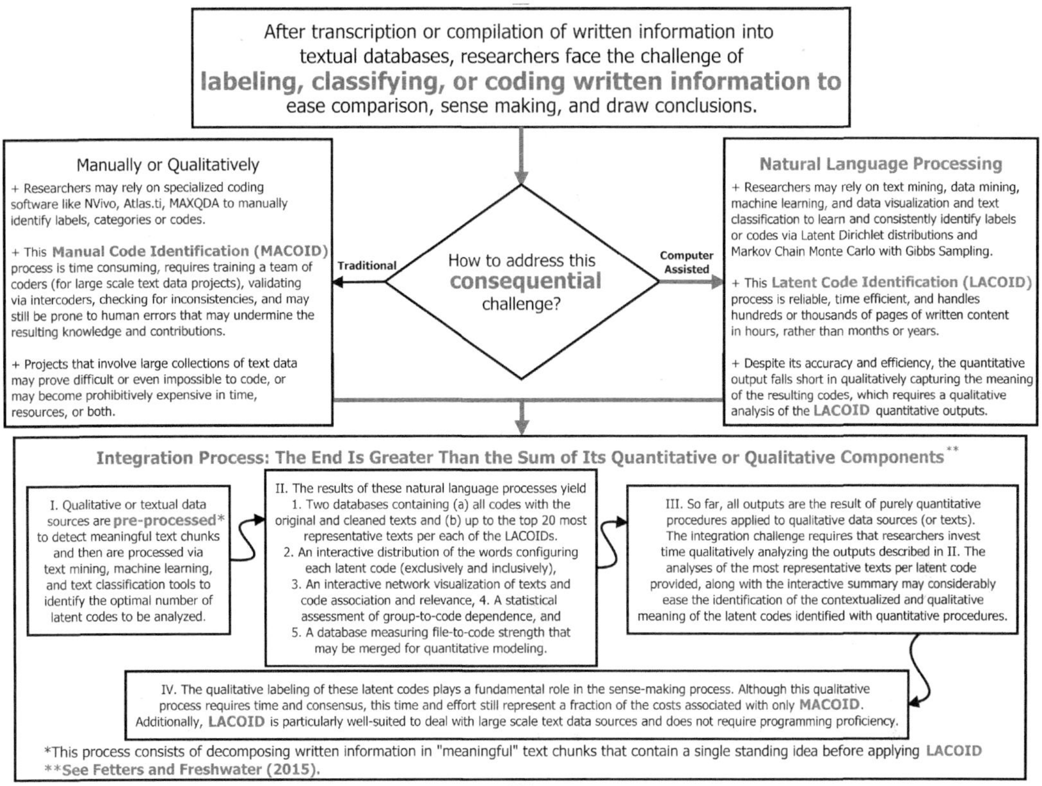

FIG. 5.2 Procedural diagram.

Practical software application and outputs

Fig. 5.3 summarizes the main components of the LACOID analytic framework. The first thread is qualitative and involves collecting qualitative evidence that must be transcribed as part of the inquiry process. For the quantitative thread, LACOID was designed to execute machine learning, text classification, and visualization procedures with just a few clicks. Indeed, the entire quantitative process entails the following steps, which we expand upon below (with additional details provided in the applied example in the "Outputs and Findings" section):

1. Select the type of text decomposition to be applied when uploading documents. Here the options are sentences or paragraphs 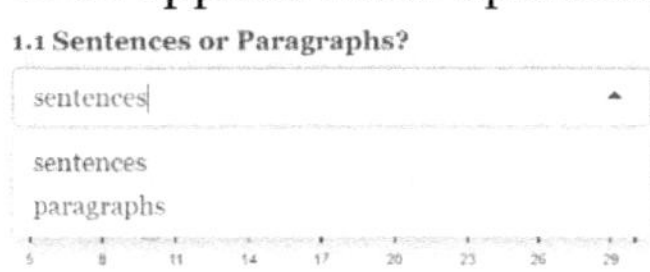

2. Upload documents 

. The text decomposition is automatically applied based on the selection in Step 1.

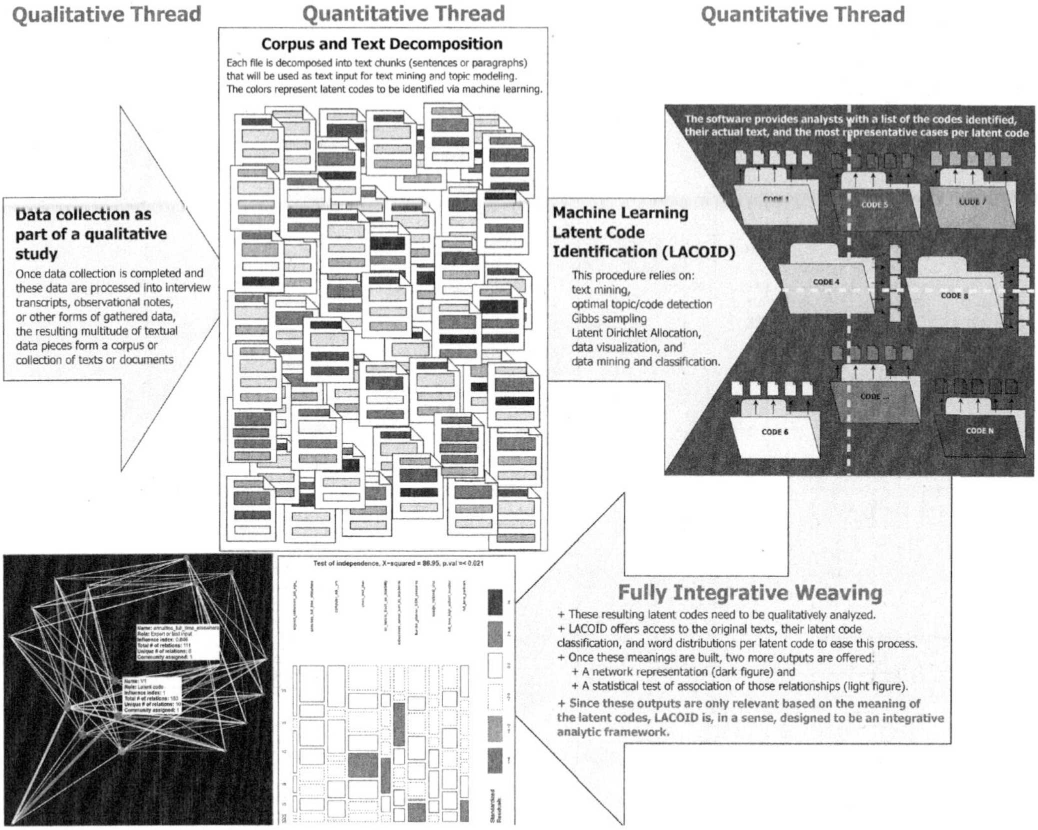

FIG. 5.3 LACOID analytic framework.

3. Execute text normalization and cleaning [Execute Text Mining] (i.e., text mining or natural language processing; see appendix).
4. Decide whether to remove or retain common words [Trim Common Words].
5. Select parameters for machine learning. Recommended values are 500 samples to establish baseline parameters (also known as burn-in samples) and a learning process based on 5000 iterations (Raftery and Lewis, 1991).
6. Execute metrics assessments [Execute Metrics] to find the optimal number of codes. See appendix and applied example, and Nikita and Chancy (2020) and González Canché (2022c).
7. Select optimal number of latent codes  based on the result of the metrics executed in Step 6. See applied example.
8. Execute the classification [Execute LACOID] that will render the Nth number of codes selected in Step 7. See applied example.

LACOID outputs

Before elaborating on these steps, we first want to discuss the set of outputs that LACOID will automatically make available.

After Step 6 is executed, LACOID will generate a plot to facilitate the detection of the optimal number of codes. Based on this plot, researchers may then proceed to classify the entire textual database, as illustrated below.

After Step 8 is executed, the following outputs will be generated:

- An interactive distribution of the words configuring each latent code
- Two databases containing
- All codes with the original and cleaned texts
- Up to the top 20 most representative texts for each of the latent codes
- A statistical test of group-to-code association (see the "Hypothesis Testing of Group and Latent Code Association" section below)
- An interactive network visualization of texts and code associations and relevance (related to the statistical test)
- A database measuring file-to-code strength that may be merged for posterior quantitative modeling. This final output provided by LACOID adheres to hypothesis generation based on integration (Fetters, 2020). In this case, the output consists of a database (see Fig. 5.4) that contains the strength of the distribution of each file (at the individual level) to each latent code. As explained in our applied results section described next, this database may be merged with quantitative databases to use latent code strength (that is, the percentage of a given participant's sentences that were ascribed to code i, for example, and how this prevalence may explain a given outcome y). From this view, the fully integrative weaving nature of all statistical tests derived from LACOID is based on the fact that these tests and outputs are only

	A	B	C	D	E	F	G	H
1	ID	Tot_number_LACOID	Weight.V1	pct.V1	Weight.V2	pct.V2	Weight.V3	pct.V3
2	boy_sixteen_1.doc	26	8	30.8	10	38.5	8	30.8
3	boy_sixteen_2.doc	24	8	33.3	9	37.5	7	29.2
4	boy_sixteen_11.doc	9	7	77.8	0	0	2	22.2
5	girl_seventeen+_153.doc	42	26	61.9	12	28.6	4	9.5
6	girl_seventeen+_154.doc	8	2	25	5	62.5	1	12.5
7	girl_sixteen_87.doc	12	5	41.7	0	0	7	58.3
8	girl_sixteen_88.doc	11	1	9.1	6	54.5	4	36.4
9	girl_sixteen_90.doc	26	7	26.9	7	26.9	12	46.2
10	girl_sixteen_91.doc	17	7	41.2	5	29.4	5	29.4
11	girl_sixteen_92.doc	28	12	42.9	6	21.4	10	35.7
144	boy_sixteen_10.doc	3	1	33.3	0	0	2	66.7
145	girl_seventeen+_143.doc	11	2	18.2	6	54.5	3	27.3
146	girl_seventeen+_144.doc	12	4	33.3	7	58.3	1	8.3
147	girl_seventeen+_145.doc	11	4	36.4	6	54.5	1	9.1
148	girl_seventeen+_146.doc	13	5	38.5	6	46.2	2	15.4
149	girl_seventeen+_147.doc	9	5	55.6	1	11.1	3	33.3
150	girl_seventeen+_148.doc	10	3	30	4	40	3	30
151	girl_seventeen+_149.doc	4	2	50	2	50	0	0
152	girl_seventeen+_150.doc	12	8	66.7	4	33.3	0	0
153	girl_seventeen+_151.doc	12	4	33.3	5	41.7	3	25
154	girl_seventeen+_152.doc	7	2	28.6	3	42.9	2	28.6

FIG. 5.4 Example of Classified Data that Can be Used in Quantitative Analyses Such as Regression Models' Outcomes or Predictors

relevant based on the qualitative understanding of the contextualized meaning of these latent codes. In other words, none of the outputs are useful or make sense based only on their quantitative descriptions; instead, their meanings need to be decoded from the original voices of participants or texts, as captured by their original sentences or paragraphs.

In addition to these outputs, all texts that did not meet the inclusion criteria (by being too vague to convey meaning, as explained below) are also available for download and analysis. This database is added for transparency; it is not necessarily considered an output of LACOID but rather a byproduct of the steps required to conduct LACOID.

All these outputs are also portable across analytic stages. The interactive word distributions, the downloadable original-and-cleaned-text tables, the most representative text summaries, the statistical tests of group-to-code association, the network visualizations, and the file-to-code strength database are useful not only for identifying latent structure, but also for supporting subsequent interpretive work. In the broader ecosystem developed in this book, researchers may bring these outputs into ISARI by uploading screenshots of the interactive plots, copied text from the summary panels, exported figures, or notes derived from the classified datasets. Because ISARI can operate locally and work with both textual and image-based artifacts, these LACOID outputs can become the basis for scholarly brainstorming, memoing, comparison of alternative interpretations, and evidence-grounded writing without moving sensitive materials outside the researcher's control.

Data and methods

In this section, we briefly describe the data employed in our example, along with the data preparation steps for hypothesis testing and network visualization, which are computed automatically by LACOID based on the resulting classified collection of texts.

Data

The textual dataset analyzed is publicly available. These data correspond to the School Leavers Study (Pahl, 2012), a collection of 154 essays "written by school children from the Isle of Sheppey in 1978... collected as part of a wider ethnographic study," where Pahl "investigat[ed] the changing patterns of work, the division of labor, unemployment, deindustrialization, and the informal economy" (para. 3). Specifically, participants were asked, 10 days before leaving school (most at age 16), "to imagine that they were nearing the end of their life, and that something made them think back to the time when they left school. They were then asked to write an account of their life over the next 30 or 40 years" (para. 4).

Note that these essays are publicly available here https://cutt.ly/8EBKXjm. This collection of files is in rich text format (".rtf"), but for researchers' convenience LACOID was designed to handle Microsoft Word ("*.doc" not "*.docx") files (*.docx files contain HTML format that conflicts with text cleaning across operative system platforms. Accordingly, if transcripts are only available in *.docx files, they should be saved as *.doc files before uploading them to LACOID). Accordingly, before uploading these files to LACOID, we made two changes: the first was the transformation from rich text format to "*.doc" Word format. These Word files are available here https://cutt.ly/nENMH6P or here González Canché (2022b) (to use, simply unzip and upload to

LACOID). The second change consisted of modifying the file names to allow for group comparisons. Specifically, following the documentation provided by Pahl (2012), here https://cutt.ly/NEBKLKC, and the rationale presented in Fig. 5.5, explained next, we created input file names that capture gender, age, and original essay ID. That is, with the combination of Essay ID 1, Gender male, and Age 16, [Essay ID | Gender | Age: 1 | male | 16] ⇨ [ID: boy_sixteen_1.doc] we created a name file called became boy_sixteen_1.doc. Similarly, when the combination was [90 | female | 16], the resulting file name became "girl_sixteen_90.doc." Finally, when the combination was [154 | female | 17+], the file name became "girl_seventeen +_154.doc."

The use of "sixteen" or "seventeen" to account for ages (as opposed to "16" or "17") is required because LACOID removes numerals from file names in order to group responses and test hypotheses of association. Specifically, *as we explain in more detail below*, all responses of "boy_sixteen_#.doc", where "#" is a placeholder for *an actual number*, as in "boy_sixteen_1.doc," will become "boy_sixteen_.doc" after initial LACOID processing, thus effectively removing the number "#" or "1" from this file name. From this perspective, for LACOID, all documents called "boy_sixteen_.doc," all documents called "girl_sixteen_.doc," and "girl_seventeen +_.doc" represent three distinctive groups accounting for the intersection of the same gender and age to be

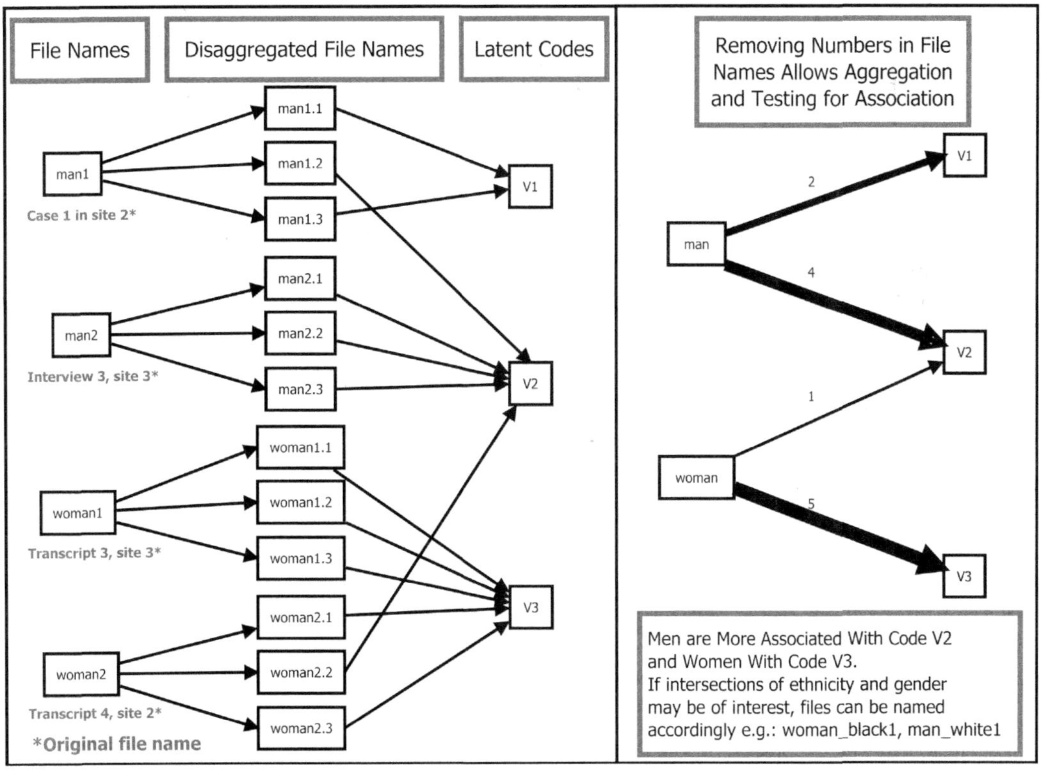

FIG. 5.5 Hypotheses testing framework and file name change rationale.

compared against other group intersectionalities (we further elaborate on this naming rationale in Fig. 5.5 below). This is achieved to test for potential associations across groups and their classified text contributions—as discussed in the next sub-section.

With these changes, the resulting groups across the 154 essays made available by Pahl (2012) were distributed as follows: "boy_sixteen_.doc" (n = 89), "girl_sixteen_.doc" (n = 51), "girl_seventeen+_.doc" (n = 13), and "girl_fifteen_.doc" (n = 1). Note that, of the 154 essays, one respondent was 15 years old. If left in the analytic sample, this single case and its associated latent codes would be considered a single group to be compared against others. Accordingly, we removed this respondent from the analytic sample. Her essay, however, can be read in the file called "girl_fifteen_125.txt," located in the same folder shared above (https://cutt.ly/nENMH6P). LACOID ignores this file when uploading all Word files, given the "*.txt" extension of this document.

Hypothesis testing of group and latent code associations

Before explaining in more detail the analytic steps described in the "Practical Software Application & Outputs" section above, let us note that the LACOID software was also automatically programmed to conduct a hypothesis test that estimates whether groups of participants (or individual participants) are more or less likely to be associated with certain latent codes. We emphasize the change in file names because this hypothesis test is conducted based on the names of the input documents. Given the relevance of this process for our integrative purpose, we next elaborate on the rationale, followed by an explanation of the hypothesis test.

Hypothesis of group and latent code association

First, let us clarify that if no names are changed, LACOID will test whether certain individual actors are more or less prone to being associated with some latent codes than with others. This statistical test aligns with the hypothesis generation integration goal (Fetters, 2020) and, by default, tests the relationship of all uploaded documents with the identified latent codes. Nonetheless, inspired by Ho et al. (2021), who tested for differences across groups rather than across individuals, if researchers prefer to test whether certain groups are more or less associated with certain latent codes, they can test these hypotheses by following this simple rule: *name each file (or Word document) to be uploaded to LACOID based on the group that will be compared in the association tests, followed by a number.*

To describe this individual or group test, we turn to Fig. 5.5. This figure contains two panels describing how LACOID handles the input documents. To begin, let us assume that we have four documents (or files) as shown in the left panel of Fig. 5.5; the original names of these files are shown in magenta. If we leave these names unchanged and load them into LACOID, each original file will be decomposed into sentences or paragraphs to avoid LCAB (as described in its own subsection below). Taking the first file as an example, "Case 1 in site 2" will become "Case 1 in site 2.**1**", "Case 1 in site 2.**2**", and "Case 1 in site 2.**3**", with the .**1**, .**2**, and .**3** in bold font indicating the respective sentence or paragraph configuring the content of this "Case 1 in site 2" input document. If these names are left unchanged, the configuring texts associated with this document will be classified with either Code 1 (V1) or Code 2 (V2). In the case of the second file

("Interview 3, case 2"), all three texts are associated with Code 2. In the two remaining cases, all but one response was associated with Code 3 (V3). At the individual level, we would conclude that Case 2 ("Interview 3, case 2") and Case 3 ("Transcript 3, case 3") were more closely associated with Codes 2 and 3, respectively, than with the other two cases. As can be seen from this example, these analyses are done at the individual level, and, although manageable with few cases, when our datasets include dozens or even hundreds of input documents, individual tests of associations between documents and latent codes become less useful. As an alternative, next we present the rationale followed to move this test to the group level.

In terms of group comparison (right panel of Fig. 5.5), assume that we have identified that Case 1 and Interview 3 are men, and Transcripts 3 and 4 are women. Accordingly, if we wanted to test for differences between men and women in their associations with the latent codes, we could simply name their respective Word files as follows: man1, man2, woman1, and woman2. However, if instead we wanted to make comparisons across sites, note that *Case 1* and *Transcript 4* belong to *Site 2,* and *Interview 3* and *Transcript 3* belong to *Site 3*. Accordingly, we could also have named these files as site_two1, site_two2, site_three1, site_three2. This latter naming approach resembles the example we present in our applied section below. Note that in this latter case, although we are still capturing the site number, we used words to represent this number rather than numerals. This is based on LACOID's need to remove numerals to create group comparisons. That is, if we named our files site_2.1, site_2.2, site_3.1, site_3.2, when LACOID removes numerals, we would retain one single group called *site_*. This is not the case when naming files man1 or woman2 because, in these cases, removing numerals retains the group terms *man* and *woman*.

For the sake of clarity, let us note that LACOID was programmed to add another numeral to each decomposed text chunk (in both the sentence and paragraph decompositions) to identify each classified text's location in its original file and to ease the retrieval of context for our participants' meanings (see our explanation associated with Table 5.1 below). For example, the text associated with "woman2.2" in Fig. 5.5 indicates that this is the second text chunk (i.e., sentence or paragraph) for the file input named "woman2." Furthermore, recall that as part of LACOID, and as depicted on the left side of Fig. 5.5, each of these decomposed texts is to be assigned to a given latent code (represented by V1, V2, and V3 in Fig. 5.5). Following our woman2 example, Fig. 5.5 indicates that one of her texts (woman2.2) was identified as belonging to the latent code V2, and the other two texts were classified as V3.

Programmatically, LACOID removes all numerals in the "**Disaggregated File Names**" section shown in the left panel in Fig. 5.5. Moreover, based on network aggregation rules, when file names are changed, the statistical tests account for group associations rather than individual input texts (see the right side of Fig. 5.5). With this information, a formal test of independence (based on Chi-squared distributions with probability values obtained via Monte Carlo simulations) is conducted, and a PDF output is generated to assess the strength of these group-to-latent code relationships (see Fig. 5.6).

In sum, when the original Word document names are changed, as depicted in Fig. 5.5, this test of independence will assess whether some groups are more likely to be ascribed to a given latent code or set of latent codes than other groups. However, if no text names are changed based on group

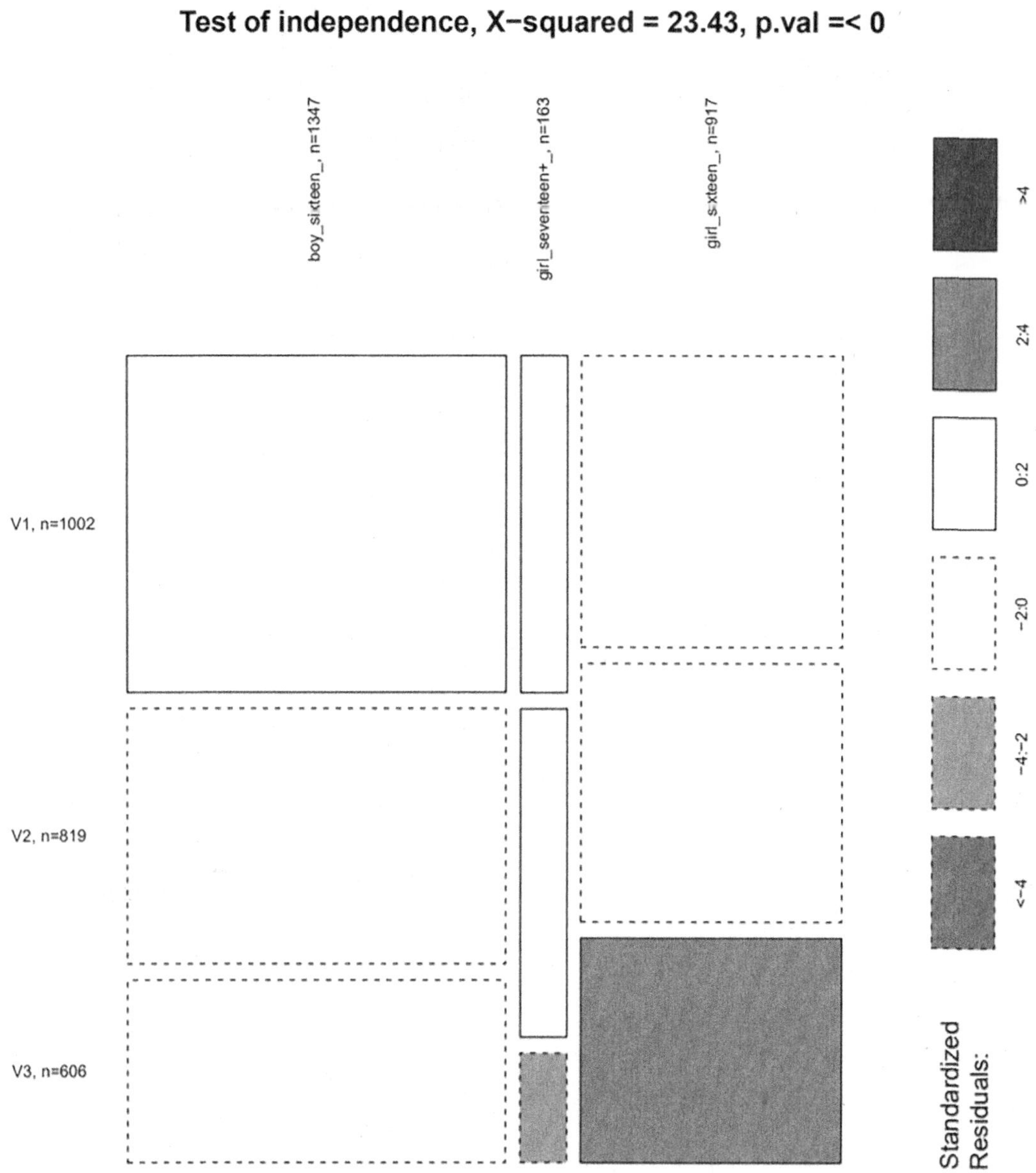

FIG. 5.6 Hypothesis testing output based on chi-square (red and blue rectangles are most influential).

ascription (i.e., man vs. woman), then the resulting test of independence will be conducted at the individual-file level (based on the original Word document that was uploaded). That is, when no document names are changed to assess group analyses, the number of comparisons is based on the number of documents uploaded rather than on the potential groups of interest. Finally, as depicted in Fig. 5.5, the intersection of two or more attributes at the document-name level can also be added to include, for example, the intersection of gender and ethnicity, following the same rationale (e.g., woman_black1, woman_black2, ..., man_white3) or site of interest like "man_site_three1" or "woman_site_two1" to conduct even more aggregated analyses. As stated above, in our application example, we tested whether the intersection of gender and age yielded significant results.

Network representation of group and latent code association

Regardless of whether we change file names to represent groups or capture individual documents with latent code associations, the relationships tested via Chi-squared tests are automatically plotted in a sociogram or network representation. Specifically, as shown in the bottom-left section of Fig. 5.3, participants' or groups' decomposed texts may be heterogeneously distributed across the collection of latent codes identified. The network presentation (shown in the bottom-left section of Fig. 5.3 and in Fig. 5.7) also highlights the strongest connection of

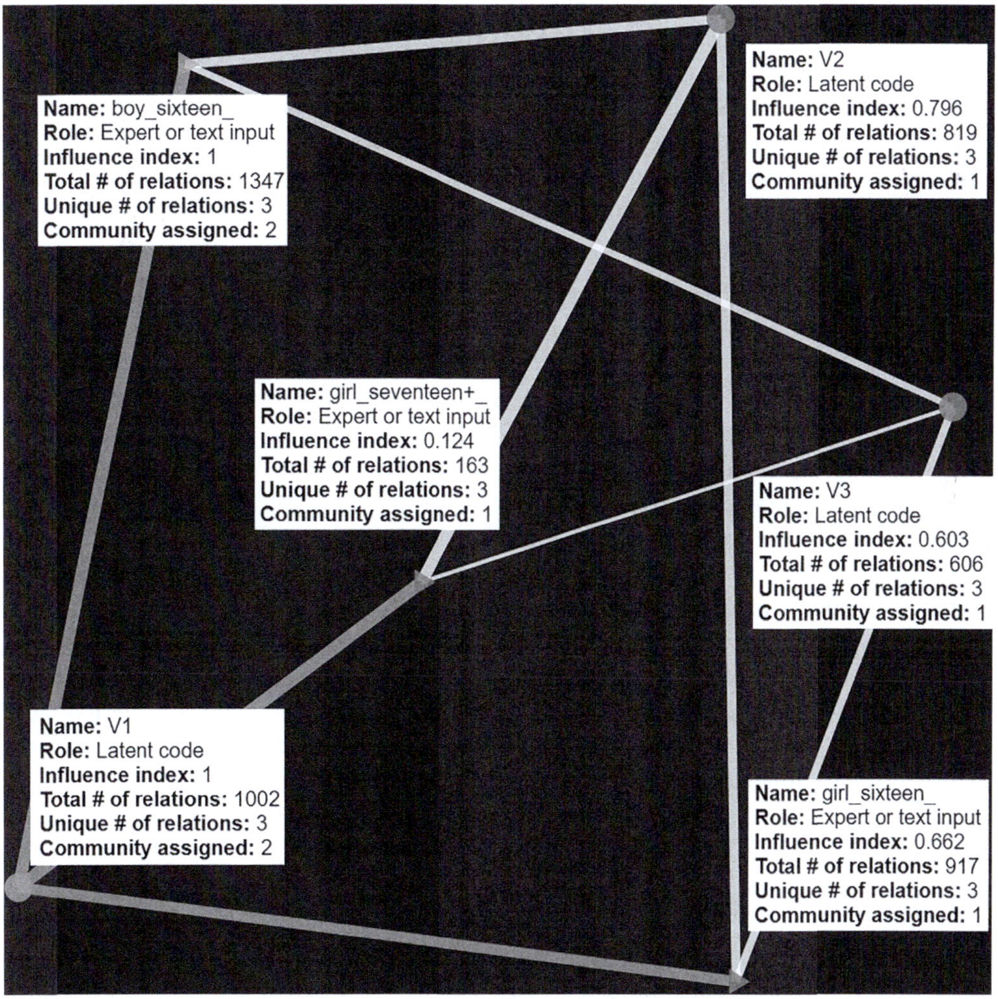

FIG. 5.7 Network depiction of group distributions and associations with latent codes (interactive version here https://cutt.ly/nE96btS).

participants' or groups' texts with a given latent code (indicated by a blue line/link, as opposed to white). Moreover, in the interactive HTML network rendering (see Fig. 5.7 and its interactive version here: https://cutt.ly/nE96btS), when researchers click on a latent code or actor (or group, as explained in the next paragraph), an information box will display how many text chunks this actor or group contributed to the analysis and how many unique connections each actor/group and latent code have. For example, if there were 10 latent codes identified, and a given actor or group has five unique relations, this indicates that this actor or group did not provide text chunks that were classified in half of the latent codes. In the case of latent codes, when clicking on a given latent code in this HTML representation, the resulting information box will show how many actors contributed text chunks that were classified under that latent code. That is, if there were 20 unique actors and a given latent code had 19 unique connections, this would mean that one of the actors did not provide any text chunk that was classified in that latent code. This network analytic process was inspired by a González Canché (2026) application of qualitative network analysis and is incorporated in LACOID.

In sum, both the hypothesis test and this network representation aim to showcase how individuals or groups may be heterogeneously associated with the resulting latent codes. Since the input associations are based on original documents that were decomposed into sentences or paragraphs with the resulting latent codes, it is worth elaborating on the meaning of document decomposition and its role in avoiding LCAB.

Text decomposition as a strategy to avoid latent code aggregation bias

Conceptually, the color schemes represented in the "Corpus and Text Decomposition" section of the process in Fig. 5.3 indicate that a single document is configured by a collection of latent codes rather than representing a single latent topic. This multiplicity of latent codes configuring each individual document is perhaps the biggest departure of LACOID from similar approaches that rely on topic modeling (see Eickhoff and Wieneke, 2018; Lynam et al., 2020). For example, Eickhoff and Wieneke (2018) applied topic modeling to 7356 research articles spanning 40 years, allowing each full article to be ascribed to just a single latent topic. Because this "one complete file to one latent code" approach may lead to a loss of meaning or a form of aggregation bias, LACOID first decomposes each input file or document into its more granular textual components, *at the sentence or paragraph levels*, without losing the original location of these components in their original file (see Figs. 5.3 and 5.5, for example), and then applies machine learning and text classification algorithms to identify the distribution of latent codes within a document. Given the salience of this process to avoid LCAB, the following subsection discusses this text decomposition.

Line-by-line coding goal

LACOID aims to resemble, as closely as possible, the line-by-line HUCOID processes (Charmaz, 2014, in Poth et al., 2021) while relying on NLP and machine learning tools. In qualitative or manual analytic processes for identifying codes, a single participant's response to an interview question or a single paragraph in an essay or document, for example, is composed of *sentences* that may be coded into similar or different *codes* or *labels* (see Saldaña, 2021). In the following example, Saldaña (2021) demonstrates that a one-paragraph essay is composed of five codes, with each code accounting for a unique sentence:

[1] My son, Barry, went through a really tough time about, probably started the end of fifth grade and went into sixth grade.	[1] MIDDLE-SCHOOL HELL
[2] When he was growing up young in school he was a people-pleaser and his teachers loved him to death.	[2] TEACHER'S PET
[3] Two boys in particular that he chose to try to emulate, wouldn't, were not very good for him.	[3] BAD INFLUENCES
[4] They were very critical of him, they put him down all the time, and he kind of just took that and really kind of internalized it, I think, for a long time.	[4] TWEEN ANGST
[5] In that time period, in the fifth grade, early sixth grade, they really just kind of shunned him all together, and so his network as he knew it was gone.	[5] THE LOST BOY

Although not every sentence in a document needs to be strongly identified with a code (see our explanation of the most representative texts in Fig. 5.1), if no purposeful text decomposition is implemented, the multiplicity of meanings configuring full documents will be aggregated into a single topic, which may lead to the loss of these multiple meanings due to aggregation—or LCAB. On the other hand, following Saldaña (2021) example, along with our additional analyses of previously manually coded texts in our own work and the work of some of our colleagues, *it follows that* when full documents are decomposed before being subjected to NLP (involving text cleaning, mining, and topic modeling), the possibility of LCAB will be minimized or even completely avoided. In line with this analytic goal, LACOID offers researchers the option of decomposing their documents at either the sentence or paragraph level.

Document-to-sentence decomposition

Sentence-level decomposition follows the line-by-line HUCOID rationale and decomposes all documents into their constituent sentences; each sentence then becomes a valid text input to be classified via LACOID. This text decomposition is clear, methodical, and objective. From machine-learning and text-classification perspectives, not only would this approach minimize or avoid LCAB, but it would also increase the statistical power across Gibbs sampling and MCMC iterations (see the online appendix) by adding more text inputs to the machine learning process—resulting in large textual datasets (i.e., thousands of text inputs). In addition, the LACOID end product will still group sentences conveying similar or strongly related meanings into the same latent codes, and this may even help detect potential plagiarism issues—see our discussion of the discovery of an identical sentence in the findings section.

One potential challenge of this approach, however, is that this nuanced text delimitation is more computationally expensive; that is, a single interview transcript may become hundreds or thousands of LACOID text inputs, and to the extent that more documents are uploaded, the resulting text inputs may reach tens or even hundreds of thousands.

Sentence-Length-Inclusion Threshold. A related challenge is that, despite increasing the sample size by including more text inputs, decomposing a document into sentences increases the chance that a proportion of these sentences may not be linguistically meaningful. Accordingly, a strategy to trim text that may not be useful in capturing meaningful pieces of information (similar to stop words removal and word standardization procedures employed in text mining that are implemented in LACOID; see appendix)—and, in doing so, reducing the computing power required to implement LACOID—consists of applying a sentence-length-inclusion threshold.

In this respect, linguists (see Cutts, 2009; Myhill, 2008) have recommended that the appropriate length of words per complete and clear sentences in the English language is 18.2 words on average, with a typical range of 9.71–28.15 words (Myhill, 2008). Similarly, Cutts (2009) recommended that, for plain English documents, the average sentence range be 15–20 words. Based on this information, to be conservative and minimize the risk of dropping meaningful sentences, LACOID was programmed to retain *original sentences* with at least 10 words by default, or the lower range found by Myhill (2008) (However, we also allow researchers to adjust this length with a single click. The allowed range is 5–30 words: 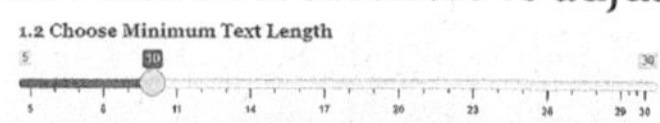

Note that this threshold includes stop words (words that by themselves convey no meaning, as described below); this means that although our threshold consists of original sentences with at least 10 words, our actual LACOID analyses may include text chunk inputs with fewer than 10 words. To illustrate what we mean, consider the following 13-word sentence, which meets the inclusion criterion:

"Keep the average sentence length of your document around 20 to 25 words."

After applying standard text-cleaning preparation techniques, such as transforming words to lowercase and removing stop words (see "Text Mining" below), the original sentence to be included as a text input in LACOID becomes:

Keep the average sentence length in the document around 20–25 words.

From this example, the "cleaned" sentence included in the LACOID analysis contains nine words. Based on our threshold and also on tens of thousands of sentences we employed to test LACOID, in no instance did the cleaning process retain single-word sentences. This latter point is noteworthy because text input word variation is relevant for topic modeling, given that this process assesses both the probability distributions of words configuring a text to a given latent code and the probability distributions of texts to latent codes throughout thousands of iterations in the machine learning process (Griffiths and Steyvers, 2004).

Finally, in the pursuit of complete transparency, LACOID allows researchers to download a database containing all the excluded sentences (those with original sentence lengths of fewer than 10 words, as shown in the applied section below), along with their file names and their positions (sentence numbers) in the original documents. This database is automatically generated after all input files have been uploaded, and it can be accessed by saving it to a local hard drive. For details, see the "Data Upload and Text Decomposition" section below.

Document-to-paragraph decomposition

The second type of text decomposition supported by LACOID requires a more conscious and strategic delimitation process by researchers. In this case, researchers may preprocess their documents to "manually" prepare the text inputs. That is, researchers may go over their transcripts and decide that a *10-page* document, for example, contains 50 text chunks (which can be paragraphs or simply text blocks delimited by a blank line) to be classified into a given number of codes based on NLP, and then feed this document to LACOID. They may replicate this process with all their documents before uploading them to LACOID. To be clear, in this instance, researchers simply need to separate any meaningful text chunk they have identified with *a blank line* before uploading it to LACOID. Once uploaded, LACOID will automatically decompose each file into *n* text chunks as predelimited by researchers.

Returning to the *10-page* example, if 50 text chunks are provided in a given document, LACOID assesses the optimal number of latent codes to be identified for these 50 text inputs, which is an integrative exercise in and of itself. That is, researchers strategically predelimit text chunks as text inputs, and then LACOID identifies their latent codes. Just as with the sentence delimitation, this process also assesses text length and removes text chunks that do not meet the minimum word threshold discussed above. Note that, based on hundreds of documents delimited by blank lines that we have used to test LACOID, in no instance was a text chunk removed due to failing to meet the 10-word minimum length threshold.

Although LACOID's *paragraph decomposition* process is straightforward, the possibility of LCAB is again present because there is a nonzero chance that text chunks potentially accounting for more than one latent code will be aggregated into a single latent code (Based on the possibility of LCAB, we do not recommend delimiting text chunks based on answers to semistructured or structured interview questions because this would limit the use of these answers as text inputs in topic modeling. That is, if our interviews are configured by 20 questions, for example, and we set the preferred text delimitation by each answer to these questions, our text input per interview would be 20, even if each text chunk (i.e., answer) is several pages long, which may surely mask a multiplicity of meanings that will be condensed in a single latent code). Relatedly, researchers' decision processes when delimiting these texts may potentially be affected by inconsistencies in text delimitation—that is, this delimitation process requires applying decision rules, which may be challenging when dealing with a large collection of texts or teams of text delimitators. On a positive note, paragraph delimitation is less computationally expensive and, to the extent that the delimitation process is rigorous and consistent, the remaining integration process may be even more straightforward given researchers' familiarity with the text based on the predelimitation exercise (see also Fetters and Freshwater, 2015; Fetters and Molina-Azorin, 2019 for discussions on integrative strategies, best practices, and modern conventions). Having said this, the default selection that LACOID implements is sentence decomposition, but researchers may update this choice with a single click in LACOID's user interface.

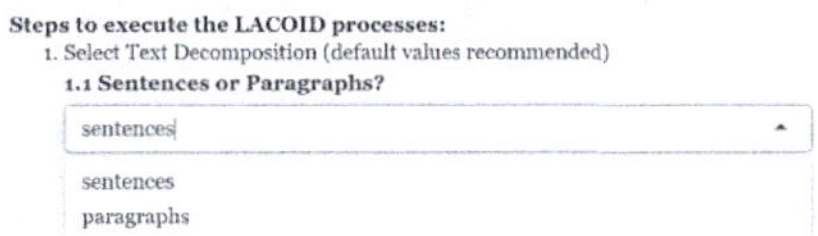

We also recommend removing the interviewers' questions and comments from the texts before uploading them to the LACOID interface. At times, it might be beneficial to add information from the interviewer's input to the response of the interviewee to ensure that the texts to be analyzed have complete and standalone thoughts, as in the following example: (I.) In general, how worried are you about your ability to retire? (R.) I am not. I mean, you know, like, I don't think I'm going to be able to. I mean, I'd like a full disclosure. I don't think any of us are gonna be able to retire. This interaction can be modified as follows (modification in italics ""): I am not "*worried about my ability to retire*". I mean, you know, like, I don't think I'm going to be able to. I mean, I'd like a full disclosure. I don't think any of us are gonna be able to retire. This modification allows the text to be uploaded to LACOID to be self-standing. In the next example, one can simply remove the interviewer's input without having to modify the respondent's information. (I.) And how do you currently save for retirement in general? (R.) I've been my sort of retirement. In that, um I've been saving and investing my money, and I don't have much of it, but I'm hoping over time I am able to have more of it. Modified text (only excludes interviewer's input): I've been my sort of retirement. In that, um I've been saving and investing my money, and I don't have much of it, but I'm hoping over time I am able to have more of it. Note that in both modifications, all references to respondents' identification (or "R." in our examples) were also removed.

Meaning retrieval postlatent code identification

Given that machine learning requires the analysis of words outside their original context (Eickhoff and Wieneke, 2018), this learning and classification process per se is useless in preserving the richness of the original and contextualized written data or information. To address this limitation, LACOID provides researchers with tools to describe the resulting codes as a function of their word frequencies (descriptive meanings, as suggested by Miles and Huberman, 1994), as shown in Fig. 5.8, and to decipher or rebuild their deeper meanings based on their original contexts (or even their inferential meanings, as also suggested by Miles and Huberman), as shown in Fig. 5.1.

Regarding the descriptive meanings (Miles and Huberman, 1994), LACOID provides a descriptive tool to quantify word frequency distributions per code (Fig. 5.8). Conceptually, in addition to serving as a descriptive tool, the distance of codes represented in the dark box quadrant plane in Fig. 5.3 (top right) also captures how similar (i.e., based on smaller distances) or dissimilar these latent codes are. Specifically, the top-right section in Fig. 5.3, represented with folders, shows a quadrant plane separated by dotted lines. In this representation, codes that are within the same quadrant (i.e., intraquadrant) have textual content that is more similar than codes located in different quadrants. That is, Code 5 is more similar to Code 7 than to Codes 1 or 6. Similarly, Code 4 is more neutral in its content given its central location, which implies that its respective textual content shares information with codes located in other quadrants. More details are discussed in our applied example (see "Outputs and Findings"). To access the interactive LACOID output, follow this link: https://cutt.ly/NrXhMnc3 or see Fig. 5.8.

Regarding the goal of deciphering or rebuilding deeper, unaltered meanings, LACOID was programmed to provide automatic access to the cleaned and normalized text chunks that were classified by LACOID, as well as to the original sentences or paragraphs containing the exact words provided by the research participants. For example, in building from our sentence-length example discussed above, our output may not alter meanings or original texts following this rationale. Assume that the first sentence in our first input document is: "*Keep the average sentence length of your document around 20 to 25 words.*" Assume also that this file is called doc1.doc. If we choose to decompose this file into sentences, the ID of this first sentence would be doc1.doc.**1**, with the last

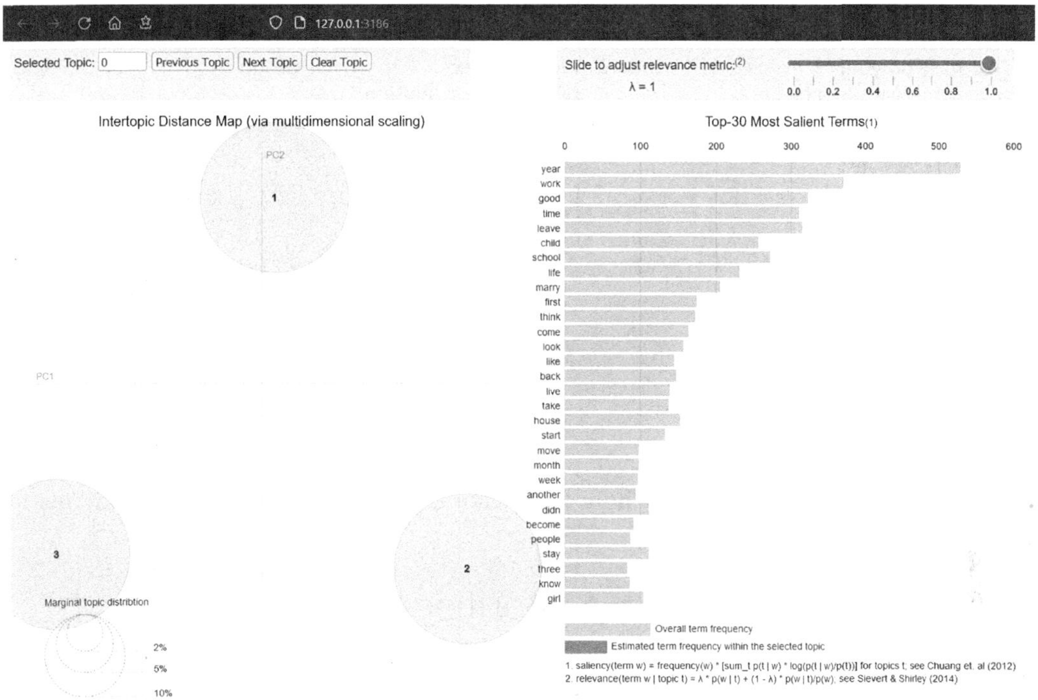

FIG. 5.8 Interactive summary of text to code distributions in the three-codes solution (interactive version here https://cutt.ly/XE9ZV71).

number indicating that this is the first sentence of that file. Finally, assume that this sentence was classified into Code 1 (or V1, as done by LACOID). With these elements, the result will be as follows:

The information in this table allows us to identify the file of origin (doc1.doc), the position of the text or sentence in that document (doc1.doc.**1**), the cleaned text for machine learning and text classification (*keep the average sentence length of the document around 20 to 25 words*), and the original text as provided in the uploaded file (*keep the average sentence length of your document around 20 to 25 words*). Note that contextualization is enhanced by the possibility of detecting the actual location of this text in the original document. That is, if instead of only observing the first sentence, we have access to "doc1.doc.**34**", we can then go and read texts or sentences that preceded sentence number 34 along with surrounding sentences, with the goal of retrieving context and observing unaltered information.

As part of the standard LACOID output, a table like Table 5.1 shown in this section, in addition will contain an assessment of how representative each text is for each latent code. As we discuss below, this is done to ease the understanding of each latent code. Although by default we show the five most representative texts, this number can be updated in the LACOID user interface by simply adjusting the selection of these representative texts from 1 to 20 (see Fig. 5.1 and Table 5.2).

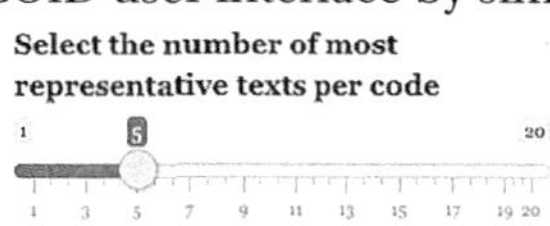

Table 5.2 Example of the classification process, text-to-code contribution, and group fit.

Q. Elaborate on your family and career balance	Code 1	Code 2	LACOID result	Text contribution	Group fit	Labeling
My career is the most important aspect of my life.	90%	10%	Code 1	1 = (.90/.90)†	1 = (.90/.90)††	Career_ Oriented
My family is as important as, or more important than, my career.	45%	55%	Code 2	.6875 = (.55/.80)†	0.675 = (0.75/.90)††	Family_ Oriented
For me, the best day is a family day, not an office day.	20%	80%	Code 2	1 = (.80/.80)†	0.75 = (0.675/.90)††	Family_ Oriented

†Divided by the highest probability within that code. ††Average contribution of the group divided by the highest contribution by the group.

Outputs and findings

This procedural section showcases how to execute LACOID and discusses the outputs for analysis and integration.

Data upload and text decomposition

In Steps 1 and 2 of the LACOID execution process, we uploaded 153 essays that were decomposed into 3222 sentences with a minimum length of 10 words. Of these 3222 sentences, 768 had fewer than 10 words and were thus excluded. LACOID allows users to download a dataset containing these excluded cases for exploration. Specifically, this dataset file includes two columns: "doc_id" and "text". The former matches the name of the original uploaded file. The last number in "doc_id" indicates the location of the text in that file. In the example below, for the file "boy_sixteen_1.doc," the first sentence, or "text," was "Reflections." Because this text is shorter than 10 words, it was excluded. LACOID identifies empty lines with a "." as in the case of the text "boy_sixteen_2.doc.1," which had an empty line before this participant started his essay with the text "Reflection." (Finally, LACOID also creates a downloadable database with the frequencies of groups to be tested. In our data, there were 89 essays completed by boys aged16, 13 essays by girls aged 17 or older, and 51 essays by girls aged 16.

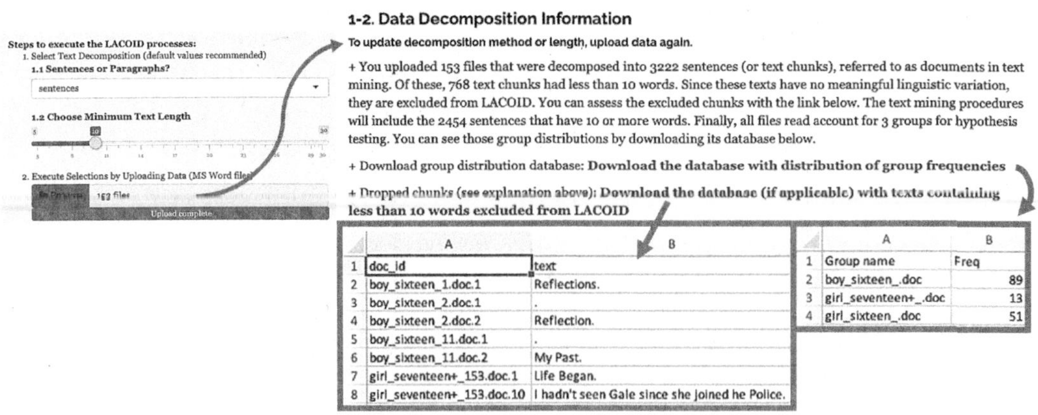

Text mining

Step 3 of the LACOID execution process consists of text data cleaning and normalization, following standard text mining procedures (see Feinerer, 2020; Silge and Robinson, 2017). During this process, some texts may be dropped due to being too infrequent in the document-term matrix to meaningfully contribute to classification. In our analyses, 27 sentences were excluded, but for

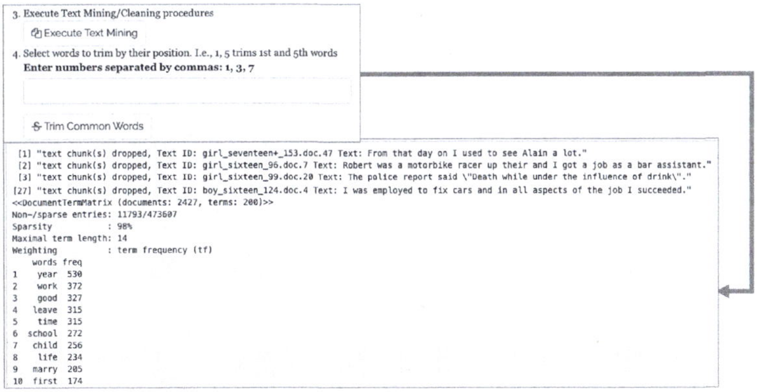

For the sake of space, we present only four of them in the figure shown next.

The words configuring these texts are outliers (and hence removed from LACOID) because they appear in less than 2% of all other texts (based on the sparsity level of 98%). These texts can also be found in their original documents based on their Text IDs. Finally, this output also shows the 10 most frequent words. Some words may be too common to be useful in classification (e.g., essays on COVID-19 having "COVID" as a frequent word) and can be excluded from the analyses (Schütze et al., 2008). We did not encounter this problem and thus did not remove any words. To remove words, simply identify them by their position. For example, to remove the words "school" and "first," which are in the sixth and 10th positions, respectively, type "6, 10" in the box corresponding to Step 4 (without quotations but with the comma separator) and click "Trim Common Words." To reinstate words after they have been removed, simply execute Step 3 again; there is no need to reupload the data in Step 2.

Metrics assessment

Step 5 includes the burn-in and MCMC/Gibbs sampling with the default recommended values of 500 and 5000, respectively (Raftery and Lewis, 1991). This step generates a plot (Fig. 5.9) displaying all four metrics (see Arun et al., 2010; Cao et al., 2009; Deveaud et al., 2014; Griffiths and Steyvers, 2004) simultaneously to select the optimal number of topics, ranging from 2 to 60 latent codes.

The plot indicates that across 5000 iterations, there are two candidates—three and seven—for the optimal number of codes based on dissimilarity (Deveaud et al., 2014) and low correlation (Cao et al., 2009); because there is more than *one candidate for the optimal number* of codes, further assessment of the pros and cons of these candidacies is depicted next.

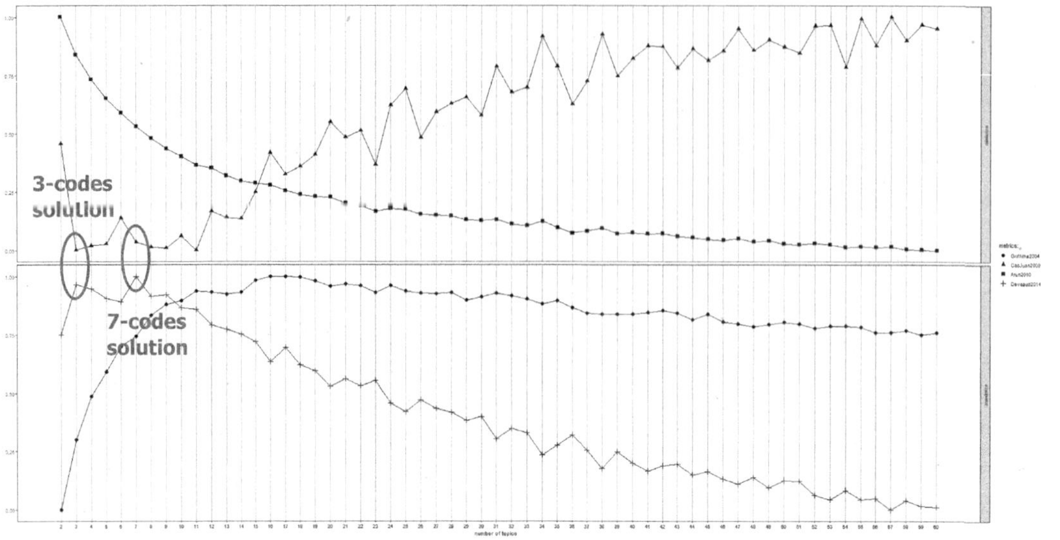

FIG. 5.9 Metrics assessment signaling two feasible solutions (circles added by research team).

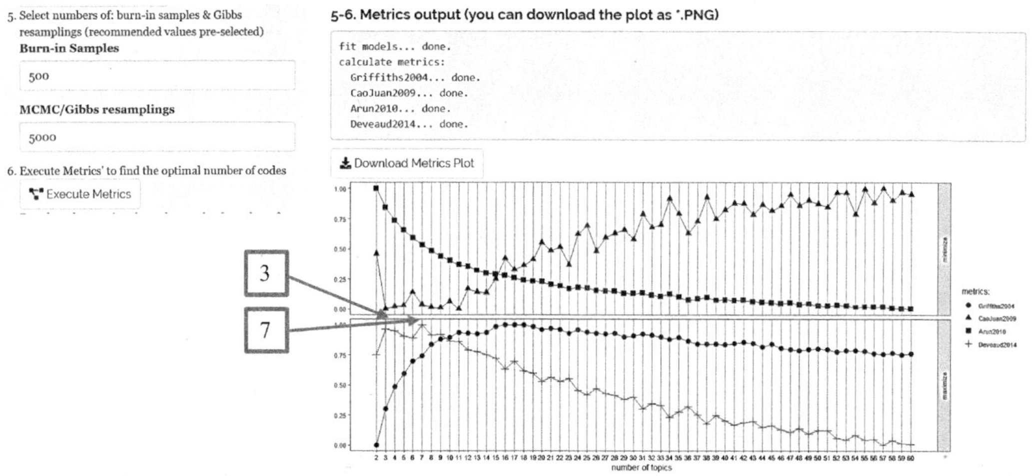

LACOID execution and selecting the optimal number of codes

To further evaluate the optimal number of codes, we executed LACOID twice: first with three codes in Step 7 and then with seven codes. The results indicate that using three codes (available here: https://cutt.ly/NrXhMnc3) renders the most differentiable latent codes (see Fig. 5.8). Codes 3 and 2 are located in the two lower quadrants, and Code 1 is in the upper quadrant. When we ran LACOID with seven codes (available here: https://cutt.ly/frXhNJmA), Code 6 fell in the center of the quadrants, meaning that its content was shared among the other codes.

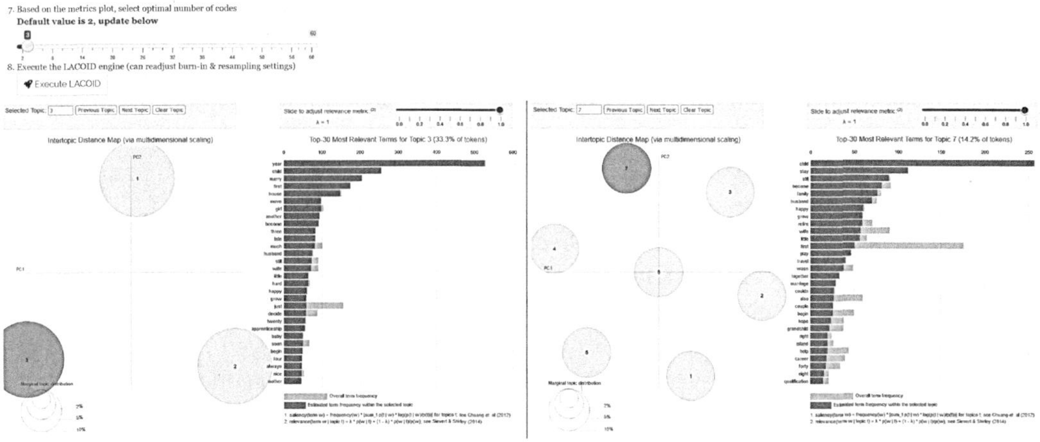

To aid with the meaning-building process, the interactive plots automatically generated by LACOID dynamically update the word frequencies in two ways: by clicking on a code (Codes 3 and 7 were selected in the plots above—not to be confused with the three-code and seven-code solutions) and by clicking on a word. For example, in the graphs shown below, the word "child" was clicked; the disappearance of all remaining latent codes indicates that only Codes 3 and 7 contain this word in their corresponding texts. As further discussed below, these codes capture meanings related to starting a family, having kids, and being married.

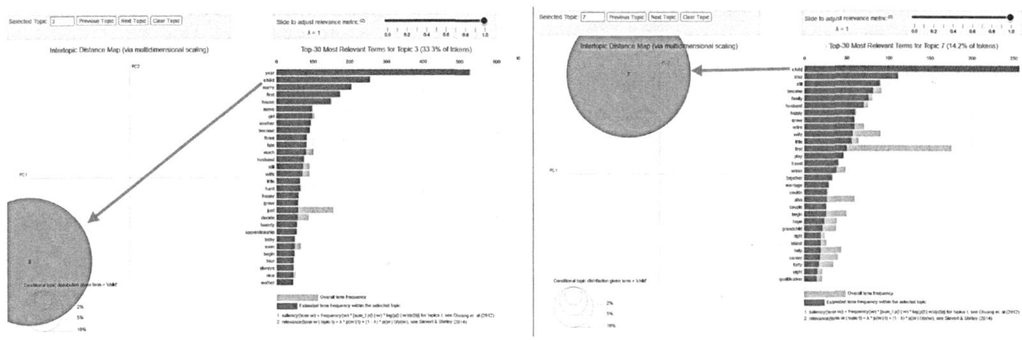

Although this analysis is useful, in cases where there is no consensus regarding the optimal number of codes, as in this study, choosing the final number of latent codes should not be based exclusively on analyzing these plots. Instead, a deeper analysis of their most representative texts should be conducted. Due to space limitations, we cannot show the entirety of this process, but, quite similar to our analyses shown next (i.e., after a careful analysis of the original texts associated with each solution), and also by noting the clear lack of latent code overlap in the plot quadrant rendered by the three-code solution, we selected this three-code option to conduct our integrative analysis.

Meaning-building and meaning-retrieval processes

A closer inspection of the interactive three-code solution (see Fig. 5.8 or https://cutt.ly/NrXhMnc3) indicates that the following descriptions were primarily or uniquely representative of each code:

Code 1 (V1) refers to leaving school, start looking for a job, or continuing with their training, either by beginning a career or transitioning to college.

Code 2 (V2) refers to remembering or reflecting on their past lives and retiring from their jobs.

Code 3 (V3) refers to starting a family, having children, getting married, moving into their first home, and the difficulties or challenges associated with transitioning to adulthood.

LACOID was designed to strengthen these understandings—retrieved from code-to-word distributions—by making it easy to retrieve and analyze the most representative texts per code. As can be seen in the graph below, researchers have the option to download the fully classified dataset or the dataset containing the 20 (or fewer) most representative original texts per code. These databases also include the text inputs that were used in the actual topic modeling (normalized and text-mined/cleaned words) in case these actual text inputs help with understanding the latent code meanings. In our analyses, we present the three most representative texts per code. Within the broader analytic architecture of this book, those same outputs may also function as prompts for local AI-supported scholarly reasoning in ISARI, where researchers may compare possible code interpretations, memo explanatory contrasts, and begin translating these distributions into evidence-grounded prose while preserving the primacy of participants' original voices.

The "relative text contribution" column in this table (and in Fig. 5.1) identifies texts that best capture the meaning of each latent code. The "relative group fit" column measures the extent to which the texts configuring each latent code coherently or exclusively represent that code. We programmed LACOID so that both measures have a maximum value of 1—see González Canché (2022c) for a more technical explanation of these measures.

Following González Canché (2022c), let us elaborate on the process LACOID follows to identify text-to-code relevance. For convenience, note that we are including only three texts and that the results obtained from the metrics assessment indicate that a two-code solution is associated with the most optimal classification outcome.

The texts contained in Table 5.2 show that the first response is clearly aligned with a career-focused standpoint, as this participant mentioned, "*My career is the most important aspect of my life.*" The columns labeled "Code 1" and "Code 2" represent the two-code solution. Note that LACOID's output does not include these two columns; instead, it shows a column called "Code" (see Fig. 5.1) that identifies the resulting classification of each text (shown as LACOID Result in Table 5.2) based on the following rationale. Each text is given a probabilistic value of representing each of the optimal codes selected in the metrics assessment (see Table 5.2 under columns "Code 1" and "Code 2"). Accordingly, assuming a two-code solution in Table 5.2, this translates into the following scenario: although the first text was classified as Code 1, in reality, after thousands of iterations in the learning process, its probability of being associated with this code, given its word configuration, was 90%, with the remaining 10% being associated with the second code. That is, this text had a 10% chance of representing Code 2 as well.

(Texts-to-code classifications are not discrete but rather represent a Dirichlet distribution of these probabilities. We programmed LACOID to identify the collection of texts with the highest probabilities of code ascription as the more likely representation or depiction of such a code.) In the case of the second open-ended response or text, which is clearly more ambiguous, as it reads, "*My family is as or more important than my career*, these probability distributions were 45% and 55% for being classified in Code 1 and Code 2, respectively. Finally, the third sentence, which reads, "*For me, the best day is a family day, not an office day*, had an 80% probability of being representative of Code 2. Taking all this information into account, we then know that, although the second response was more ambiguous than response 3, and this second sentence was also close to being classified as Code 1, it was classified under Code 2 given its higher probability value (i.e., 55% vs. 45%). In this respect, note that topic modeling does not traditionally offer a measure of how strong the text-to-code associations are. That is, although our example clearly shows that texts 1 and 3 are more strongly associated with Codes 1 and 2, respectively, than text 2 is with Code 2, topic modeling would not provide this information as part of standard outputs. To address this limitation, typical of topic modeling, LACOID's main outputs (both in the fully classified dataset and the dataset with the most representative responses) offer two measures to assess text-to-code strength and detect the most representative texts per code (see Fig. 5.8).

In the LACOID output shown in Fig. 5.1, The columns that allow the identification of text representativeness are "Text Contribution" and "Group Fit."The text contribution is populated by dividing the probability of each text to be classified in the code where such a text was actually classified over the maximum probability of a text being classified in that same code. That is, taking text 2 in Table 5.2, for example, we have that the maximum estimated probability of a text being classified under Code 2 was 80% (text 3); however, the probability of this text 2 to be classified in Code 2 was 55%. Accordingly, the text contribution or representativeness of text 2 for Code 2 is the result of dividing 0.55/0.88=0.6875. In contrast, the third text will be divided by its own probability, which is the highest observed, rendering a value of 1. This indicates then that text 3 is the best representative of Code 2 across all texts classified under Code 2. This same rationale is applied to text 1 and Code 1. This is how we can request that LACOID's interactive UI displays only the most representative texts per code (see Figure 5.1).

The second assessment is group fit. This assessment takes the average probability values obtained by each code given the probability of their configuring texts to being classified in such a code and then divides the resulting code average probabilities by the maximum average value observed across all codes. Specifically, in the case of Code 2, this average value is estimated from 0.55 and 0.80 (the probabilities of its configuring texts 2 and 3 to be classified in Code 2). This average value is 0.675. In the case of Code 1, this average value is 0.90. Since 0.90 is the maximum average value observed across all codes, Code 1 will have the maximum group fit for its average is divided by the maximum average across codes. In the case of Code 2, its average is divided by the maximum code group average (or 0.675/0.90) rendering an estimated value of 0.75. This implies that Code 1 had the maximum group fit (similar to an internal consistency indicator). This indicator, which does not change within codes, may be interpreted as a tool to assess the extent to which the probabilities of texts configuring each code were consistently high (or not).

In a sense, then, these two fit measures allow us to assess individual text-to-code contributions while also estimating how consistently these texts are being classified under their corresponding codes. Since these fit measures are a function of the optimal code number,

7-8. Code modeling and most representative texts per code

• After LACOID completion you can download the Chi-squared test of text or group association with latent codes

Hypothesis test: **Click here to Open the Chi-squared test of independence**

• You will also be able to download the resulting LACOID datasets

Download Full LACOID data (all text chunks) | Download File-to-LACOID distribution (quant. modeling) | Download Most Representative cases (shown below)

Select the number of most representative texts per code

Show 9 entries

Search:

Most representative texts per latent code and coherence assessments

	code	text	relative_text_contribution	relative_group_fit	quote	text_input
5	V1	girl_sixteen_113.doc.2	0.831	0.974	I left school in the summer of 1979 to work in an office as a secretary it was a big place is London and I didn't know many people, after I didn't know many people, after about a year and a half I left to work for a bigger and better company where I could earn more money as I was planning to marry within the next year or two and we had planned to save to buy our own house near his parents.	company didn didn earn good half house know know leave leave london many many marry money near next office parent people people place save school secretary work work year year
7	V1	boy_sixteen_50.doc.2	0.866	0.974	I could build my own house to have something to have ever since I had left school what I had in mind and that is what happen and once I had finish that house I would work myself up for a foreman on a building site to design new place in different town all across the country make my firm very famous in what I had achieved since working at a shop.	different ever finish firm happen house house leave make mind place school shop since since something work work
9	V1	boy_sixteen_83.doc.2	1	0.974	My career started when I joined the army I decided that this was what I wanted to do when I was fifteen so I went to the careers office to find out more about the tests I would have to take they told me to apply after Christmas, when I told my parents this, my mother said to me that seems very late, so she phoned up the army careers office at Chatham and they told her that there were only two places left for testing of the next day, so they invited me to go over and take the entrance test.	apply army army career career career decide find join late leave mother next office office parent place seem start take take tell tell tell test test test want
1	V2	boy_sixteen_1.doc.27	0.922	0.984	Life is very tedious now and I spend most of my time just sitting around thinking of the past.	around just life past spend think time
6	V2	boy_sixteen_39.doc.7	0.944	0.984	My life has had its good and bad times but altogether a good life, my mum lived to 88 and dad to 79 they both had good times.	good good good life life live time time
8	V2	boy_sixteen_50.doc.7	1	0.984	We all use to help one another out of trouble when we were not at work like repair our car etc and in my spear time I like garden a lot growing think and all at repair my car of my own I have had the odd moments when we had trouble at work when there was a Strike everyone hat to stop for a fortnight so we lost a lot of time because we was on contract work so (within) 3 month we made-up that time by doing overtime and then we was back on our feet as normal I think that I had a good life with the good think and the back then what happen I would love to have my life all over again what a interest life I have had.	another back back good good grow happen help life life life like like lose love make month stop think think think time time time work work work
11	V2	girl_seventeen+_148.doc.12	0.922	0.984	Life is very tedious now and I spend most of my time just sitting around thinking of the past.	around just life past spend think time
2	V3	girl_sixteen_97.doc.27	0.971	1	My husband felt reluctant to let my daughter grow up as he needed to mother her but when he saw her in her bridal gown he realised how much life we had both seen together and we concentrated on living life for us and when we watched our grandchildren grow up we both felt contented and continued to grow old still as much in love as the day we got engaged, or as I feel when we left school and grew up.	daughter engage feel feel feel grandchild grow grow grow grow husband leave life life live love mother much much need school still together watch
3	V3	girl_sixteen_99.doc.11	0.971	1	He was a fine little baby and we loved him dearly, but when there was three mouths to feed life became a little harder, the money didn't go far and we missed the money that I used to bring in, but we got over the little crisis, Steve soon got another job and the baby grew up to be a healthy boy.	another baby baby become bring didn grow hard life little little little love money money soon three
4	V3	girl_sixteen_112.doc.12	0.975	1	Then we waited another six years when our third child was born, this time a boy, another two years later our fourth and last child was born, another girl.	another another another child child girl last late time wait year year
10	V3	boy_sixteen_124.doc.5	1	1	At the age of twenty I finished my four year apprenticeship and continued at the garage for another four years until I saved up enough money to buy a small holding about ten miles from my first house.	another apprenticeship enough finish first four four garage house money save small twenty year year

Identified with the metrics assessments, all LACOID steps are interconnected across the entire classification process.

With this explanation in mind, each latent code will then have at least one text with the highest value, whereas there will be only one latent code with a value of 1. In the figure above, for example, the third code has the highest relative group fit, and within each code, there is one text that best captures the latent meaning of that code.

In the case of Code 1, the text "boy_sixteen_83.doc.2" is its most representative text. Congruent with our previous descriptive analysis, this text captures the start of a career in the army. The "cleaned" version under the column "text input" shows that this actor used "career" three times—hence measuring word frequency per text. The other two most representative texts also mention leaving school and starting to work. Accordingly, we named this latent code *transitions from school to work/career*.

In the case of Code 2, its most representative text was provided by "boy_sixteen_50.doc.7," and this text captures thoughts about past good times and things they used to do. The second most representative text (boy_sixteen_39.doc.7) describes when his father and mother died and how they "had good times." The third most representative text appeared twice in the essays (the implications of this duplication are discussed below). This content corresponded to "boy_sixteen_1.doc.27" and "girl_seventeen+_148.doc.12" and indicated that their lives are tedious, and they constantly remember the past. Based on this content, we are naming this code *remembering past lives nostalgically.*

The analytic power of text mining, sentence decomposition, and LACOID is demonstrated with the identification of this duplicate case, which resulted in the same "relative_text_contribution" value (0.922) and prompted us to go back to the original essay files (as downloaded from the original study) to corroborate that LACOID did not alter texts. We found that both essays differ in content but close with the exact same sentence. From a methodological perspective, the implications of this duplicate detection are that (1) since both texts have the same word distributions, they have the same relative text contribution, (2) LACOID is useful to detect potential plagiarism at the sentence level—or even potential issues in data preparation, and (3) when there is a "tie" in the relative text contribution, we programmed LACOID to deploy all instances (configuring IDs and text chunks), even if the deployment of this information resulted in showing more than the expected number of most representative text chunks. That is, we selected to show the top three most representative cases, but LACOID deployed four instances, since one instance had the same relative text contribution.

The third latent code has the highest group fit. Code 3 also had *two texts tied in their representativeness,* but the input *texts were not exact copies.* In the most representative text, "boy_sixteen_124.doc.5" described finishing his apprenticeship and saving to buy more properties. In the second most representative text, "girl_sixteen_112.doc.12" described how long they waited for the arrival of their third and fourth children. The tied cases also referred to their children growing up and their family lives. Based on this content, we named this code *home andfamily life family growth.*

Hypotheses generation as part of integration

Although a deeper qualitative analysis may be strengthened by analyzing the context in which texts were provided, for our practical demonstrative purposes, let us rely on our initial qualitative understanding of the meaning of these codes to showcase how we can test if there is an association between the intersection of gender and age and these three latent codes. To assess these relationships, LACOID offers two outputs. One is the network depiction of the distribution of these group-code relationships (see interactive plot here: https://cutt.ly/nE96btS, where clicking on lines and circles [latent codes] or triangles [groups] provides more information about these relationships—see Fig. 5.7). The second is the statistical test (using Chi-square) for these relationships—Fig. 5.6. Similar to the other interactive plot, the network depiction is launched automatically. In the case of the statistical test, researchers have the option of opening the resulting PDF and saving it for inclusion in their papers.

The network depiction shows that for all three groups, the code *transitions from school to work/career* (Code 1) had the highest frequency. This highest frequency is automatically detected and highlighted with a blue link (rather than white, as shown in the network graph below). The groups and latent code colors in this network are based on a community detection algorithm, wherein groups and codes with the strongest ties/links between them are classified in the same community (Reichardt and Bornholdt, 2006). As can be seen in the network depiction, boys and *transitions from school to work/career* (Code 1 or V1 in the network) are classified as part of the same community (both in orange), whereas girls 16 and 17 + are classified under Codes 2 and 3 or V2 and V3 (*remembering past lives nostalgicallyand home andfamily life family growth*, respectively), all in magenta.

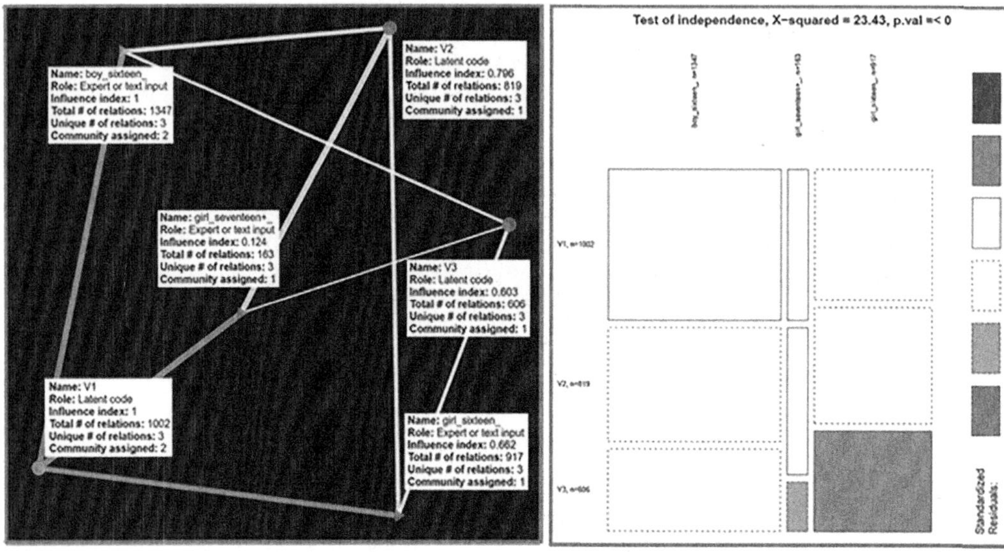

Although the network depiction is useful, it does not formally test whether these observed relationships are stronger or weaker than expected under the hypothesis of no association. The Chi-square test (Fig. 5.6) shows statistically significant results (p-values computed by Monte Carlo simulations are not zero but smaller than three decimal places), indicating associations across attributes and classified responses that are not the result of random chance. Specifically, note that the responses provided by boys aged 16 and girls aged 17+ are slightly more frequently associated with *transitions from school to work/career* (Code 1 or V1) than what would be expected under random chance; however, this hollow solid rectangle does not depart from expected values in a statistically significant way. That is, only blue and red rectangles are considered to be statistically driving these associations. Specifically, in the case of girls aged 16, their responses are statistically significantly classified more frequently under the code *home andfamily life family growth* (Code 3 or V3, as indicated by the blue box). Finally, in the case of girls aged 17+, their responses are statistically significantly less likely to be classified under this same code, *home and family life/family growth* (as depicted by the red box). This same group (girls aged 17+) is also *slightly more likely* to be classified under the code *remembering past lives nostalgically* (Code 2 or V2).

LACOID's output integrates with quantitative databases

The latent codes generated at the document level may be merged with more traditional quantitative analyses, such as regression models. That is, in the same way that we captured the most prevalent associations of *groups* with latent codes (blue links in the network depiction), we programmed LACOID to measure the strongest relation of each essay author (i.e., Word file) to these latent codes (or the distribution of these classifications). Specifically, one of the three downloadable datasets resulting from executing Steps 7–8 is called "File-to-LACOID distribution." This database contains the distribution of all latent code IDs (LACOID) across all input files—see figure below. As discussed earlier, our analytic sample includes 153 essays; hence, this database contains 153 rows (the first row in the Excel file is "column names") (Fig. 5.4).

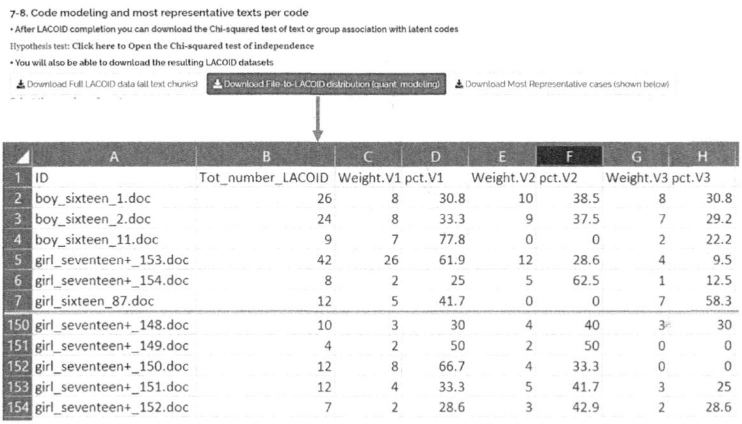

	A	B	C	D	E	F	G	H
1	ID	Tot_number_LACOID	Weight.V1	pct.V1	Weight.V2	pct.V2	Weight.V3	pct.V3
2	boy_sixteen_1.doc	26	8	30.8	10	38.5	8	30.8
3	boy_sixteen_2.doc	24	8	33.3	9	37.5	7	29.2
4	boy_sixteen_11.doc	9	7	77.8	0	0	2	22.2
5	girl_seventeen+_153.doc	42	26	61.9	12	28.6	4	9.5
6	girl_seventeen+_154.doc	8	2	25	5	62.5	1	12.5
7	girl_sixteen_87.doc	12	5	41.7	0	0	7	58.3
150	girl_seventeen+_148.doc	10	3	30	4	40	3	30
151	girl_seventeen+_149.doc	4	2	50	2	50	0	0
152	girl_seventeen+_150.doc	12	8	66.7	4	33.3	0	0
153	girl_seventeen+_151.doc	12	4	33.3	5	41.7	3	25
154	girl_seventeen+_152.doc	7	2	28.6	3	42.9	2	28.6

The column "Tot_number_LACOID" accounts for how many LACOIDs each file contained and the columns starting with "Weight" indicate the total number of texts for each participants that were classified under each of the resulting codes. For example, "boy_sixteen_11.doc" and "girl_seventeen+_152.doc" had nine and seven LACOIDs, respectively. Accordingly, given that for this boy seven of these text contributions were classified under Code 1 or V1, 77.8% (or 7/9) of his text inputs (sentences in our analyses) were classified in *transitions from school to work/career*(Code 1or V1), whereas for this girl, 42.9% (or) of her sentences were classified under *remembering past lives nostalgically* (Code 2 or V2).

This resulting table has an inherent quantitative nature. If researchers have other individual- or input text-level-related attributes of interest, such as socioeconomic status, college enrollment, or a site's contextual attributes of these text documents (e.g., interview transcripts, observation notes), these classified responses may be merged with such attributes to address quantitative questions. For example, assume that our participants' transcripts have attributes indicating college enrollment. Considering this attribute, we can ask: "Are participants strongly associated with the latent code *transitions from school to work/career also* more likely to pursue a college education? Or are participants more related to the latent code *home and family life family growth* less likely to attend college?" The same may apply to socioeconomic attributes or any other indicator at the individual- or text-input-levels. These questions also align with hypothesis generation via integration (Fetters, 2020).

Discussion of the findings

These relationships, shown in Figs. 5.7 and 5.6, suggest that the foci of the essays written by boys aged 16 and girls aged 17+ are slightly more concentrated on transitioning to the job market, careers, and training (Code 1 or V1 in Fig. 5.6), whereas the essays of girls aged 16 tend to be more focused on home and family life and family growth (Code 3 or V3 in Fig. 5.6) with the opposite being true for girls aged 17+. Although these associations were retrieved from the integration of computer or statistically-assisted machine learning and NLP algorithms, as well as the initial qualitative analyses of these resulting classifications, neither these categories nor these associations should be taken as the only form of valid knowledge or the only understandings to be retrieved from these essays. To the contrary, there are many more nuanced and important responses in the textual database that deserve further attention. For example, some essays clearly denote depression and suicidal thoughts. Participant "boy_sixteen_4.doc" mentions being constantly depressed, not being happy, and simply waiting to die. Participant "boy_sixteen_7.doc" more blatantly mentions suicide:

I am now fifty-nine, and I don't want to live any longer. Therefore, I will leave all my money and property to my wife and children (including the one at university), retire with them, and on my 60th birthday, I will kill myself (I hope).

Although some instances of these feelings can be inferred from Code 2 or V2, "*remembering past lives nostalgically*, especially in the duplicate sentence, "Life is very tedious now, and I spend most of my time just sitting around thinking of the past" (participants "boy_sixteen_1.doc.27" and "girl_seventeen+_148.doc.12"), these instances do not quite capture the severity and potential consequences of depressive and suicidal thoughts. The main implication from this discussion is that, although the codes and our understandings retrieved from machine learning are useful, they should not replace our human capabilities of grasping more nuanced understandings. As Poth et al., 2021 said, the main limitation of these techniques is precisely the "absence of human reasoning around classification" (Poth et al., 2021, p. 351). Relatedly, Ho et al., 2021 demonstrated that a single text may be ascribed to two codes via HUCOID. This is not possible in LACOID because each text will be ascribed to the code it contributes to the most based on the probability distributions generated during the machine-learning process (see also González Canché (2022c) for a more detailed explanation). In this respect, although "relative text contribution" to a given code and "relative group fit" of a given code aid in assessing the coherence of input sentences or texts to a given topic, human nuance remains the most powerful meaning-building tool at the disposal of qualitative and mixed methods researchers.

Based on this discussion, although we appreciate the consistency and time-efficiency of applying computer rules and machine learning in classification, we should never forget the analytical power of human nuance that can be gained from reading the original essays and finding more balanced understandings. In other words, the classificatory LACOID outputs should be treated as guides to interpretation rather than as interpretations in themselves, and their full analytic value emerges only when they are revisited qualitatively and, when useful, carried into local environments such as ISARI for further memoing, comparison, and evidence-grounded synthesis.

##Integrating LACOID Outputs into Local Scholarly Brainstorming

One of the strengths of LACOID is that its outputs are portable across analytic stages. The interactive word-to-code distributions, the original-and-cleaned text tables, the most representative text summaries, the group-association tests, the sociograms, and the file-to-code strength databases are useful not only for identifying latent structure in textual corpora, but also for supporting subsequent interpretive work. In the broader methodological ecosystem developed throughout this book, these outputs can be brought into ISARI, a local, fully offline, open-source, and multimodal brainstorming environment designed to support scholarly memoing, comparison, synthesis, and evidence-grounded writing.

This possibility matters because machine classification, however rigorous, does not determine its own meaning. Researchers still need to decide how latent codes should be named, how competing code solutions should be compared, what more nuanced understandings remain visible only in the original texts, and how the resulting coded structure should be translated into qualitative or mixed-methods argument. Because ISARI can work with textual notes, screenshots, exported figures, scanned pages, and other image-based or document-based artifacts, researchers can upload LACOID outputs in multiple forms—for example, screenshots of the interactive quadrant plots, copied text from representative-passage tables, downloaded code-association figures, or notes derived from the file-to-code distribution database—and use them as the basis for local scholarly reasoning.

This workflow should not be mistaken for automation of interpretation. On the contrary, it extends LACOID's equal-status mixed-method design. The outputs remain guides, not conclusions. ISARI can help researchers compare latent code interpretations, generate analytic memos, test explanatory contrasts, and begin drafting evidence-grounded prose, but the authority to decide what the latent codes mean remains with the research team. In that sense, LACOID and ISARI are methodologically aligned: one classifies and organizes large-scale textual evidence while preserving original voices and contextual location; the other helps researchers think with those outputs under conditions of privacy, interpretive accountability, and local control.

Contribution to qualitative and mixed methods literature

Compared to previous methodological and applied contributions to the mixed methods literature, LACOID offers two major points of departure. The first is LACOID's explicit effort to democratize access to these rigorous and sophisticated analytical and statistical tools by removing all computer and statistical programming requirements (see also González Canché, 2022c,

2022d, 2022e, for related no-code data science applications to analyze qualitative evidence). We programmed the software infrastructure of LACOID and provided this applied example to be fully transparent, reproducible, and accessible locally and cost-free. At the same time, transparency should not be confused with unrestricted dissemination of outputs. Because LACOID's standard products preserve original and unaltered textual material together with contextual indicators of their location, researchers should ensure that any screenshots, exported tables, or interactive outputs shared beyond the research team do not weaken confidentiality, especially in small, distinctive, institutionally sensitive, or otherwise vulnerable samples (see González Canché, 2025). In many cases, local storage, selective excerpting, or restricted sharing may be methodologically and ethically preferable to broad release of complete outputs. Our emphasis on transparency, reprodicibility, and accessibility is methodologically significant. Because LACOID is available as a local, no-code application together with public source code and Code Ocean replication pathways (González Canché, 2022a), the framework supports reproducibility, inspectability, and long-term usability. These same qualities also make LACOID especially compatible with the broader local analytic ecosystem advanced in this book, in which outputs can remain under the researcher's control

Local access is important in the design phase to minimize the risks associated with uploading data to servers. This local architecture is not only a technical feature; it is also an ethical one. Because LACOID preserves original wording, contextual location, and highly interpretable passages, its outputs may contain information that is unusually rich and therefore unusually revealing (González Canché, 2025). For that reason, both the classification stage and any later interpretive work conducted with those outputs are methodologically safer when they remain under the researcher's local control, including when those outputs are later examined in ISARI as part of an offline scholarly workflow. That is, although we purposefully analyzed a collection of publicly available data to motivate the replication of our analytic procedures, we recognize that most typical projects cannot share data and that data protection and confidentiality should be prioritized, which is why we designed LACOID to run locally and it now can be integrated with a local intelligent system (ISARI) as we discuss in Chapter 9.

The second point of differentiation of LACOID is that we developed it to mirror (to the best of our abilities) the qualitative process of coding data (or HUCOID) in a line-by-line fashion. Our main motivation for decomposing texts was to avoid or minimize LCAB, where latent meaning may be lost due to using a multiplicity of sentences as a single text input in the machine learning and classification processes. As discussed earlier, our application example empirically corroborated the relevance of decomposing texts before applying NLP and text classification procedures.

Another contribution of LACOID is hypothesis generation (and testing) through an integrative framework. As depicted in our procedural diagram, the only way for this hypothesis-testing process to be meaningful is by having a clear understanding of latent codes. As depicted in our example, this understanding is possible only via the weaving of qualitative and quantitative threads. From this perspective, LACOID is designed to "force" this integrative process as a fundamental part of the meaning-building, hypothesis-testing process.

##Software Access, Local Architecture, and Reproducibility

Mac users can download LACOID here https://cutt.ly/jc7n3OT, PC users here https://cutt.ly/wc7nNKF. In addition, Code Ocean peer reviewed replication procedures can be found in González Canché (2022a), which executes the software on the Code Ocean server. To execute the application locally with the Mac or Windows versions, you simply need to: (1) extract the zip folder, (2) open the folder starting with "LACOID-App-" and ending with Win or Mac, and (3) double click on the file called "LACOID-App-(Win or Mac)." The first time LACOID is executed, you may need to select the menu "View" and "Force Reload" to generate all paths to this newly extracted folder. If, after forcing reload, LACOID does not display, close the window, execute LACOID again, and, if needed, force reload once more. The interface will now load. This process is only needed the first time LACOID is launched on a computer. The software applications have been comprehensively tested for viruses and are housed in a secure server location. Accordingly, all executable files are virus-free but may require administrative permission to be deployed. For the execution of the metrics assessment, Windows requires access to parallelization to speed up the process. This process will also request administrative permission the first time LACOID is executed on a Windows machine (this process is not required on Mac). Finally, each of these executable files for the Mac and Windows operating systems contains all the source code required to modify and/or expand the application's capabilities in the file called "app.R." Finally, Mac users are also recommended (and in some instances they do need) to place the LACOID software (and all software products discussed in this book) in their applications folder for newer versions can only read and write files from this location.

Moving forward

To conclude this methodological guidance study, we offer a set of questions that may help researchers and practitioners guide their research projects using LACOID. These questions are not prescriptive but are instead designed to spark scientific curiosity during the integration of the quantitative and qualitative analytic threads required by LACOID as a mixed-methods analytic integrative framework. The information in square brackets is provided to further contextualize these questions based on understandings gained in our analyses. These questions are:

1. What is the optimal number of latent codes embedded in our textual database?
   a. Is there a unique or clear solution, or are there competing, feasible solutions? [In this study, we found two solutions—one with three codes and the other with seven codes.]
   b. If there is more than one feasible solution, what approaches and decision-making processes can researchers use to select one solution over another?
   c. Are there any limitations associated with the final solution selected? For example, does selecting the lower number increase the chance of LCAB, wherein texts that may be associated with other latent codes are aggregated into a less optimal number of codes? The answer to this question can only be reached through qualitative analysis of these solutions, following the approach discussed in this study, and with the added nuance gained from reading the original texts and using the competing LACOID solutions as

guides. In addition to our qualitative analyses, the clear, nonoverlapping delineation of the three-code solution allowed us to discard the competing seven-code choice.

2. What is the contextualized meaning of these codes?
   a. How do the word-to-code distribution and the codes' quadrant locations aid in the meaning-building process?
   b. Can similarities and differences be observed in competing optimal solutions? [In our renderings, Code 3 and Code 7 overlapped in family formation, as highlighted by the selection of the word "child" as an example.]
   c. How do the word-to-code distribution and quadrant positions fall short in capturing these contextualized meanings? [Although these quantitative techniques are useful for gaining an initial understanding, the original texts should still be analyzed.]
   d. Does the analysis of original texts lead to understandings that are congruent with those gained from word-to-code distributions?
   e. Does the analysis of the text-mined data contribute to a better understanding of the meaning of these latent codes?
   f. Is the analysis of the most representative text, along with measures of within-code coherence and within-group coherence, useful in improving our qualitative understanding?
   g. Is there any evidence of plagiarism or potential text preparation issues, such as the use of the exact same sentence we found in our example?
3. Is the analysis of group association relevant to the research project? If so, which groups or intersections across groups (e.g., gender and age in our case) are more meaningful?
   a. Is there any statistical association? If so, how is this statistical association explained through qualitative understandings?
   b. What were the strongest associations found?
   c. Are these depictions useful in strengthening qualitative understandings? If so, how, or in what ways?
   d. How useful is it to observe that each group is not exclusively or uniformly linked to only one latent code, but rather that there is a distribution of groups to latent codes?
   e. How does the community detection process of LACOID compare to the statistical analyses? [In the example, girls were identified as belonging to Code 2 and Code 3 as communities, but the statistical tests indicated differences wherein the age 16 group was more frequently associated with Code 3 than expected, and the age 17+ group was less associated with this same code than expected—see Fig. 5.6. What are the meanings and implications of these variations?]
6. Finally, how can the integrative analyses of the associations of the answers to these questions help improve our understanding of the inherent structures embedded in these texts?
   a. Can these latent codes be used in more traditionally quantitative analyses, such as regression models?

Overall, and to conclude, LACOID's epistemological goals may be realized only to the extent that these latent codes are decoded with the contextualized understandings embedded in the original texts. The power of human nuance and knowledge of the topic of study is the best asset that qualitative and mixed methods researchers have to leverage the machine-learning and artificial intelligence analytic power that NLP and LACOID bring to the mixed methods table. The time has come to fully remove the computer and statistical programming barriers to these powerful analytic tools. We hope LACOID may truly serve to expand access and strengthen mixed methods in qualitative research via efficient and effective classification and hypothesis testing, based on the effective, unaltered, and contextualized retrieval of participants' meanings. Within the broader infrastructure developed in this book, these same outputs may also serve as structured prompts for local scholarly reasoning in ISARI, where latent code patterns, representative passages, and file-to-code distributions can be compared, memoed, and translated into evidence-grounded qualitative and mixed-methods explanation. Please see also González Canché (2022c), González Canché (2022d), and (2022e), for related no-code data science applications to analyze qualitative data dynamically.

References

Alexander, E., Eppler, M.J., Bresciani, S., 2019. Visual replay methodology: a mixed methods approach for group discussion analysis. J. Mixed Methods Res. 13 (1), 33–51.

Arun, R., Suresh, V., Veni Madhavan, C.E., Narasimha Murthy, M.N., 2010. On finding the natural number of topics with latent Dirichlet allocation: some observations. In: Zaki, M.J., Yu, J.X., Ravindran, B., Pudi, V. (Eds.), Advances in Knowledge Discovery and Data Mining. Springer, Berlin Heidelberg. https://link.springer.com/chapter/10.1007/978-3-642-13657-3%5C_43, pp. 391–402.

Belotto, M.J., 2018. Data analysis methods for qualitative research: managing the challenges of coding, interrater reliability, and thematic analysis. Qual. Rep. 23(11), 2622–2633. Available from: https://nsuworks.nova.edu/tqr/vol23/iss11/2/. doi:10.46743/2160-3715/2018.3492.

Cao, J., Xia, T., Li, J., Zhang, Y., Tang, S., 2009. A density-based method for adaptive lda model selection. Neurocomputing 72 (7-9), 1775–1781. doi:10.1016/j.neucom.2008.06.011.

Chang, T., DeJonckheere, M., Vydiswaran, V.V., Li, J., Buis, L.R., Guetterman, T.C., 2021. Accelerating mixed methods research with natural language processing of big text data. J. Mixed Methods Res. 15 (3). doi:10.1177/15586898211021196.

Charmaz, K., (2014). Constructing Grounded Theory, 2nd. Sage.

Coffey, A., Atkinson, P., 1996. Making Sense of Qualitative Data: Complementary Research Strategies, Sage Publications, Inc.

Cutts, M., 2009. Oxford Guide to Plain English, 3rd. Oxford University Press, New York, NY.

Deveaud, R., SanJuan, E., Bellot, P., 2014. Accurate and effective latent concept modeling for ad hoc information retrieval. Document Numérique 17 (1), 61–84. doi:10.3166/dn.17.1.61-84.

Eickhoff, M., Wieneke, R. (2018). Understanding topic models in context: a mixed-methods approach to the meaningful analysis of large document collections. *Proceedings of the 51st Hawaii International Conference on System Sciences*. https://pdfs.semanticscholar.org/9663/20028a68a5dd3d044f2bf1b40239aa2fcf58.pdf

Feinerer, I. (2020). Introduction to the tm package text mining in r. R CRAN. https://cran.r-project.org/web/packages/tm/vignettes/tm.pdf

Fetters, M., Curry, L., Creswell, J., 2013. Achieving integration in mixed methods designs: principles and practices. Health Serv. Res. 48 (6 Pt. 2), 2134–2156.

Fetters, M.D., 2020. The Mixed Methods Research Workbook: Activities for Designing, Implementing, and Publishing Projects, Sage.

Fetters, M.D., Freshwater, D., 2015. Publishing a methodological mixed methods research article. J. Mixed Methods Res. 9 (3), 203–213. doi:10.1177/1558689815594687

Fetters, M.D., Molina-Azorin, J.F. (2019). A checklist of mixed methods elements in a submission for advancing the methodology of mixed methods research 13 (4), 414–423. doi:10.1177/1558689819875832.

Fetters, M.D., Molina-Azorin, J.F. (2021). Special Issue on COVID-19 and Novel Mixed Methods Methodological Approaches During Catastrophic Social Changes.

González Canché, M.S. (2022a). Latent Code Identification [LACOID]: a Machine Learning-Based Integrative Framework [and Open-Source Software] to Classify Big Textual Data, Rebuild Contextualized/Unaltered Meanings, and Avoid Aggregation Bias [Source Code]. 10.24433/CO.6576994.v1

González Canché, M.S. (2022b), "School Leavers Study for Latent Code Identification Replication", Mendeley Data, V1, doi: 10.17632/gzhfdtmhcm.1

González Canché, M.S., 2022c. Machine driven classification of open-ended responses (MDCOR): an analytic framework and free software application to classify longitudinal and cross-sectional text responses in survey and social media research. Expert Syst. Appl. doi:10.1016/j.eswa.2022.119265.

González Canché, M.S., 2022d. Mapping, organizing, and visualizing interdependent events (MOVIE): a rigorous analytic framework and cost-free software application designed to model temporal and dynamic complex realist structures in social research settings. Methodol. Innov. 15 (3), 263–288. doi:10.1177/20597991221119012.

González Canché, M.S., 2022e. Network analysis of qualitative data: an integrative software application to visualize and assess similarities in participants' qualitative contributions. J. Mixed Methods Res. 16 (3), 373–377. doi:10.1177/15586898211051584.

González Canché, M.S., 2025. Ethics and trust in social network analysis: how data science and interactive visualizations impact anonymity, confidentiality, consent, and IRB procedures?. In: Handbook of Social Network Analysis and Education, (pp. 114–131). Edward Elgar Publishing. doi:10.4337/9781035312191.00013.

González Canché, M.S., 2026. Network analysis of qualitative data (NAQD): An analytic framework and software for qualitative data science visualization and hypothesis testing. Int. J. Qual. Methods. 25. doi:10.1177/16094069261439942.

Griffiths, T.L., Steyvers, M., 2004. Finding scientific topics. Proc. Natl. Acad. Sci. U.S.A. 101 (suppl 1), 5228–5235. https://www.pnas.org/content/101/suppl%5C_1/5228.

Grün, B., Hornik, K., 2011. Topicmodels: an r package for fitting topic models. J. Stat. Softw. 40 (13), 1–30. doi:10.18637/jss.v040.i13.

Ho, P., Chen, K., Shao, A., Bao, L., Ai, A., Tarfa, A., Brossard, D., Brown, L., Brauer, M., 2021. A Mixed methods study of public perception of social distancing: integrating qualitative and computational analyses for text data. J. Mixed Methods Res. 15 (3), 374–397. doi:10.1177/15586898211020862.

Humble, Á.M. (2019). *Computer-aided Qualitative Analysis Software*. Atkinson, P., Delamont, S., Cernat, A., Sakshaug, J.W., Williams, R.A. (Eds.). Sage Publications Limited.

Janasik, N., Honkela, T., Bruun, H., 2009. Text mining in qualitative research: application of an unsupervised learning method. Organ. Res. Methods 12 (3), 436–460.

Jason, L., Glenwick, D. (Eds.), 2015. Handbook of Methodological Approaches to Community-based Research: Qualitative, Quantitative, and Mixed Methods. Oxford University Press, New York, NY.

Lynam, T., Damayanti, R., Rialine Titaley, C., Suharno, N., Bradley, M., Krentel, A., 2020. Reframing integration for mixed methods research. J. Mixed Methods Res. 14 (3), 336–357.

Miles, M.B., Huberman, A.M., 1994. Qualitative Data Analysis: An Expanded Sourcebook, Sage.

Myhill, D., 2008. Towards a linguistic model of sentence development in writing. Lang. Educ. 22 (5), 271–288.

Nikita, M., Chaney, N. (2020). ldatuning: tuning of the latent Dirichlet allocation models parameters. R Project. https://CRAN.R-project.org/package=ldatuning

Pahl, R. (2012). *School leavers study, 1978. [data collection]* (tech. rep.). UK Data Service. SN: 4867. https://beta.ukdataservice.ac.uk/datacatalogue/studies/study?id=4867

Poth, C.N., Bulut, O., Aquilina, A.M., Otto, S.J., 2021. Using data mining for rapid complex case study descriptions: example of public health briefings during the onset of the COVID-19 Pandemic. J. Mixed Methods Res. 15 (4). doi:10.1177/15586898211013925.

Raftery, A.E., Lewis, S. (1991). *How many iterations in the Gibbs sampler?* (Tech. rep.). https://apps.dtic.mil/sti/pdfs/ADA640705.pdf

Reichardt, J., Bornholdt, S., 2006. Statistical mechanics of community detection. Phys. Rev. E 74 (1), 14. 016110.

Saldaña, J., 2021. The Coding Manual for Qualitative Researchers, sage.

Schütze, H., Manning, C.D., Raghavan, P., 2008. Introduction to Information Retrieval. Cambridge University Press. https://nlp.stanford.edu/IR-book/.

Silge, J., Robinson, D., 2017. Text Mining with R: A Tidy Approach. O'Reilly Media, Inc. https://www.tidytextmining.com/.

Stroll, A., Martinich, A.P. (2022). Epistemology. Encyclopedia Britannica. https://www.britannica.com/topic/epistemology

Teddlie, C., Tashakkori, A. (Eds.), 2009. Foundations of Mixed Methods Research: Integrating Quantitative and Qualitative Approaches in the Social and Behavioral Sciences. Sage, Thousand Oaks, CA.

Tesch, R., 2013. Qualitative Research: Analysis Types and Software, Routledge.

6

Machine Driven Classification of Open-ended Responses (MDCOR)

Introduction

Open-ended questions, or open-response questions, in questionnaires or survey research do not provide respondents with a set of predetermined answers (Geer, 1988; Kintzer, 1977; Kunz et al., 2020) but instead allow participants to respond freely using their own words (R. Popping, 2015; Reja et al., 2003a). Because respondents may offer insights about processes or reasons (Geer, 1988; Lazarsfeld, 1944) or may simply express opinions and ideas that do not conform to the options provided in closed-response questions (i.e., where options are predetermined), open-ended questions may help researchers expand the scope of their understanding about the topic of study (Reja et al., 2003a; Schuman and Presser, 1979). From this perspective, open-ended questions are particularly relevant when exploring a complex issue or new topic, as opinions may be heterogeneous, and even long lists of options may likely omit important nuances and points (Geer, 1988; Kelley, 1983; Lichtman and Kelley, 1984).

Despite these knowledge-based, value-added benefits, open-ended questions have two important costs. The first is that, given the burden associated with responding openly (McColl et al., 2001a), respondents might be more likely to leave open-ended questions unanswered than they would closed questions (Friborg and Rosenvinge, 2013; Popping, 2015; Kunz et al., 2020). In this respect, concerned with factors that may explain missing data in open-ended responses, Geer (1988) explored whether nonresponse was driven by participants' lack of ability to articulate a response rather than by their lack of interest in responding. This differentiation is important, for if the former is true, then our insights gained would be based on having the responses of those with the "right" skill articulation level to produce such a response. Perhaps more concerning is that, if this is true, our insights would systematically omit the responses of those with valid concerns but a lower ability level required to provide a response. Notably, Geer (1988) found that "individuals who fail to respond appear uninterested, not unable to respond" (Geer, 1988, p. 2). Although we discuss how nonresponses may be analyzed in our application section, our study predominantly focuses on reducing the second cost, which we discuss next.

The second cost associated with the analysis of open-ended responses is that this analytic process has traditionally been much more time-consuming and involved compared to the analysis of closed responses (Reja et al., 2003a; Kunz et al., 2020). Considering that open-ended responses are textual or qualitative data, these responses need to be labeled into categories or codes (Miles and Huberman, 1994) to ease their comparison and ultimately reach meaningful understandings (Holsti, 1969; Parker and Holsti, 1970; Kunz et al., 2020). Depending on the number of responses and the complexity of the questions, not only could this process take

Data Science, Interactive Visualizations, and Generative AI Tools for the Analysis of Qualitative, Mixed-Methods, and Multimodal Evidence. https://doi.org/10.1016/B978-0-443-21961-0.00004-8

weeks or months if conducted "manually," but this labeling or classification task also constitutes a truly consequential challenge (Coffey and Atkinson, 1996a; Miles and Huberman, 1994): the rigor and consistency behind the construction of these labels may ultimately shape research findings and conclusions (Belotto, 2018).

Guided by the relevance of this consequential challenge, this paper seeks to address the following overarching question:

How can vast amounts of qualitative evidence be classify effectively and efficiently without losing context or the original voices of our research participants, while also leveraging the nuances that human reasoning brings to qualitative and mixed-methods analytic processes?

How to close open-ended responses?

As we showcase in Fig. 6.1, we can currently employ two approaches to analyze open-ended responses: (1) human-driven via manual code identification, and (2) machine learning-driven based on statistical and probabilistic text classification.

Manual or human coding approach

With respect to human or manual classification, its main advantage is the deeper and more nuanced understanding gained based on human reasoning (Poth et al., 2021a). This same manual or human classification, however, comes with the well-documented increase in

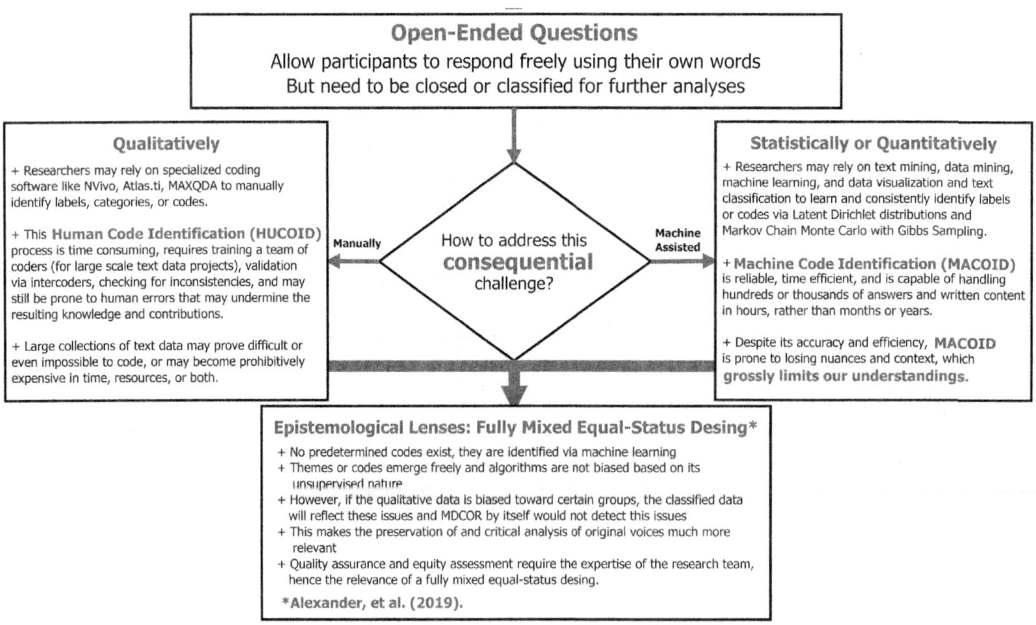

FIG. 6.1 Conceptual and Epistemological Framework.

classification inconsistencies and errors, particularly when dealing with vast amounts of documents and teams of coders (Belotto, 2018; Chang et al., 2021a; Poth et al., 2021a).

Although specialized software for qualitative analysis, such as ATLAS.ti, NVivo, or MAXQDA, may ease the labeling challenge (Humble, 2019; Tesch, 2013b), the coding process remains time-consuming and labor-intensive (Popping, 2015; Kunz et al., 2020). That is, this type of software is not yet capable of being trained to learn and identify text pieces that may be good candidates for specific codes or labels; consequently, the final product remains fully dependent on researchers' individual inputs and coding decisions (Chang et al., 2021a; Poth et al., 2021a).

A common strategy for minimizing subjectivity and room for interpretation in human code assignment is through intercoder agreement (Belotto, 2018; Eickhoff and Wieneke, 2018a; Miles and Huberman, 1994). In this validation process, "team members separately code (a sample of) the given data set [and then] compare and discuss the assigned codes and applied coding rules and recode the data according to the agreed solution [to] correct for discrepancies in individual judgment, and for joint mistakes that become apparent during analysis" (Eickhoff and Wieneke, 2018a, p. 904). From this view, the intercoder agreement test/training may increase coding reliability, but, once more, depending on the scale of the project, the costs in both time and resources may become unaffordable, particularly when dealing with time-sensitive information. Relatedly, despite these training efforts, the final product is still prone to potential inconsistencies, particularly in present times when big data has become part of our lives.

Machine learning driven classification approach

On the other hand, machine learning-assisted techniques rely on data science, data visualization, and natural language processing (NLP) tools for data labeling and classification. These quantitative and probabilistic processes are cost-effective and consistent. They are probabilistic because a given text or document is classified into a topic or code based on the high probability that the words configuring such a text or document are consistently classified in that topic or code through thousands of learning iterations (more on this below) (Albalawi et al., 2020a; Deveaud et al., 2014b; Nikita and Chaney, 2020). Notably, however, despite this time-efficiency gain, this probabilistic classification approach typically comes at the cost of potentially losing or altering participants' meanings or voices for the following interrelated reasons.

First, participants' own words are normalized or altered as part of the NLP and text cleaning or preparation. Depending on the type of normalization employed, the resulting words may be closer to or farther away from their initial meanings. For example, assume we have the following set of original words provided by our respondents: "educational," "educated," "educate," "education," and "educative." If we apply a normalization technique called stemming, which only preserves the root of the words, the resulting vocabulary will yield only one term, "educ," with a frequency of 5. If, instead, we apply lemmatization, the result removes inflectional endings, leaving the base of the word (Schütze et al., 2008a). That is, the use of lemmatization will result in the following vocabulary: "educational," "educate," "education," and "educative," all with a frequency of 1, except for "educate," which will have a frequency of 2. From this view, stemming leads to a much more reduced vocabulary compared to lemmatization. The problem here is that, after these normalization techniques are applied and machine learning is implemented, researchers are typically constrained to interpret the meaning of each statistical or probabilistic topic or code via the analysis of these normalized texts (for a recent example, see Table 3 of Ho et al., 2021a, on page 384), rather than by analyzing or retrieving the original or unaltered words associated with those classified open responses.

The second important limitation associated with currently available machine learning classification tools is the possibility of losing context or aggregating a multiplicity of meanings (Eickhoff and Wieneke, 2018a). That is, when these techniques are applied to entire documents, multiple meanings may be aggregated into a single code, which can lead to a significant loss of nuance in our understanding. For example, Eickhoff and Wieneke (2018a) applied topic modeling to 7356 research articles spanning 40 years, classifying each full article into only one single latent topic. The main concern with this classification approach is that, although the content of each article may be associated with more than one topic—especially if each article were classified using human classification techniques (Ho et al., 2021a)—machine learning classification is designed to yield one topic per document. Having said this, although machine learning-driven text classification may lead to a loss of nuance when classifying entire documents, this issue becomes more problematic when each input document contains multiple pages. On the other hand, when the responses are relatively short, such as open-ended responses or social media posts, machine learning-based text classification can be quite powerful in implementing a version of the *line-by-line* coding approach (Charmaz, 2014a), similar to the one used in human or manual code identification (see Saldaña, 2021a, for example).

Integrative approach and purpose

Considering that coding results ultimately affect a study's findings and conclusions (Belotto, 2018), we should prioritize efforts to achieve the most valid, reliable, and consistent labeling or coding outcomes across the vast amount of open-ended information to be analyzed.

These efforts constitute the overarching purpose of this study.

The analytic approach and software tool we offer in this study, *machine-driven classification of open-ended responses* (MDCOR), builds on the premise that, despite the added accuracy and efficiency of using computer-assisted NLP techniques to learn and classify texts, those techniques alone may alter participants' words and could fall short in yielding nuanced understandings. Accordingly, MDCOR was programmed to preserve the original voices of our research participants while offering researchers tools to identify the optimal number of codes contained in the data, ease the retrieval of information, render interactive summaries and data visualizations, and offer group comparisons of the resulting classified responses (see Arun et al., 2010a; Cao et al., 2009b; Deveaud et al., 2014a; Griffiths and Steyvers, 2004). The ultimate goal of MDCOR is to facilitate the development of nuanced understandings of the qualitative meanings embedded in open-ended responses to strengthen our pursuit of relevant findings and valuable conclusions. In this sense, MDCOR contributes not only to efficiency and consistency, but also to interpretive transparency by making the path from unstructured text to latent code identification more inspectable, revisitable, and documentable. This means that classification should never be mistaken for interpretation itself. MDCOR can identify latent structure with consistency and efficiency, but the work of deciding what those latent codes mean, how they should be named, how they relate to one another, and what they imply for the study's broader argument remains a qualitative and interpretive responsibility of the research team. In short, then, the purpose of MDCOR is to classify vast amounts of qualitative evidence effectively and efficiently without *altering the original voices* of our research participants, while leveraging the nuances that human reasoning brings to the qualitative and mixed methods analytic tables.

MDCOR represents a comprehensive, no-code development in computer-assisted techniques to implement machine learning and text classification to retrieve latent codes in written data. Our subsequent sections summarize recent literature and describe the conceptual and methodological adequacy of using machine-learning techniques to classify open-ended survey responses, offering assessments of optimal classification metrics and initial statistical assessments of group-to-code association. Moreover, we showcase a rigorous, user-friendly, and cost-free software application developed to implement MDCOR's analytic framework. As detailed in the software description, MDCOR not only reduces subjectivity and human coding or classification errors while yielding time-efficient and reliable results, but also provides researchers with a fully classified dataset that may be merged with the original quantitative dataset to apply traditional quantitative analyses. By removing manual classification burdens, MDCOR opens the possibility of time-efficiently and rigorously reaping the benefits of open-ended questions, even when analyzing tens of thousands of responses.

The relevance of these outputs extends beyond classification alone. Once latent codes have been identified, interactive word summaries have been rendered, and code-by-group relationships have been estimated, researchers still need to determine what those patterns mean, how code distinctions should be interpreted, which contrasts deserve further memoing, and how the resulting structure should be translated into narrative explanation. In the broader analytic architecture of this book, MDCOR outputs can therefore be understood not as endpoints, but as structured guides that may also be brought into ISARI, a local, fully offline, open-source, and multimodal brainstorming environment designed to support scholarly comparison, memoing, synthesis, and evidence-grounded writing. In this way, MDCOR and ISARI are methodologically aligned: MDCOR classifies and organizes large-scale textual evidence while preserving original voices; ISARI helps researchers think with those outputs under conditions of privacy, interpretive accountability, and local control.

Related literature

This section is divided into two main parts. The first part illustrates how the closure of open-ended responses has been addressed in the literature. The second part presents recent applications of data science and NLP as time-sensitive tools for classifying vast amounts of qualitative evidence.

Closing open-ended responses

Responses to open-ended questions are free from the potential biases that result from constraining respondents to a set of predetermined options (Reja et al., 2003a). That is, to the extent that predetermined options fail to address relevant alternatives (Geer, 1988; Lichtman and Kelley, 1984), the information gathered may be limited or even misleading (Friborg and Rosenvinge, 2013; Lazarsfeld, 1944). This is why scholars like Lazarsfeld (1944) recommend beginning the questionnaire design with open-ended questions as the starting point for identifying the corresponding categories that can be used in the closed versions of these questions.

A longstanding concern regarding the use of open-ended questions is the difficulty of analyzing these questions due to the need to conduct extensive coding or classification (Popping, 2015; Reja et al., 2003a; Schuman and Presser, 1979; Kunz et al., 2020). That is, as Popping stated, "open-ended questions must be closed [i.e., coded or classified] in the analysis process" (Popping, 2015, p. 25), and the question then becomes how to close them as rigorously and efficiently as possible. In this respect, Popping noted two main limitations in this area of research. First, methodological developments for the closing of open-ended questions "has not received that much attention in the literature" (Schuman and Presser, 1979), (Popping, 2015, p. 27). Second, Popping argued that although technological developments in this area have been important, general software applications capable of conducting text analyses "are usually written in the context of a specific study rather than for general use" (Popping, 2015, p. 36). This "one code per document" is a procedural limitation of most machine learning text classification approaches, which we are describing as latent code aggregation bias below and which our framework and software tool aim to overcome.

In this respect, MDCOR represents a contribution that addresses both of these limitations. First, the analytic framework aims to address current shortcomings in machine learning text classification. Specifically, MDCOR does not alter participants' original voices and offers easy access to the classified text and their unmodified qualitative content to strengthen the sense-making process. Second, the software application fueling MDCOR has been designed for general use rather than for specific projects, so that any researcher with a database containing open-ended responses (or opinions, such as social media posts) may use MDCOR to classify text responses and even conduct exploratory statistical analyses of relationships among these codes and participants' attributes. Moreover, although we are making the back end of MDCOR publicly available, free to share, and free to distribute (Mac version is available here: https://cutt.ly/UNyzlxW and Windows version is available here: https://cutt.ly/KNyzGUV), this computer and statistical-code-intensive option is intended for researchers interested in contributing to or expanding MDCOR's capabilities. MDCOR's user interface (UI) does not require any computer or statistical programming, and its current form is capable of replicating all outputs discussed in our application section.

Recent applications of data science and natural language processing

Computer-assisted tools involving data science, NLP, text and data mining, machine learning, and topic modeling have been used infrequently in survey and mixed methods research (Chang et al., 2021a; Ho et al., 2021a; Lynam et al., 2020a; Poth et al., 2021a). This methodological gap is notable, considering both the relevance of these analytic tools for dealing with large-scale textual data sources and their adequacy for reliably, systematically, and efficiently classifying vast amounts of unstructured information (Chang et al., 2021a; Eickhoff and Wieneke, 2018a; Janasik et al., 2009a; Poth et al., 2021a). The following paragraphs synthesize recent empirical applications of computer-assisted techniques to classify textual data rapidly and rigorously, with the goal of situating these studies within MDCOR's capabilities.

Poth et al. (2021a) applied text mining—a set of techniques, technologies, and rules for efficiently, cost-effectively, and consistently extracting information and tendencies from large collections of text data (Poth et al., 2021a, pp. 351–352)—to rapidly generate a complex case description of the COVID-19 pandemic. Specifically, the researchers relied on text mining and sentiment analysis to identify fluctuations across six time periods in the content of 72 media briefing transcripts, including word counts and counts of positive, negative, and neutral words conveyed in those briefings. After these techniques were applied, the authors qualitatively analyzed a subset of 16 media briefings (of the 72 in total included in their analytic sample). The coding process followed a line-by-line approach to develop a preliminary code list (Charmaz, 2014a). In addition, Poth et al. (2021a) applied topic modeling to detect clusters of similar word groupings across the media briefings. Together, these authors contributed to mixed methods research by offering a timely example of text mining, sentiment analysis, and topic modeling to analyze large volumes of textual data and, as a team, integrate these findings with their qualitative and more contextualized analyses. Aligned with the purpose of MDCOR, the integration conducted by Poth et al. (2021a) enabled these researchers to avoid the pitfalls of relying on one approach, instead fully integrating their multiplicity of qualitative and quantitative outputs to enhance their sense-making.

In another recent application, Chang et al. (2021a) relied on NLP—computational approaches to analyze textual data in natural language (Chang et al., 2021a, p. 399)—to rapidly analyze textual data during times of catastrophic change. These authors rightly stated that qualitative data are particularly salient during a crisis to better understand the experiences of affected individuals, arguing that NLP may help expedite the analytic process without sacrificing scientific integrity. The authors' depiction of qualitative or manual code identification (or thematic analysis), using computer-assisted qualitative data analysis software, is of particular salience for both their study and MDCOR. They noted that MAXQDA's and NVivo's text mining functionalities are limited to "count[ing] word frequencies and word combinations" (Chang et al., 2021a, p. 401). NLP, on the other hand, examines phrases, finds clusters, and may naturally account for slang and informal grammar based on word tokenization. Recent examples of these latter situations include times of crises like the COVID-19 pandemic (Chang et al., 2021a; Ho et al., 2021a; Poth et al., 2021a).

To demonstrate the usefulness of NLP, Chang et al. (2021a) provided two examples. In the first example, which relies on thousands of clinical notes documented by clinicians during virtual care visits, NLP allowed these researchers to analyze all qualitative data (rather than a subset or sample) in a reliable and time-efficient way. In their second example, they used NLP to identify themes in survey responses from nearly 5000 text messages "in 2 full days, rather than the typical 2 full months it would otherwise require for the team" (Chang et al., 2021a, p. 407). In this case, the authors conducted analyses at the word-level using unigrams (single words), bigrams, and trigrams (two words and three words, respectively) that remained in a sentence after removing stop words (words that convey no meaning by themselves; see Appendix available online at http://dx.doi.org/DOI_of_Chapter6) and punctuation. These authors then summarized data with wordclouds, a visual summary that is similar to MDCOR's interactive distribution of latent codes and their configuring words (see our example of the interactive plot resulting from MDCOR discussed in our applied section here: https://cutt.ly/lrXhCh4e and in Fig. 6.2).

Another relevant application of computer-assisted analyses of qualitative data is offered by Ho et al. (2021a). In this study, the authors analyzed 2270 open-ended responses to holistically understand factors associated with participants' reported social distancing. Their design relied on first understanding what participants considered to be important reasons to practice social distancing, the factors that contributed to variations in these reasons, and how these reasons impacted their self-reported social-distancing behaviors. These authors relied on topic modeling, noting that its unsupervised nature makes qualitative analysis of the resulting topics necessary to identify the contextualized meaning embedded in those latent topics. Similar to MDCOR, their analyses fundamentally consisted of "closing" these open-ended responses through topic modeling. However, before proceeding with this NLP approach, they created a qualitative codebook relying on in vivo manual or human-driven code identification. As part of this process, the research team also conducted intercoder reliability, and this coding process allowed single responses to be assigned to more than one code.

Ho et al. (2021a) also provided assessments of the optimal number of topics (an approach also implemented by MDCOR; see Fig. 6.3). In addition, they provided the top five keywords (the top five most frequent words that only appear in a given latent topic) to offer a clearer idea of the meaning of each latent topic, a similar capability also provided by MDCOR in the form of an interactive topic-word frequency summary. That is, as explained in our application section,

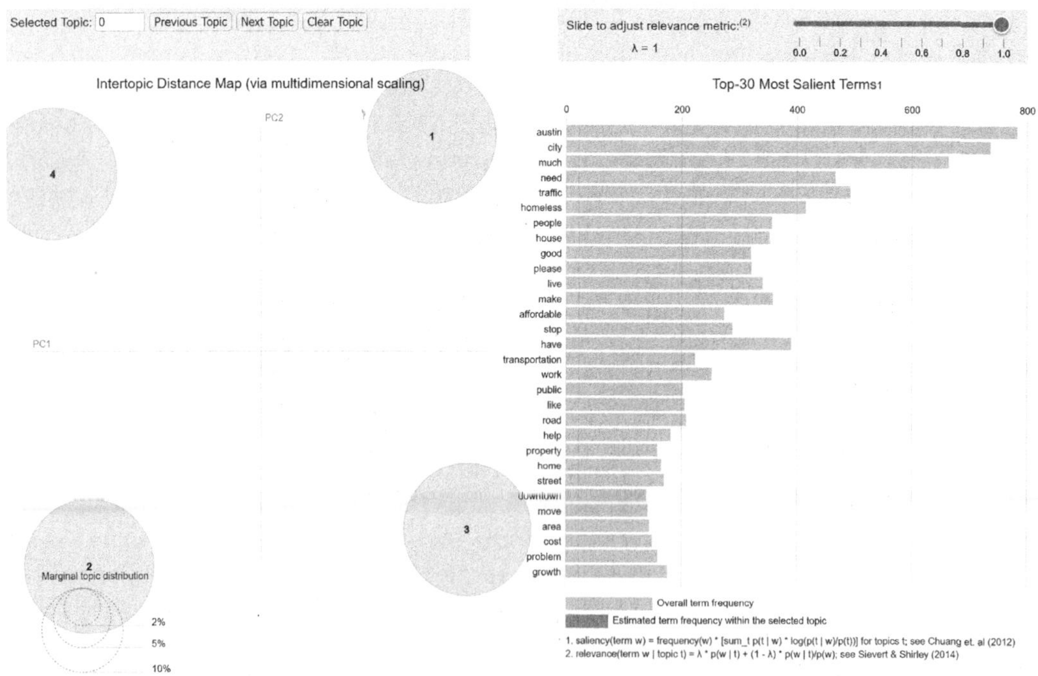

FIG. 6.2 Word summaries of each latent code—see the interactive version at https://cutt.ly/cBAITVd.

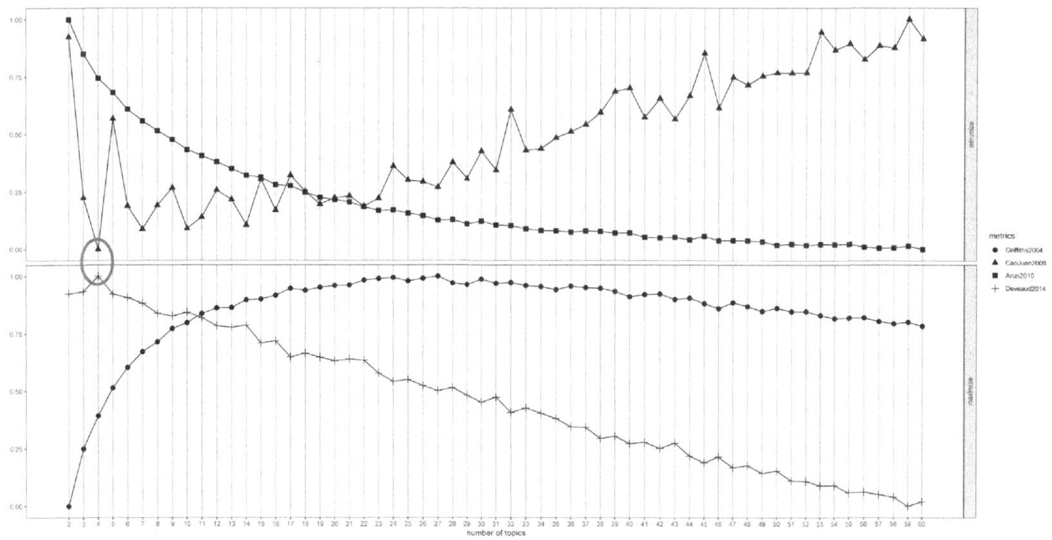

FIG. 6.3 MDCOR Assessment Metrics Showing the Convergence of Two Metrics: Cao et al. (2009a) **and** Deveaud et al. (2014a)

the interactive plot rendered by MDCOR (see https://cutt.ly/lrXhCh4e or Fig. 6.2) also allows identification of words exclusively classified in a given latent code as well as words shared across codes, as we illustrate in our applied example and in Fig. 6.4.

Despite Ho et al. (2021a)'s research team assessment of the optimal number of topics, they concluded that two topics were not meaningful based on their integrative analyses "because the highly associated examples of these topics did not provide a consistent topic" (p. 384). Another important contribution of Ho et al. (2021a)'s study is their comparison of topic modeling and qualitative results, demonstrating that "overall the two approaches yielded similar results" (Ho et al., 2021a, p. 387). They also pointed out that any differences in the percentage of responses assigned to each topic varied because the topic model approach identified topic ascription based on the highest contribution (see our appendix for more details and Table 6.1 for an applied example), hence creating mutually exclusive and exhaustive categories, whereas their qualitative analyses allowed each response to potentially be labeled in more than one code. Based on this possibility of code overlap or closeness in meanings, MDCOR offers two interactive HTML plots: one that places resulting classes or codes in quadrants (see Fig. 6.2) and a second that assesses the degree to which each pair of codes share participants' responses based on these participants' attributes, such as gender or home ownership status, for example (see Fig. 6.5), as we further explain in our applied section.

In closing this section, and based on these recent real-world applications, we are confident in stating that MDCOR represents an important methodological contribution. Not only does it package best practices in topic modeling, but it also offers these tools at no cost and without requiring computer or statistical programming expertise—which remains a barrier for most natural language processing and machine learning applications as stated

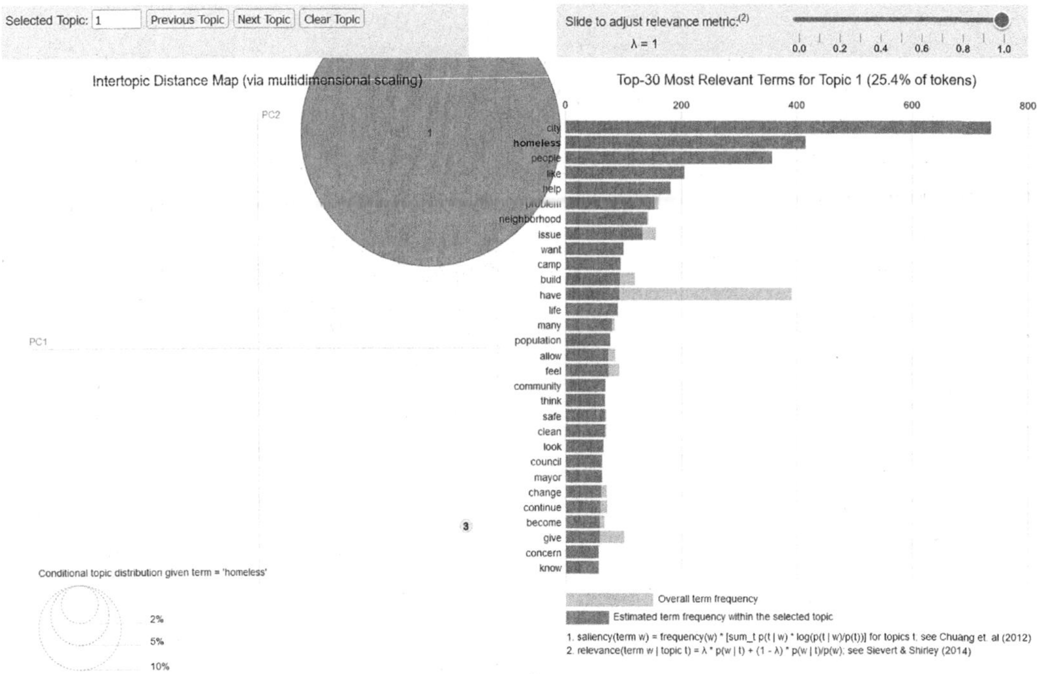

FIG. 6.4 Rationale code and word selection in the interactive summary—To replicate this figure, click Code 1 and then place your cursor on the word "homeless" in the interactive version at https://cutt.ly/cBAlTVd.

Table 6.1 Example of classification process and text to code contribution and group fit.

Q. Elaborate on your family and career balance	Code 1	Code 2	MDCOR Result	Text Contribution	Group Fit	Labeling
My career is the most important aspect of my life	90%	10%	Code 1	1 = (.90/.90)†	1 = (.90/.90)††	Career_ Oriented
My family is as or more important than my career	45%	55%	Code 2	.6875 = (.55/.80)†	0.675 = (0.75/.90)††	Family_ Oriented
For me, the best day is a family day not an office day	20%	80%	Code 2	1 = (.80/.80)†	0.75 = (0.675/.90)††	Family_ Oriented

†Divided by the highest probability within that code
††Average contribution of group divided by highest contribution by group.

by Chang et al. (2021a). That is, this latter point is important because the lack of application or use of NLP and topic modeling is, in part, explained by the fact that "deep linguistic analyses of text-based research data [like the ones employed in MDCOR] are still reserved for investigators or research teams with a computer science background" (Chang et al., 2021a, p. 400). Accordingly, with the removal of this programming hurdle, MDCOR may contribute to expanding access to these analytic benefits.

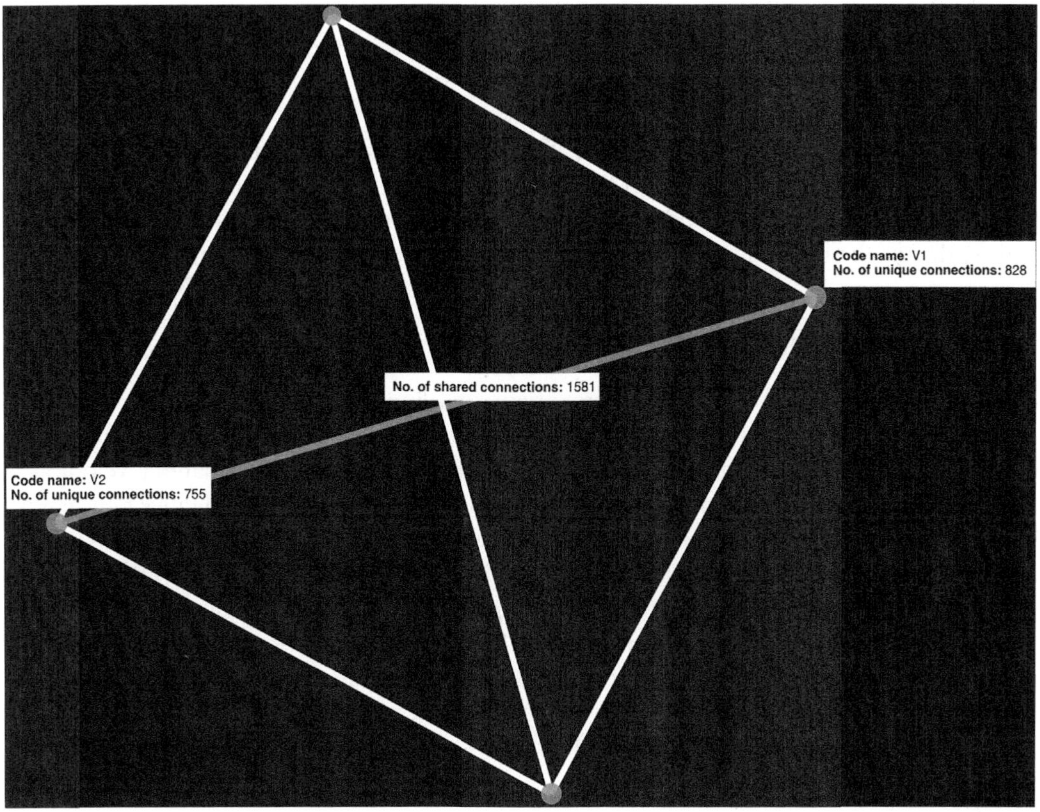

FIG. 6.5 Code relationships by group association—see the interactive version at https://cutt.ly/UBJ6Hfa.

Conceptual and epistemological lenses

Shapiro (2020) described two types of approaches to coding open-ended questions: the instrumental and the representational frameworks. In the instrumental approach, a computer program relies on a set of thematic codes provided by researchers to classify open-ended responses into those themes or codes. The limitation of this approach is that, by relying on a deductive method—wherein the open-ended responses adhere to a predetermined list of codes (Boyatzis, 1998; Eickhoff and Wieneke, 2018a; Tashakkori and Teddlie, 2008a, 2009b)—respondents' meanings that do not conform to those predetermined codes are ignored (Popping, 2015). On the other hand, in the representational approach, themes or topics emerge freely, without a predetermined list of codes, and therefore may better reflect a more comprehensive set of perspectives organically, as respondents react to the open-ended prompts or questions (Geer, 1988; Popping, 2015; Shapiro, 2020). In this representational approach, after themes or categories have been identified, researchers need to interact with these proposed themes and discover or interpret their meanings (Popping, 2015). MDCOR follows this representational approach and relies on the following conceptual and epistemological lenses. This representational commitment is important for the broader architecture of the book because it means that machine-driven classification is not treated as a substitute for human interpretation. Instead, the classification process identifies latent structure that must still be read, named, compared, and situated by researchers. In later stages of analysis, this same commitment supports the use of local AI as a thinking partner rather than as an answer authority: the role of systems such as ISARI is not to decide what the codes mean, but to support researchers as they compare, memo, and synthesize the meanings already preserved in participants' original responses.

How is knowledge produced?

Understanding epistemology as the nature, origin, and limits of human knowledge (Stroll and Martinich, 2022), MDCOR is guided by the belief that knowledge can be built from machine learning and text classification analyses of qualitative evidence, but only if this quantitative output is fully integrated with the qualitative analysis of this classified and unaltered qualitative information. This implies that, since we are relying on open-ended questions to gather unrestricted reactions, no predetermined codes may be applied during the classification process; instead, codes may emerge freely from the input data, which explains MDCOR's reliance on unsupervised learning.

More specifically, Fig. 6.1 summarizes MDCOR's epistemological premise: despite the added accuracy and efficiency of using computer-assisted NLP techniques to identify latent codes based on word-to-code (see Table A4, available online at http://dx.doi.org/DOI_of_Chapter6) and text-to-code (see Table A5, available online at http://dx.doi.org/DOI_of_Chapter6) probabilistic distributions.

Those techniques alone fall short in yielding meaningful understandings that may lead to valuable conclusions. Instead, qualitative and nuanced understandings of the contextualized meanings of those latent codes, based on the original and unaltered responses of our participants, are needed as part of these analytic processes to generate relevant and useful knowledge. This is the reason why both the statistical and the qualitative approaches depicted in Fig. 6.1 converge and constitute the epistemological lenses driving MDCOR's analytic framework and software implementation.

Accordingly, MDCOR's epistemology represents a fully mixed equal-status design (Alexander et al., 2019; Fetters et al., 2013a; Jason and Glenwick, 2016a; Teddlie and Tashakkori, 2009a), wherein rigorous, valid, and useful knowledge is only achieved through the equally important insights gained by integrating the quantitative analysis of qualitative evidence with the qualitative analysis of the resulting classified codes. From this perspective, machine-driven classification does not eliminate the need for qualitative interpretation; it intensifies the importance of it. The latent codes identified by MDCOR only become analytically useful when researchers return to the original responses, compare the most representative texts, assess overlaps and distinctions across codes, and reconstruct meaning in relation to the purposes of the study. Taken together, then, this mixed equal-status design analytic strategy implies, then, that if one of these approaches is conducted poorly, the entirety of the analysis may fall short in (re)constructing the meanings and knowledge embedded in our qualitative evidence.

A fundamental component of MDCOR's epistemological perspective is its emphasis on the preservation of our research participants' original voices. This preservation serves as an empirical quality assurance mechanism. That is, although MDCOR will always consistently identify latent codes without altering the original meanings or texts from input responses (Eickhoff and Wieneke, 2018a; Lynam et al., 2020a), this does not mean that the resulting knowledge will always be useful or unbiased for the following reasons. If the data-gathering process of these input responses lacks methodological rigor or if it is collected without internal consistency or validity, MDCOR will not be able to fix or solve this data-input quality problem—but it may help make it evident. For example, as detailed below, one of the secondary products of MDCOR is a dataset containing responses that were either nonsensical or constituted by words that appear in less than 1% of all texts in the vocabulary. Since these texts are outliers, they are excluded as valid inputs for the classification process. Nonetheless, researchers may assess their content and decide whether these texts may be worthy of inclusion—see the second point in our UI section on page 19 for more details.

In terms of bias issues, although MDCOR's algorithms do not suffer from bias toward certain groups due to their unsupervised classification nature (which means that the learning process starts without any preconceived notions or the representational approach described above), if the qualitative data do contain these biases (e.g., our participants' discourses or responses are themselves biased toward certain groups in the form of racist or misogynistic comments), the classified results will carry these issues forward, precisely because MDCOR does not alter participants' meanings or original texts.

That is, bringing these two points together, the preservation of original texts is important because MDCOR, by itself, would not know whether the input information meets a certain quality standard or whether this input information is subject to biases. Accordingly, text preservation allows for the possibility of conducting *quality and equity assessments*, both of which require the expert and qualitative analysis of the research team. This qualitative-quantitative codependence, once more, reflects the relevance of a fully mixed equal-status design, where both qualitative and quantitative perspectives are truly important in this knowledge-building process.

For that reason, MDCOR should be understood as a framework that strengthens rather than relaxes methodological responsibility. The preservation of original text, the inspection of dropped responses, the interpretation of latent codes, and the evaluation of possible bias all require disciplined qualitative judgment. Machine learning can support classification, but it cannot replace the methodological care required to decide whether the resulting categories are conceptually coherent, ethically defensible, and substantively meaningful.

Methodological foundations, flow diagram, and user interface

The implementation of our conceptual and epistemological frameworks is represented in Fig. 6.6 as a flow diagram. Note that a complete version of the methodological and statistical foundations of MDCOR is presented in the appendix. In this section, we describe these methodological elements practically and illustrate their location in MDCOR's flow diagram.

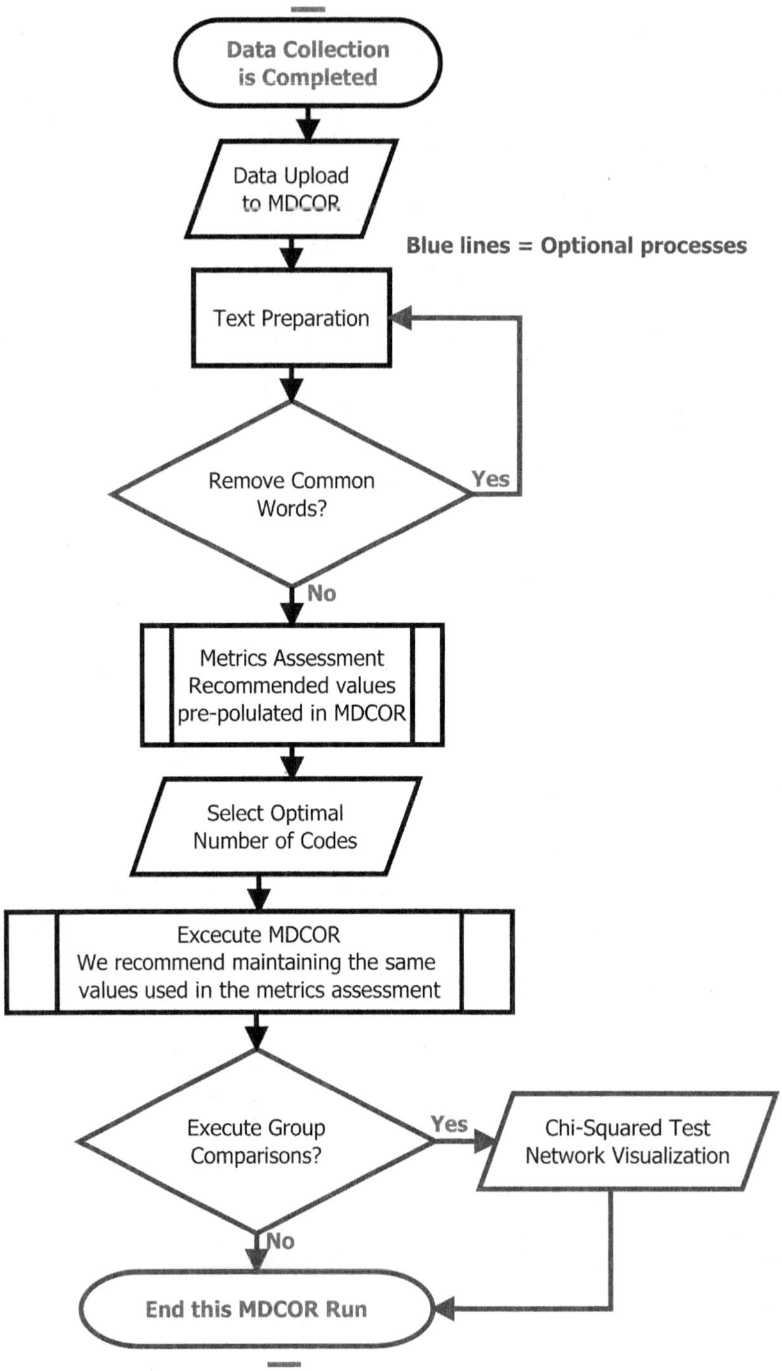

FIG. 6.6 MDCOR's software flow diagram.

Machine driven classification

Based on MDCOR's epistemology, its main analytic engine consists of an unsupervised classification method designed to find natural groups of texts based on shared information (Griffiths and Steyvers, 2004; Grün and Hornik, 2011). This classification framework consists of a training and learning algorithm that, based on Bayesian statistics and statistical physics (Griffiths and Steyvers, 2004) via Markov chain Monte Carlo (MCMC) and Gibbs sampling (Grün and Hornik, 2011), estimates.

- Word importance for a given code includes all words in the entire vocabulary (see Table A1, available online at http://dx.doi.org/DOI_of_Chapter6), irrespective of their nesting texts, that are associated with a given code (see also Table A4, available online at http://dx.doi.org/DOI_of_Chapter6)
- Proportion of words configuring (or nested within) text i that were classified in code j (see Tables A3 and A5 in the appendix).

Please note that these indicators are explained in more detail in the appendix (see Equation A3). For our applied purposes, let us describe MDCOR's flow diagram as a preamble to the full description of its UI, along with all the outputs resulting from its implementation.

Flow diagram

The flow diagram implemented by MDCOR is shown in Fig. 6.6. This diagram indicates that the analytic process starts when the data collection is completed, and these data are uploaded to the UI. Given MDCOR's goal to fully rely on free software, the data input format is a universal comma-separated-value ("∗.csv") file. In terms of types of data supported, MDCOR handles cross-sectional (i.e., participants responded once) or longitudinal (i.e., participants may have multiple responses or contributions over time) data formats. These data may be retrieved from surveys, social media posts—like YouTube commets, for example—or any other source where at least one ID column and one text column are present. Based on the intensity and computing power required, currently only one text column (or open-ended response) may be analyzed at a time.

Once the data are uploaded, the text preparation step implements NLP, text mining, and normalization (see Feinerer (2020) for best practices, guideliens, and programming tools), as we discussed earlier. Beyond the coded actor-to-text structure itself, MDCOR does not require additional feature engineering for its classification procedures, which is one of the reasons it can move efficiently from open-ended responses to interpretable latent code structures. After the initial text cleaning or preparation is concluded, MDCOR offers the possibility of eliminating words that are too common and would not be useful in the learning and classification processes (Schütze et al., 2008a)—more on this below in the UI section. This is, however, an optional step, as it is not necessary to delete these words for MDCOR to continue with the classification process.

The next step involves a metrics assessment to determine the most optimal number of codes from a statistical point of view. These metrics, further explained in the UI section (and the appendix), require MDCOR users to select the values of four components: burn-in and Gibbs samplings, β and α hyperparameters. Accordingly, our UI discussion places particular emphasis on these four statistical components. Having said this, we recommend that researchers rely on the default values we offer in the UI, as these values are supported by empirical methodological research (Griffiths and Steyvers, 2004; Grün and Hornik, 2011). Nonetheless, researchers may modify these values if they wish to conduct sensitivity and robustness checks.

Once the metrics assessments are completed, researchers will have one plot at their disposal to facilitate the identification of the optimal number of codes. With this tool, they can select this *optimal* number in the UI and execute MDCOR. This final required step will perform the final text classification. Note that this machine-learning-driven text classification also requires the values of the four elements selected in the metrics assessment. In this regard, for consistency, we recommend researchers use the same values chosen for the metrics assessment. That is, if set *m* is chosen for the input values fueling the metrics, use this same set *m* for the final MDCOR analysis. If another set, *k*, is used for robustness checks, use this same set *k* in both the metrics assessment and the MDCOR implementation.

Finally, the flow diagram in Fig. 6.6 shows that a final and optional *post-hoc* analysis is available. This analysis offers the possibility of assessing whether the resulting classification is associated with participants' attributes, based on a categorical indicator. Once more, we elaborate further on this and the other outputs in the following section. As a visual summary of this flow diagram, please see Fig. 6.7. This figure illustrates the entire process and showcases the main outputs automatically produced by MDCOR.

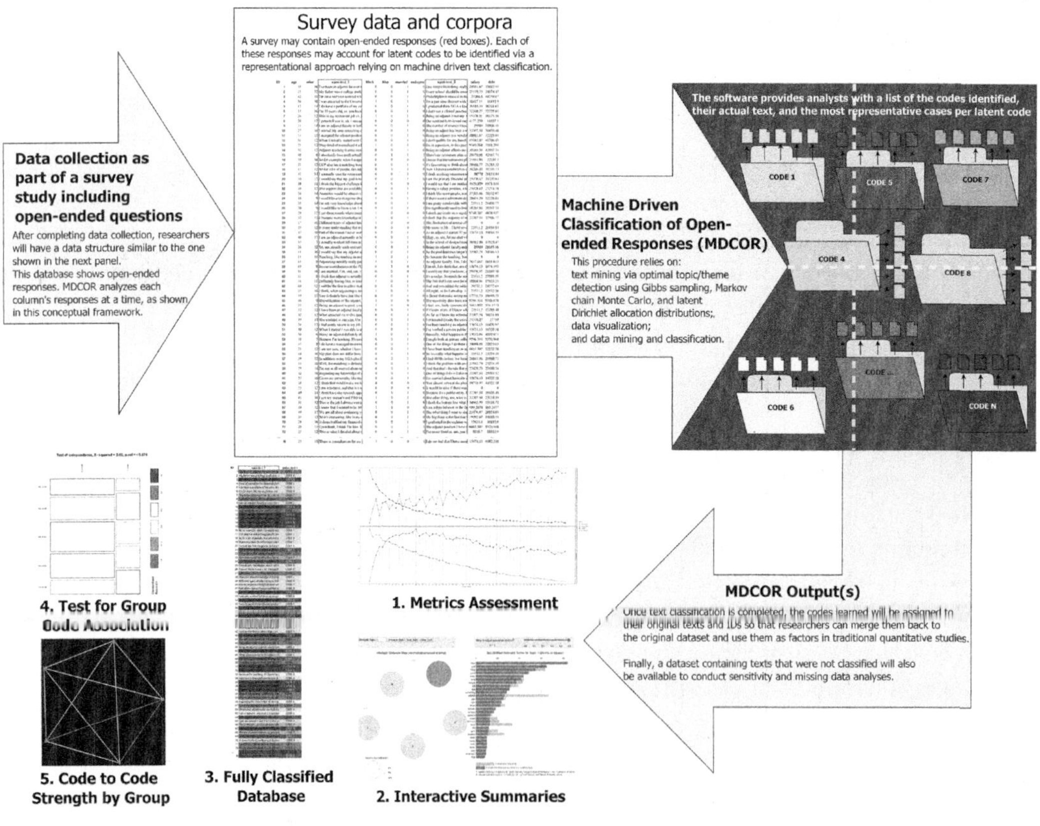

FIG. 6.7 Procedural diagram and MDCOR outputs.

User interface

Fig. 6.8 shows MDCOR's UI. The processes and analyses depicted in our flow diagram, as well as the conceptual and epistemological frameworks further described in the online appendix, can be performed with just a few clicks. The following description illustrates these steps, along with the output resulting from each selection:

1. **Load the dataset**: To start using MDCOR, upload a database under "1. Select database from local folder." This database needs two columns as the minimum components: the *unit ID* and the column containing the *open-ended responses* (or any form of text data, such as social media posts). As described below (in the eighth step), although we can test for code associations by group, this group or categorical column is optional. That is, it is not required during the learning process, which increases the flexibility of MDCOR with respect to the minimal number of elements required to execute all natural language, machine learning, and text classification procedures. Finally, note that users may decide to include cells without responses. Although MDCOR will ignore this lack of responses during

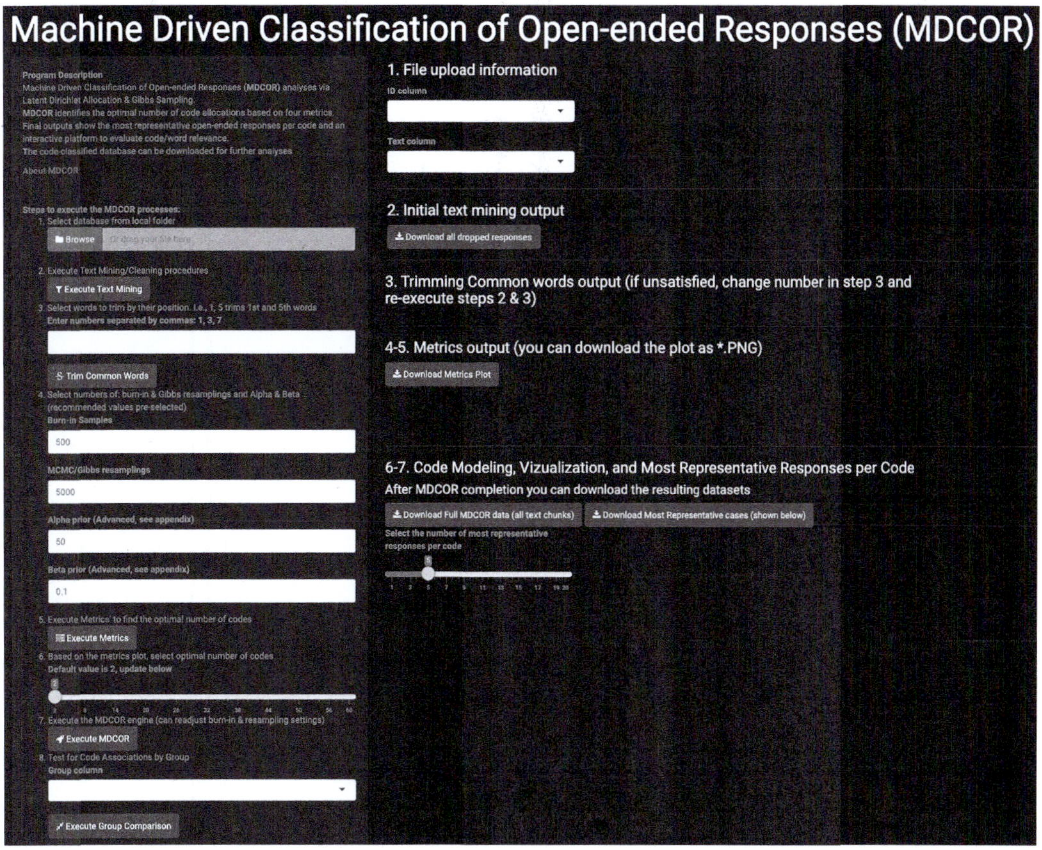

FIG. 6.8 User Interface Platform. Mac users can download MDCOR here: https://cutt.ly/xv6nnuN, and PC users can download it here: https://cutt.ly/Lv6nTLG.

the classification, the *post-hoc* group analysis will assess whether certain groups were more likely to not provide a valid response (see Fig. 6.9), which may increase our insights into the mechanisms driving our understanding.

2. **Conduct text cleaning/mining**: Once the ID and Text columns are selected, we can then execute the NLP and text-mining procedures (described in the appendix) by clicking "2.

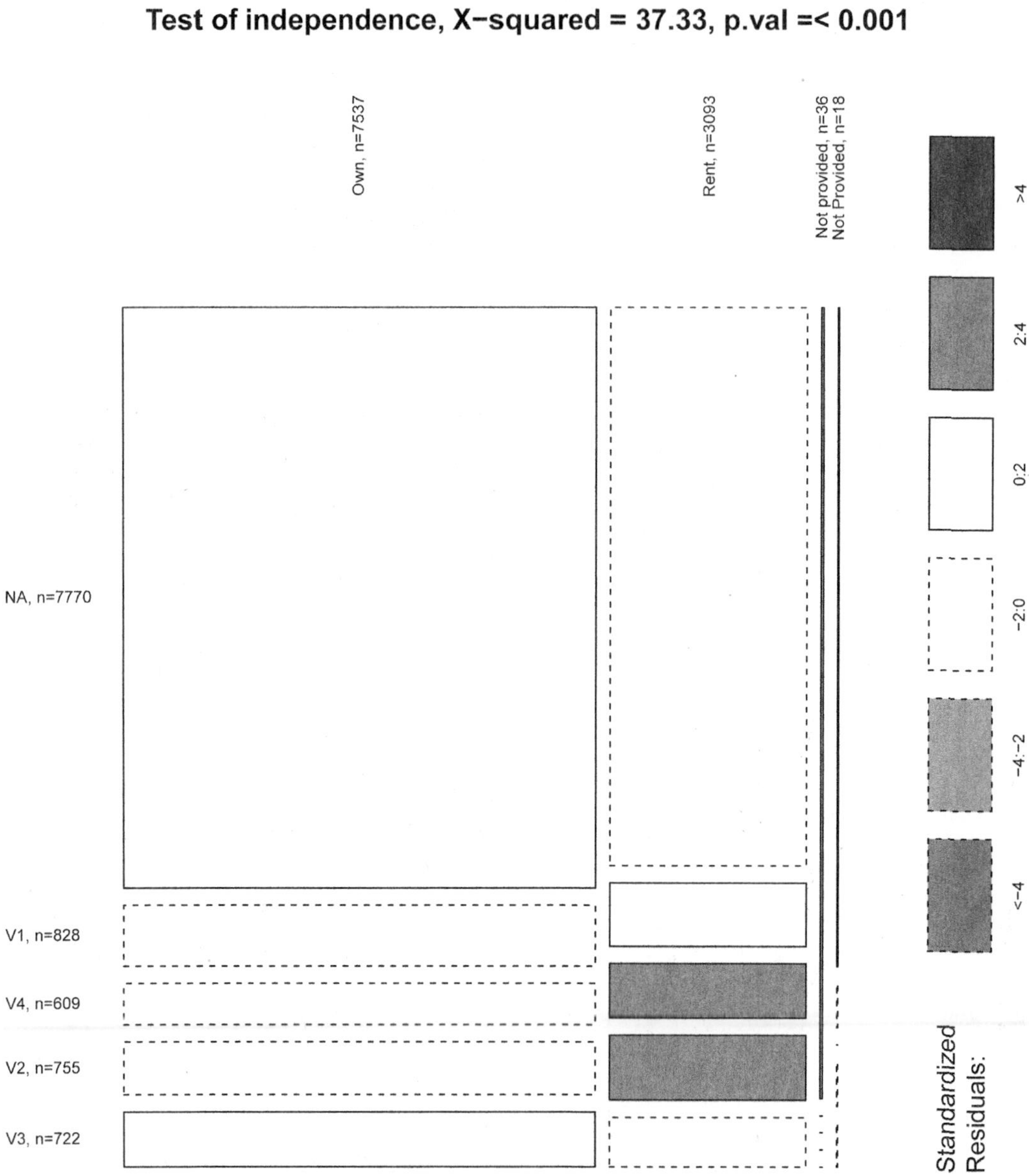

FIG. 6.9 MDCOR Assessment of Group by Code Relationships.

Execute Text Mining/Cleaning procedures." Based on machine-learning requirements, open-ended responses may be excluded from classification if their content is configured by sparse, unique, or outlier words that do not allow their probabilities to be learned and thus assigned to a latent code (see appendix). Guided by the goal of full analytic transparency, all responses that are dropped may be downloaded after Step 2 is executed by clicking a link displayed in the MDCOR's UI called "Download all dropped responses," as shown in Fig. 6.10. The goal of allowing researchers to download these excluded open-ended responses and their respondent IDs is to enhance transparency and also allow them to further qualitatively assess the content of these excluded responses.

In the example presented in Fig. 6.10, we can see that one of these excluded responses (more details on the dataset are provided in the application section below) was provided in Spanish. In this respect, let us note that this response was not dropped because it was written in Spanish, but rather because it was the only response in the entire dataset (containing over 10,600 rows) that was written in Spanish, which made the content of this response a true outlier in the dataset. This means that if more participants had responded in Spanish, both English and Spanish responses would have been retained in the analytic sample.

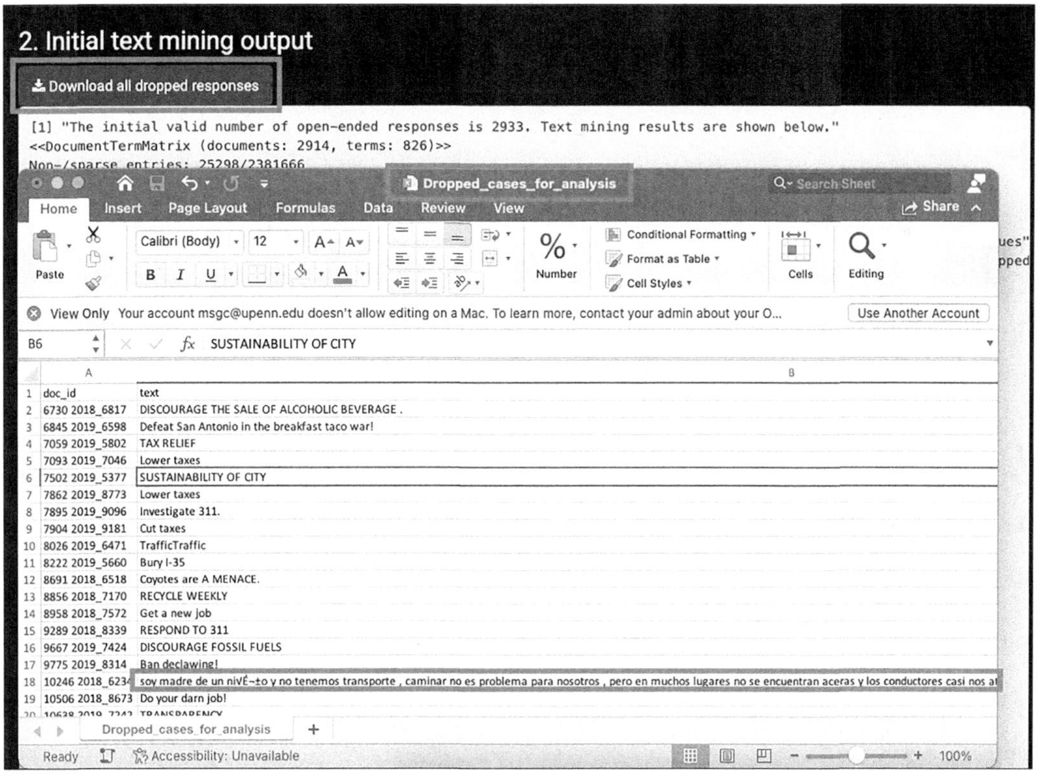

FIG. 6.10 Dropped responses.

3. **Remove most common words (optional):** As indicated in the flow diagram, this third step is optional and allows users to remove words that are too common to add value to the learning and classification processes (see appendix, H. Schütze et al., 2008a). For example, if we were analyzing open-ended responses regarding attitudes toward COVID-19, it is quite likely that most participants' responses would include "COVID-19," which decreases the classificatory usefulness of this concept or word.

To ease the identification of common words, we programmed MDCOR's text-mining output to print the 20 most frequent words across all open-ended responses as shown in Fig. 6.11. Note that in this text mining output, Austin, the city where the survey was conducted, occupied the first place with a frequency of 796. Note also that, after removing sparse terms or outliers, 19 responses were dropped, resulting in 2914 responses to be employed in the learning and classification analyses. This indicates that, assuming that the word "austin" was used once among responses that included this word, which is the "worst-case scenario" in terms of commonality

2. Initial text mining output

Download all dropped responses

```
[1] "The initial valid number of open-ended responses is 2933. Text mining results are shown below."
<<DocumentTermMatrix (documents: 2914, terms: 826)>>
Non-/sparse entries: 25298/2381666
Sparsity           : 99%
Maximal term length: 14
Weighting          : term frequency (tf)
[1] "Based on the DocumentTermMatrix information, a total of 19 responses were dropped due to sparsity (irrelevance) issues"
[1] "If applicable, to assess dropped cases and potentially use them for non-response analyses, select 'Download all dropped
[1] "If applicable, MDCOR shows the first 10 responses dropped and the top 10 most common words"
 [1] "Response(s) dropped, ID: 6730 2018_6817 Text: DISCOURAGE THE SALE OF ALCOHOLIC BEVERAGE ."
 [2] "Response(s) dropped, ID: 6845 2019_6598 Text: Defeat San Antonio in the breakfast taco war!"
 [3] "Response(s) dropped, ID: 7059 2019_5802 Text: TAX RELIEF"
 [4] "Response(s) dropped, ID: 7093 2019_7046 Text: Lower taxes"
 [5] "Response(s) dropped, ID: 7502 2019_5377 Text: SUSTAINABILITY OF CITY"
 [6] "Response(s) dropped, ID: 7862 2019_8773 Text: Lower taxes"
 [7] "Response(s) dropped, ID: 7895 2019_9096 Text: Investigate 311."
 [8] "Response(s) dropped, ID: 7904 2019_9181 Text: Cut taxes"
 [9] "Response(s) dropped, ID: 8026 2019_6471 Text: TrafficTraffic"
[10] "Response(s) dropped, ID: 8222 2019_5660 Text: Bury I-35"
            words freq
1          austin  796
2            city  734
3            much  676
4         traffic  502
5            need  475
6        homeless  414
7            have  395
8            make  363
9           house  358
10         people  355
11           live  345
12         please  327
13           good  326
14           stop  293
15     affordable  279
16           work  257
17 transportation  227
18           road  212
19           keep  207
20         public  206
```

FIG. 6.11 Text mining/cleaning and top 20 most frequent words.

of word usage among respondents (Latent refers to unobserved factors that make observed indicators (traditionally columns in databases) be correlated, like in factor analysis. In textual databases the input data are unstructured, which requires text mining and normalization procedures (presented in the appendix) to create document-term matrices. The columns in this document-term matrices are populated by the "cleaned" vocabulary in our texts (which are sentences or paragraphs) and the goal of topic modeling via Latent Dirichlet Allocation is to identify the probability of a given sentence or paragraph (represented in the rows of this document-term matrix) to be part of a latent topic or code. As in factor analyses, these latent codes have no meaning for this meaning needs to be retrieved from the sentences or paragraphs that have been statistically identified as belonging to such a latent code), the percentage of respondents using this word would have been 27.32% (or $\frac{796}{2914}$). From this view, and although no rule of thumb exists regarding the assessment of the common presence of a term, we consider that Austin's prevalence in the corpora is not "overwhelmingly" common to merit its removal from this dataset for only 3 out of 10 respondents used this term at the most. Having said this, researchers could decide to exclude some frequent terms to conduct sensitivity analyses. Accordingly, we next describe the process to drop terms from the analytic database.

To drop any number of words, simply type their order from the 20 most common words output. That is, as shown in Fig. 6.11, we have the following order: "1. austin," "2. city," "3. much," ..., "20. public." If we wanted to remove "austin" and "much", for example, we simply need to type "1, 3" (without quotations) in the box shown in Step 3 in Fig. 6.8, and then click "Trim Common Words" in the UI to indicate that the first and third words should be removed. Note, however, that although these words are presented in isolation in Fig. 6.11, they form part of open-ended responses full of context. Accordingly, we reiterate that words should be removed only if there is a strong rationale for doing so. Before moving to the next step, note that deleted words may be reinstated by re-executing the text-mining process in Step 2 and moving on to Step 4. MDCOR is conducted with the matrix resulting from the most recent call of Steps 2 or 3.

4. **Select machine learning sampling parameters**: As described in the appendix, the learning parameters are based on 500 burn-in and 5000 posterior resamples as the recommended values for assessing the optimal number of codes (see Raftery and Lewis, 1992), but these values can be changed before executing these metrics in Step 5. Depending on the size of the corpus and the number of burn-in and resamples, this process may take minutes, hours, or even days. For example, the text-mining and classification of the 2914 open-ended responses presented in the appendix were completed in 9 minutes with a 12-core Windows machine and in 13 minutes with an 8-core Mac machine. MDCOR has also been tested to classify 500,000 Twitter comments, and the classification time took about 43 hours on the same Windows machine—the Mac procedures are yet to be timed.

The other two elements that researchers may update before executing these metrics assessments are β and α, two hyperparameters of the prior distribution for the term distribution of the topics and the topic distribution of documents (Grün and Hornik, 2011, p. 6), respectively. As described in the appendix, these starting values are updated as part of the learning process, feeding from our selection of burn-in and Gibbs resamples over the entire classification process. Note that although our software application allows users to manually set the starting values of β and α, we highly recommend relying on the default values (0.1 and 50 for β and α, respectively), as these values have been identified in the literature based on rigorous evidence (Griffiths and Steyvers, 2004; Grün and Hornik, 2011). Nonetheless, if researchers wish to conduct sensitivity tests with other values, they can easily update those inputs in MDCOR's UI.

5. **Execute metrics for optimal number of code selection**: After the recommended values are maintained or updated, researchers may press "5. Execute Metrics" to find the optimal number of codes. Once these metrics are executed (more on this in our applied section below), a metrics plot that facilitates the identification of the optimal number of topics will appear under "4–5 Metrics Output" (see online appendix). This plot can be downloaded as a PDF and should be used for publication purposes to justify the selection of the optimal number of codes, as we further discuss in our applied section below.

With respect to these metrics assessments, note that, as indicated by Nikita and Chaney (2020), the higher the agreement across metrics is, the more trust we may have in the selection of the optimal number of codes. In our example (see Fig. 6.3), we are showing the convergence of two metrics (Cao et al., 2009a; Deveaud et al., 2014b) out of the four metrics. In this case, these two metrics indicated that four codes constituted the most optimal solution (see the number 4 on the x-axis of Fig. 6.3—we further detail the meaning of these metrics in the appendix). Accordingly, this convergence enabled us to justify the selection of four latent codes as capturing the optimal number of codes across these 2914 open-ended responses.

6. **Select the optimal number of codes**: The final step required before executing the final classification consists of selecting the number of codes that the machine learning process will identify across the collection of open-ended responses or texts. This value is set to 2 by default in the UI (see Step 6 in Fig. 6.8), but researchers should adjust this value based on the insights obtained from the metrics plot. Once more, as we discussed above, the results shown in Fig. 6.3 indicate that we should select 4 as the optimal number of codes.
7. **Execute MDCOR**: With this four-code solution, we are ready to click "7. Execute MDCOR." The burn-in and MCMC/Gibbs resampling values, as well as the α and β hyperparameters used in this process, may be adjusted in Step 4 in Fig. 6.8. For consistency, however, we recommend using the same input values when executing both the metrics approach (Step 5) and MDCOR (Step 7).

There are two main outputs associated with this step. The first is an interactive descriptive tool (see Fig. 6.2, with its interactive version located at https://cutt.ly/lrXhCh4e) designed to quantify word frequency distributions per code. This interactive summary also indicates the distribution of the top 30 most important words configuring each code. In addition to serving as a descriptive tool, the distances among codes in the box quadrant plane also convey similarities or differences in the underlying meanings of these codes. For example, let us observe the conceptual representation of this quadrant in the dark square in Fig. 6.7. Specifically, in our conceptual representation shown in Fig. 6.7, Code 7 is more similar to Code 5 than to Codes 1 or 6. Similarly, Code 4 is more neutral, meaning that its content shares similarities with other codes given its central location. This implies that it may be more difficult to clearly differentiate the content of Code 4 from the content of other codes. On the other hand, codes located at the corners of this plane have content that is more unique and discernible.

In returning to our four-code solution, which we will qualitatively discuss in our applied section, note that each code is in its own quadrant and at its uppermost corresponding corner. This finding, in essence, indicates that the content of each of these codes should be relatively clearly discernible from the other three codes.

Although this interactive summary is a useful tool for reconstructing the meaning of each code, the most important output offered by MDCOR (separated into two databases) is the fully classified dataset containing all codes and all texts. To ease the qualitative understanding of this database, a subset of these data, containing up to the top 20 most representative texts per code, is also offered for analysis, and this dataset is actually displayed in MDCOR's UI (see Fig. 6.12). Based on the relevance of this output for MDCOR's epistemological lenses, let us illustrate its components.

Conceptually speaking, the dark box in Fig. 6.7 showcases these main outputs, which are a set of *N* codes that are more or less correlated (i.e., the closer they are in the quadrants, the more information in common they share). Note also that within each code, we may identify texts that, *among all the collections of responses*, best represent each code—represented as files stored in each folder in this dark box in Fig. 6.7. To elaborate on the process MDCOR follows to identify text-to-code relevance, let us discuss Table 6.1. For convenience, note that we are only including three responses in the following example; however, this same rationale applies to thousands or millions of cases. Let us also assume that we asked an open-ended question on "family and career balance" and that the results obtained from the metrics assessment indicated that a two-code solution is associated with the most optimal classification outcome.

The texts contained in Table 6.1 show that the first response is clearly aligned with a career-focused standpoint for this participant, who mentioned, "*My career is the most important*

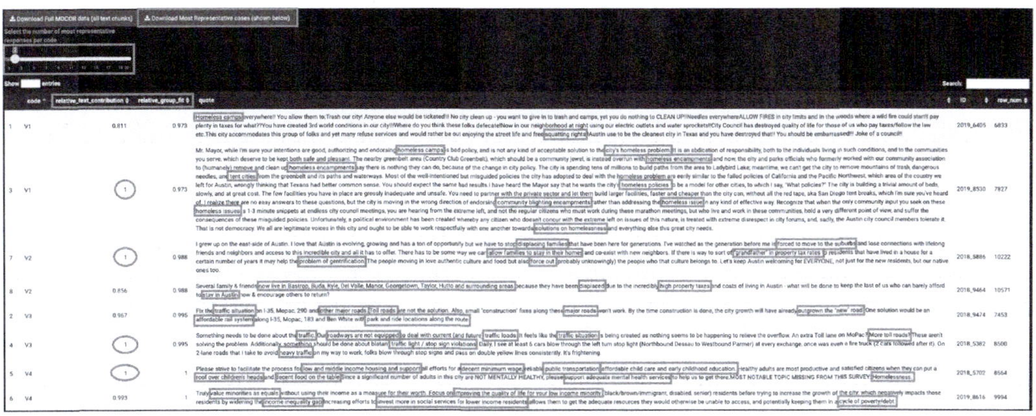

FIG. 6.12 Example of code classification, original text inputs, and measures of fit.

aspect of my life." The columns labeled "Code 1" and "Code 2" represent the two-code solution. Note that MDCOR's output does not include these two columns; instead, it shows a column called "Code" (see Fig. 6.12) that identifies the resulting classification of each text (shown as MDCOR Result in Table 6.1) based on the following rationale. This final output provided by MDCOR supports hypothesis generation through integration. (Fetters, 2020). In this case, the output consists of a database (see Fig. 6.4) containing the strength of each text's contribution to its classified code, along with indicators of fit that can be interpreted only in relation to the qualitative meaning of those codes as reconstructed from participants' unaltered responses. As explained in our applied results section described next, this database may be merged with quantitative databases to use latent code strength (that is, the percentage of a given participant's sentences that were ascribed to Code i, for example, and how this prevalence may explain a given outcome y). From this perspective, the fully integrative weaving nature of all statistical tests derived from MDCOR is based on the fact that these tests and outputs are only relevant when grounded in the qualitative understanding of the contextualized meaning of these latent codes. In other words, none of the outputs are useful or make sense based solely on their quantitative descriptions; instead, their meanings need to be decoded from the original voices of participants or texts, as captured by their original sentences or paragraphs. As explained in the appendix, each text is given a probabilistic value of representing each of the optimal codes selected in the metrics assessment (see Table A5, available online at http://dx.doi.org/DOI_of_Chapter6). Accordingly, assuming a two-code solution in Table 6.1, this translates into the following scenario: although the first text was classified as Code 1, in reality, after thousands of iterations in the learning process, its probability of being associated with this code given its word configuration was 90%, with the remaining 10% being associated with the second code. That is, this text had a 10% chance of representing Code 2 as well. In the case of the second open-ended response or text, which is clearly more ambiguous as it reads, "*My family is as or more important than my career*, these probability distributions were 45% and 55% for being classified in Code 1 and Code 2, respectively. Finally, the third sentence, which reads, "*For me, the best day is a family day, not an office day*, had an 80% probability of being representative of Code 2. Taking all this information into account, we then know that, although the second response was more ambiguous than response 3, and this second sentence was also close to being classified as Code 1, it was classified under Code 2 given its higher probability value (i.e., 55% vs. 45%). In this respect, note that topic modeling does not traditionally offer a measure of how strong the text-to-code associations are. That is, although our example clearly shows that texts 1 and 3 are more strongly associated with Codes 1 and 2, respectively, than text 2 is with Code 2, topic modeling would not provide this information as part of standard outputs. To address this limitation typical of topic modeling, MDCOR's main outputs (both in the fully classified dataset and the dataset with the most representative responses) offer two measures to assess text-to-code strength and detect the most representative texts per code.

Aligned with our conceptual representation shown in the dark square in Fig. 6.7, MDCOR's UI, by default, displays the top five most representative responses per code; however, users may select to display up to the 20 most representative texts, as we show in Fig. 6.12. Following our example contained in Table 6.1, we next discuss the process followed to detect the most representative text per code. This rationale is the same as that applied to detect the top N most representative texts per code.

Similar to Table 6.1, MDCOR's main output is a database that contains code, ID (and row number to allow longitudinal responses), original text, and two measures of fit. A conceptual representation of this output is shown in Table 6.1 and the actual MDCOR output is shown in Fig. 6.12. These columns that allow the identification of text representativeness are "Text Contribution" and "Group Fit." (*.docx files contain HTML format that conflicts with text cleaning across operative system platforms. Accordingly, if transcripts are only available in *.docx files, they should be saved as *.doc files before uploading them to LACOID). The text contribution is populated by dividing the probability of each text to be classified in the code where such a text was actually classified over the maximum probability of a text being classified in that same code. That is, taking text 2 in Table 6.1, for example, we have that the maximum estimated probability of a text being classified under Code 2 was 80% (text 3); however, the probability of this text 2 to be classified in Code 2 was 55%. Accordingly, the text contribution or representativeness of text 2 for Code 2 is the result of dividing $\frac{0.55}{0.88} = .6875$. In contrast, the third text will be divided by its own probability, which is the highest observed, rendering a value of 1. This indicates then that text 3 is the best representative of Code 2 across all texts classified under Code 2. This same rationale is applied to text 1 and Code 1. This is how we can request that MDCOR's interactive UI displays only the most representative texts per code (see Fig. 6.12).

The second assessment is group fit. This assessment takes the average probability values obtained by each code given the probability of their configuring texts to being classified in such a code and then divides the resulting code average probabilities by the maximum average value observed across all codes. Specifically, in the case of Code 2, this average value is estimated from 0.55 and 0.80 (the probabilities of its configuring texts 2 and 3 to be classified in Code 2). This average value is 0.675. In the case of Code 1, this average value is 0.90. Since 0.90 is the maximum average value observed across all codes, Code 1 will have the maximum group fit for its average is divided by the maximum average across codes. In the case of Code 2, its average is divided by the maximum code group average (or $\frac{0.675}{0.90}$) rendering an estimated value of 0.75. This implies that Code 1 had the maximum group fit (similar to an internal consistency indicator). This indicator, which does not change within codes, may be interpreted as a tool to assess the extent to which the probabilities of texts configuring each code were consistently high (or not). In a sense then, these two fit measures allow us to assess individual text to code contribution while also estimating how consistently were these texts to being classified under their corresponding codes. Since these fit measures are a function of the optimal code number identified with the metrics assessments, all MDCOR steps are interconnected across the entire classification process.

8. **Test for code associations by group (optional):** As indicated in our flow diagram, this last step is optional and serves to test whether there is evidence that respondents from specific groups in our datasets are more likely to select some codes over others (i.e., group and code association). Moreover, in the case of nonresponse, this test also assesses whether some of these groups were less likely to respond to this open-ended prompt.

Conceptually, for this test to make "sense," we need to have a clear understanding of the qualitative meaning of these latent codes, which may be achieved through the analysis of the unaltered words of our participants, as we will illustrate in our applied section below. *Procedurally*, to conduct this test, researchers simply need to select a categorical or factor indicator from the dataset uploaded to MDCOR. In our applied example, to be discussed next, we conducted analyses based on home ownership status in the city of Austin. This indicator may also contain nonresponses, even if these participants did answer the open-ended question of interest (e.g., 54 participants did not provide a valid home ownership status response). Accordingly, both nonresponses to the open-ended question and nonresponses to the categorical indicators may also be analyzed with this test. *Statistically*, this test is conducted via a Chi-squared test of association with MCMC p-values. The result of this test is shown in Fig. 6.9. This figure can be downloaded as a PDF by clicking the message "Click here to Open the Chi-squared test of independence," which will be visible in MDCOR's UI once this *post-hoc* test is executed.

Finally, another interactive output resulting from executing group comparisons also assesses the degree to which codes are interrelated. This interactive output is presented as a network of these relationships. In this network depiction (see Fig. 6.13, with its interactive

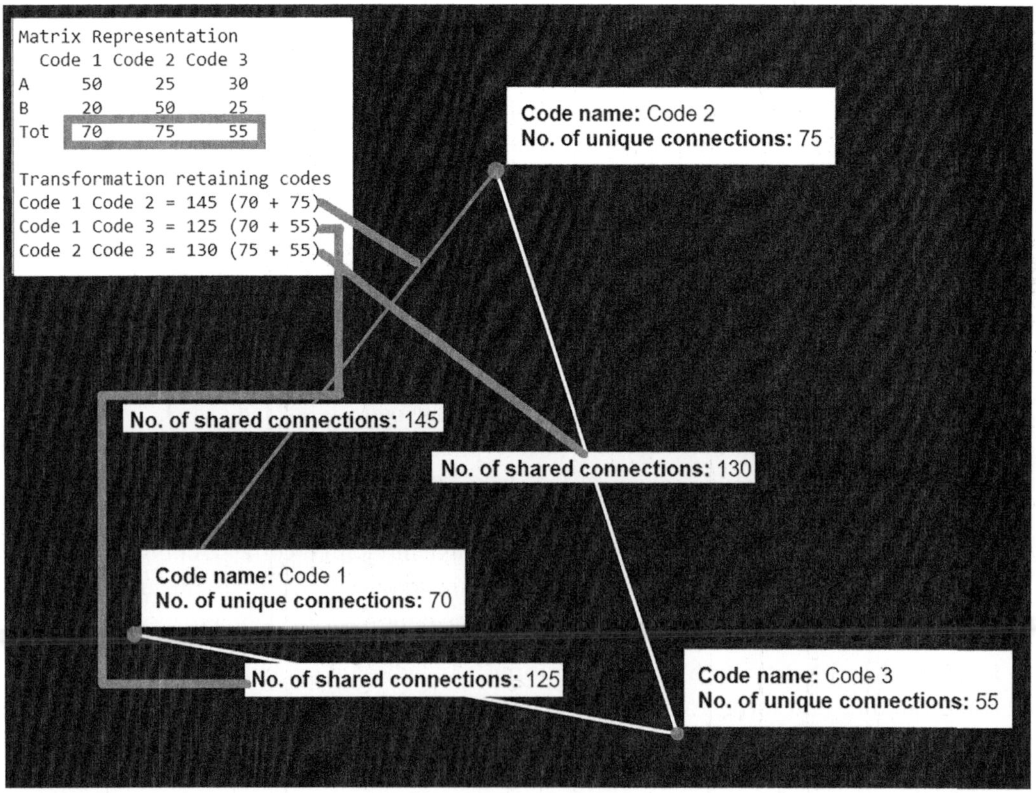

FIG. 6.13 Example Code Relationships by Group Association as presented on page 27—see the interactive version at https://cutt.ly/oB6XGh2.

version available at https://cutt.ly/oB6XGh2), the strength of each relationship is measured by the number of participants within a given group who selected both codes. For the sake of clarity, let us assume we have three codes: 1, 2, and 3, and two groups: A and B. Members of group A had 50 text classifications in Code 1, 25 text classifications in Code 2, and 30 text classifications in Code 3. In the case of group B, the distribution is Code 1 = 20, Code 2 = 50, and Code 3 = 25. This information is represented in matrix form in the top-left section of Fig. 6.13. From these group-to-code distributions, we can estimate the total number of texts classified in each code. The same matrix shows that Code 2 had the highest frequency of text allocation, with 75 responses being classified under Code 2. These values are available in the interactive network depiction when clicking on a circle, as also shown in Fig. 6.13.

Nonetheless, this network also shows that the lines connecting each code are weighted by the added responses of the groups in the network. For example, as indicated in the transformations shown below the matrix in Fig. 6.13, we can see that the strength of the relation or line connecting Codes 1 and 2 is 145. This value is a function of the number of texts from Groups A and B that were connected with both Codes 1 and 2 (50 + 25 + 20 + 50 = 145 = 70 + 75). As can be seen in Fig. 6.13 and in its interactive version (https://cutt.ly/oB6XGh2), MDCOR was programmed to detect the highest connectivity among codes with a red connection. In this case, this indicates that Codes 1 and 2 are the most highly connected codes in the network as a function of the number of texts that respondents from the two groups analyzed provided to both codes. Since this network is deployed as an interactive HTML object, researchers may click on the lines or the nodes (i.e., codes) to retrieve the number of shared connections in the case of the lines or the name and number of unique connections for the codes. The output corresponding to our example is shown in Fig. 6.5, with its interactive version available at https://cutt.ly/UBJ6Hfa, as we discuss in our applied section next.

This concludes our presentation of the UI and the outputs provided by MDCOR. As previously stated, all these outputs are only useful to the extent that the meanings of codes are enriched with participants' voices. At the same time, the use of original responses in interactive outputs requires careful ethical judgment (González Canché, 2025). When classified texts, representative responses, or code-to-group visualizations are shared beyond the research team, analysts should ensure that no combination of textual content, contextual clues, or group attributes weakens confidentiality, especially in restricted, institutionally sensitive, or small-sample studies. In many cases, selected excerpts, screenshots, or locally retained outputs may be preferable to broad distribution of full interactive materials.

Taken together, then, MDCOR's fully mixed equal-status design implies that the end product is greater than the sum of its parts (Fetters and Freshwater, 2015; Lynam et al., 2020a) and that rigorous and time-efficient knowledge can be built through the qualitative and quantitative integration of MDCOR's outputs.

These outputs are also portable across analytic stages. The interactive code summaries, the most representative response tables, the downloadable classified datasets, and the HTML visualizations of code-to-group relationships are useful not only for identifying latent structure, but also for supporting subsequent interpretive work. In the broader ecosystem developed in this book, researchers may bring these outputs into ISARI by uploading screenshots of the interactive plots, copied text from the summary panels, exported figures, or notes derived from the classified datasets. Because ISARI can operate locally and work with both text and image-based artifacts, these MDCOR outputs can become the basis for scholarly brainstorming, memoing, code comparison, and evidence-grounded writing without moving sensitive materials outside the researcher's control.

To illustrate this meaning-building process, we offer an applied example using a publicly available dataset. We hope that researchers are encouraged to use this database to replicate the analyses and interact first-hand with MDCOR.

Application

MDCOR is available for Windows and Mac operating systems (see access and installation information in the closing section of this chapter). Because MDCOR runs locally, it can be used to analyze restricted data, such as surveys conducted by the National Science Foundation and the National Center for Education Statistics. Indeed, the appendix presents an analytic exercise using one of these federally protected datasets: The High School Longitudinal Study (HSLS) 2009–2018. For replication purposes, our main analysis in this applied section relies on a publicly available dataset called the "Community Survey."

The city of Austin has administered a community survey for the years 2015, 2016, 2017, 2018, and 2019 (see https://data.austintexas.gov/stories/s/wn8p-f4h4 for an interactive dashboard) to "assess satisfaction with the delivery of major City Services and to help determine priorities for the community as part of the City's ongoing planning process." This network analytic process was inspired by an application of qualitative network analysis (González Canché, 2022c, 2026) and is incorporated in LACOID. Although we downloaded the dataset analyzed in this study from the former link, given that the city of Austin is interested in continuing to administer this survey, there is a chance that the data we used for this analysis and the data hosted on the city of Austin's website may differ in the following years. Moreover, as of 2026, the city has paused data release. Accordingly, to ensure the replication of our findings, we recommend researchers download and analyze the dataset we employed in our analyses, which can be accessed at the following link: https://github.com/democratizing-data-science/MDCOR/blob/main/Community_Survey.csv. This data can also be accessed via Mendeley at González Canché (2022a).

Before proceeding with the analyses, note that we decided not to modify any aspect of this community survey dataset and conducted the analyses with the information provided *as is* by the City of Austin. Our goal in doing so was to ensure transparency in the process and to test how MDCOR may handle relatively *uncleaned* data. For example, one of our indicators of interest is home ownership status. The data, as provided by the City of Austin, contains the following categories: "Not provided" and "Not Provided," which are treated as two different options—in addition to "Own" and "Rent." Although we could have addressed this inconsistency, we chose not to, in order to observe what the impact may be in the post-hoc outputs (i.e., this does not impact the classification process). Having said this, these types of processing errors should be corrected and documented when employing MDCOR for publication purposes.

The community survey data contains 10,684 rows and 251 columns. Of these columns, our analyses will rely on the following three indicators, which are taken verbatim from the survey:

"ID", "Q25—If there was one thing you could share with the Mayor regarding the City of Austin (any comment, suggestion, etc.), what would it be?", and "Do you own or rent your home?"

Findings

In this section, we offer an example of how to read and analyze the outputs provided by MDCOR. When using MDCOR in your own projects, we recommend following the order depicted in our flow diagram and UI to present your findings, as shown next.

NLP and text mining

As described above, once the text mining was conducted (see Fig. 6.10), 19 responses were removed from the analytic dataset and were therefore not included in the classification process. One of these responses was written in Spanish, which made its configuring words outliers in the dataset, thus explaining the exclusion of this response from the analytic sample. Regarding this exclusion, note that it was not the Spanish language itself that was the reason for dropping this response; rather, it was the fact that this response was the only text provided in Spanish in the entire survey. As stated above, NLP allows the inclusion of slang, informal grammar, and even other languages, as long as these texts are not considered outliers for the classification process.

Other dropped responses were too short or even too vague. For example, one response referred to coyotes ("Coyotes are A MENACE," participant ID 8691 2018_6518). A search among all the responses in this survey indicated that the words "coyote" and "menace" only appeared once in the survey, which justified their exclusion. To expand on the rationale for text exclusion, as implemented by MDCOR, note that Fig. 6.11 shows that after NLP and text mining/preparation, the maximum sparsity of the matrix for the text classification is 99%. This sparsity level indicates that all remaining 2914 open-ended responses, with a total of 826 terms (i.e., its vocabulary, as shown in Table A1 available online at http://dx.doi.org/DOI_of_Chapter6), are configured by terms appearing in at least 1% of all responses.

This NLP and text mining output (Fig. 6.11) also shows the top 20 most frequent words after applying lemmatization (see Appendix). In our analyses, because the most prevalent word ("Austin") is expected to appear in at most 27.32% of the responses, we did not remove this word, as we are not concerned that its prevalence in the responses may hinder our classification procedures. Indeed, our analyses relying on interactive word summaries (see Fig. 6.2) corroborated that this word is overwhelmingly present in only one of our four identified codes (see Fig. 6.14), which further justified our decision to not remove this word from the analyses. As such, when in doubt, we recommend keeping the word, conducting the metrics assessment to select the optimal code number, conducting MDCOR, and analyzing the resulting word summary as we illustrate below.

Metrics assessment

Fig. 6.3 shows the results of Steps 4 and 5 in Fig. 6.8. Specifically, the resulting metrics plot displays the four assessments of the optimal number of codes associated with all 2914 open-ended responses. Because Step 3 in Fig. 6.8 is optional, the metrics plot may be based on either the complete or the trimmed document term matrix. As an analytic note, to assess variation in optimal latent code identification with the complete and trimmed document term matrices, follow these steps: 1. Execute text mining in Step 2 and then go directly to Step 5 to perform the metrics assessments and save the resulting plot. 2. Trim the document term matrix in Step 3, re-execute Step 5, and save the resulting plot. If the result with the trimmed dataset is preferred, simply adjust the number of codes in Step 6 and execute Step 7 to qualitatively analyze the output. However, if the result associated with the nontrimmed matrix is preferred, re-execute Step 2, adjust the number of codes in Step 6, and re-execute Step 7.

We executed the metrics with the recommended 500 burn-in samples and 5000 MCMC/Gibbs resamplings (Raftery and Lewis, 1992). Ideally, all four metrics would converge into a unique solution for the optimal number of codes; however, based on text complexity, the convergence of at least two metrics in detecting an optimal number of codes may suffice to make an informed decision regarding the optimal number of codes that capture meaningful information (see appendix), and (Nikita and Chaney, 2020). The resulting plot indicated that, based on the metrics proposed by Deveaud et al. (2014b) and Cao et al. (2009a), the optimal number of topics is four, as can be seen by the maximization and minimization indicators (discussed in the appendix), respectively. We added the red oval for clarity. With this information in mind, we executed the final text classification.

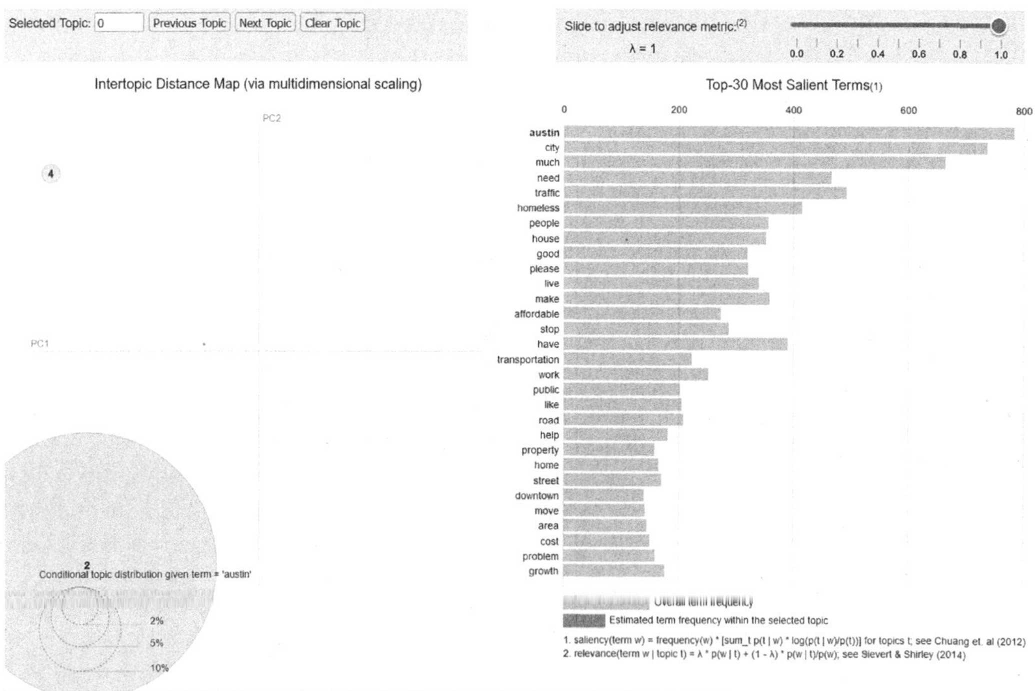

FIG. 6.14 Assessing the presence of Austin Across Codes—To replicate this figure, place your cursor on the word "austin" in the interactive version at https://cutt.ly/cBAlTVd.

MDCOR execution

After executing MDCOR, there are two main outputs. One is the interactive word and code summary, and the second is the classified responses. This latter output is separated into a subset of the most representative responses displayed in the UI and another set containing all classified responses available for download. In this section, we focus on the interactive word and code summary and the subset of the most representative responses.

Interactive summary statistics per latent code

As part of the integrative goal, MDCOR also provides a visual and interactive representation of the most frequent words configuring each code. This interactive description is a useful tool for starting to build our understanding of the qualitative meanings of these codes.

To begin, recall that the open-ended question asked participants to share anything with "the Mayor regarding the City of Austin." Aligned with our selection of a four-code solution, Fig. 6.2 shows four codes, each located in its respective quadrant. In this respect, note that, as described earlier (see Fig. 6.7), the distance among codes indicates the level of similarity or dissimilarity of their latent meanings. Given the position of these four codes in Fig. 6.2, we are confident that their meanings may not be too difficult to differentiate.

The interactive version, available at https://cutt.ly/lrXhCh4e, allows us to select codes to display their configuring words, and within codes, to select these configuring words to assess the extent to which those words are present in other topics. To illustrate this process, we show Fig. 6.4. In this figure, we clicked Code 1 and then placed the cursor on the word "homeless." This process changes the size of all remaining codes in the following ways: Codes 1 and 2 dissipated, meaning that the word "homeless" is not part of those codes. Code 3 did not disappear but became quite small. Finally, we can see that the size of Code 1 increased, which means that the word "homeless" is a key configuring attribute of this code. A similar exercise may be conducted with any other code and word combinations. Nonetheless, even if no code is selected, we can still select a word. For example, without clicking any code, "austin" is the most frequent word (see Fig. 6.2). In this context, if we place our mouse cursor on the term "austin," the result will show that only Codes 2 and 4 contain this word (see Fig. 6.14), with Code 2 having the highest concentration of this word—this solidified our decision to not remove this term from the analyses.

What meanings can we start (re)building?

To briefly illustrate how this interactive graph may be used to build meanings, let us start by selecting Code 1. This selection indicates that the most important words configuring this code describe a problem with homelessness in the city and its neighborhoods. In the case of Code 2, the word frequencies indicate some form of concern with home property costs and affordability issues in the city of Austin. In the case of Code 3, its configuring words clearly indicate traffic concerns. These words include roads, streets, lanes, and time. Finally, Code 4's most frequent terms highlight the cost of living in general, including affordability, public services, housing, and health care needs. Once more, although these interactive summaries are informative, these words are taken out of their original context (Eickhoff and Wieneke, 2018a). For that reason, the summary and classificatory outputs should be treated as guides to interpretation rather than as interpretations in themselves. That is, although summaries may give us a hit of most important workds per code, it is neccesary to retrieve or build meaning using complete responses, rather than word-to-code frequency per code. In other words, the meanings of latent codes still need to be rebuilt from participants' unaltered responses and from the qualitative judgment of the research team. Within the broader analytic architecture of this book, these same summaries can also serve as prompts for local AI-supported scholarly reasoning in ISARI, where researchers may compare alternative readings of latent code structure, memo possible interpretations, and begin translating those patterns into evidence-grounded prose while preserving the primacy of participants' original voices. Although the integration of MDCOR with ISARI can support all classified texts, the use of the top 20 most representative texts per code has proven effective in retrieving the "human meaning" of these codes using intelligent systems, like ISARI. Having said this, for this exercise, we are presenting the top two texts per code, but we recommend researchers select a higher number in the pursuit of more nuanced understandings.

Contextualized meanings of the most representative responses

To create Fig. 6.12, we adjusted the selection of the most representative texts per code from five to two, as shown in the top-left corner of this figure. To ease the analyses, we sorted these texts by their code so that the first code is displayed in the first two rows, the second code in the subsequent two rows, and so on and so forth.

Code 1: Homelessness. As can be observed in Fig. 6.12, the classification performed remarkably well. Code 1 consistently reflected concerns associated with homelessness in the city. The most representative text (second row) was quite clear regarding this issue, which was also corroborated by the content of the second most representative response. Note, however, that this second most representative open-ended response, although also concerned with homelessness, focused more on the consequences of this issue regarding trash, fires, and the overall quality of life. This broader set of concerns reflected in this response was associated with a lower text contribution to Code 1, even though both responses are clearly aligned with the topic of "homelessness."

Code 2: Displacement, property taxes, and gentrification. The top two most representative responses indicated concerns about the displacement of families due to high property taxes, which are forcing families to move to the suburbs. Participant ID 2018_5886, whose response was the best representation of Code 2, highlighted an issue of gentrification. In the case of the second most representative response, the original text also mentioned displacement due to high property taxes; however, instead of referencing suburbs, this respondent listed the names of the neighborhoods where Austin's inhabitants have been displaced. Once again, the differential emphasis of this response, which focused on naming neighborhoods, may help explain its lower text contribution to Code 2, even though its content also clearly underscores issues with displacement and property taxes.

Code 3: Traffic, roads, and tolls. The third code clearly reflects issues with traffic in the city of Austin. These unaltered responses reflect participants' frustrations with roadways, which "may not be equipped to deal with current and future traffic loads" (Participant ID 2018_5382). In this case, both responses mention concerns with toll roads. The second most representative response also deviates a bit from simply highlighting concerns to focus on possible solutions, such as rail systems and park-and-ride locations. Although this deviation impacted the overall contribution of this response to this code, given its higher adherence to the topic, its relative text contribution was higher than that of the previous two codes (i.e., 96.7 vs. 0.856 or below).

Code 4: Quality of life & cost of living. This fourth code had both the highest relative text contributions and the highest relative group fit. This indicates that the configuring responses should more coherently highlight a common theme. The two most representative responses raised concerns based on socioeconomic grounds. That is, both responses indicate that everyone (minorities and nonminorities) should receive a decent minimum wage, affordable public transportation, childcare, early childhood education, and mental health services. The second most representative response also included immigrants in his or her response, further arguing for the relevance of closing income gaps and potentially helping minorities to get out of "a cycle of poverty/debt" (participant ID 2019_8616).

Post-hoc analyses of code by group association

Our description of each code, based on participants' voices, illustrates how the combination of the quantitative and the qualitative, unaltered texts serves to make sense of the meaning of those latent codes. Note that MDCOR also includes an empirical assessment of whether some groups are more likely to select some codes over others. To conduct these analyses, MDCOR requires a categorical indicator. In full transparency, we selected home ownership status before qualitatively rebuilding the meaning of these codes. That is, we did not first rebuild the qualitative meaning of these codes and then select home ownership status. In retrospect, however, given the qualitative meaning of these codes, it is evident that two of these codes (Codes 1 and 2) are more closely aligned with issues of housing insecurity, and Code 4 reflects concerns with affordability. Accordingly, the following exercise, in a sense, may help further validate our topic modeling findings.

The main results of this test are shown in Fig. 6.9. The MCMC probability values associated with the estimated Chi-squared indicate that there is enough evidence to conclude that some group responses are more or less similar than what one should have expected to observe under random chance. The colors in Fig. 6.9 signal the strength of this association, with negative relationships in red (i.e., we expected higher concentrations than observed in reality) and positive associations in blue (i.e., we expected fewer concentrations than observed). Dotted and solid hollow rectangles do not indicate statistically relevant associations. With this description in mind, note that only respondents who were renting (instead of being the owners of their homes) were more likely to be classified in Codes 2 and 4. Once more, to be able to better understand this relationship, we should go back to the actual qualitative meaning of these codes (Fig. 6.13).

Code 2 is related to housing insecurity, and Code 4 is associated with affordability and the cost of living. Specifically, Code 2 indicates concerns about home displacement, including rising costs and gentrification, while Code 4 encompasses concerns about quality of life and the cost of living. From this perspective, it is noteworthy that only participants who rent, rather than own a home, were significantly associated with these two codes. This finding, in essence, serves to further validate our machine learning classification and the interpretation of our codes based on participants' own voices (once more, note that more qualitative texts should be included, rather than only two, as we did for illustrative purposes).

Our network representation, shown in Fig. 6.5, serves to assess code strength given the group of participants who coselected those codes. Although we kept nonresponse as a category in the Chi-squared test to assess whether this nonresponse category was associated with any of the groups of interest, the network representation does not include this nonresponse category. This exclusion allows us to more clearly focus on how our identified codes with valid open-ended responses were related to one another. Notably, in this case, we found a slight yet important change. Our findings show that Code 1 (homelessness) and Code 2 (displacement/gentrification) have the strongest links across the network. That is, in essence, this highlights another important pattern: "homelessness" and "displacement" concerns are more closely related in terms of being selected by participants, regardless of their group of ascription (or categorical indicators), than the other two codes.

This finding complements the results shown in Fig. 6.9 and does so while following a different mechanism. The Chi-squared depiction showed that participants who rented were more concerned with these codes compared to owners (and nonrespondents). In this network depiction, we observed that Codes 1 and 2 had the highest level of association; however, this association level came from all groups of respondents instead of only the group of renters. That is, when combining all responses, the concerns of participants across groups converged into signaling these code categories as the most related in the network. In other words, when we take all four codes into account, it is clear that these two codes touch on similar issues, namely housing insecurity. Accordingly, this network result serves to further highlight relationships across codes that may help strengthen our understanding.

Discussion and next steps

Our application example corroborated MDCOR's epistemology: useful knowledge can be built from machine learning and text classification analyses of qualitative evidence, but only if this quantitative output is fully integrated with the qualitative analysis of this classified and unaltered qualitative information. That is, although we knew that the four-code solution was the most optimal solution from a probabilistic point of view, we still needed to re-build the meaning of these codes with the visual inputs provided by MDCOR while relying on our participant's unaltered voices. Moreover, as indicated above, although MDCOR's algorithms are not biased, the responses may still signal or contain biases toward certain groups. This latter possibility makes text preservation fundamental for allowing us to conduct *quality and equity assessments*. That is, by observing the responses as provided by participants, we may not only analyze whether these responses adhere to the open-ended question (a form of response validity) but also whether these responses may reflect biases against certain groups or issues. In both instances, the expert and qualitative analysis of the research team is necessary to conduct these assessments.

From this view, our study showcases an integrative framework and its open-source software application to classify open-ended responses via text mining, machine learning, and text classification and visualization techniques. The analyses are run locally, and the input data analyzed can be downloaded and merged back with the original survey data for subsequent use in quantitative methods. This local architecture is not only a practical feature; it is also an ethical one. Because open-ended responses may contain personally revealing, sensitive, institutionally restricted, or otherwise context-rich information, keeping classification and subsequent reasoning workflows on the researcher's own machine can be methodologically preferable to relying on external cloud systems (see González Canché (2025) for an ample discussion of ethical concerns regarding the integration of data science into qualitative and mixed-methods studies). That same logic extends to downstream interpretation: when MDCOR outputs are later examined in ISARI, local and fully offline execution helps preserve data stewardship, institutional trust, and the researcher's responsibility to govern how sensitive textual evidence is used.

From these integrative perspectives, the study offers a rigorous and comprehensive approach to processing open-ended responses rigorously and efficiently. As illustrated in our literature review, this time-efficiency is particularly relevant when dealing with textual data gathered in times of crises, wherein fast actions based on nuanced understandings are needed. Nonetheless, even in the absence of a crisis, we argue that sophisticated, no-code, free-to-use and distributable tools like MDCOR should be readily available to all researchers, not only those proficient in computer and statistical programming. From this view, our no-code approach aims to open access to these machine learning and interactive visualization tools for researchers without any computer or statistical programming proficiency requirements.

As indicated in the introduction, MDCOR reduces the cost of closing open-ended questions, and it also offers a post-hoc assessment that not only tests group-to-code association but also assesses whether nonresponse may be associated with certain attributes of survey participants. That is, although dealing with nonresponses was not the primary focus of MDCOR, our post-hoc assessment offers a quick evaluation of nonresponse association with certain attributes of interest. In our case, we did not find any relationship between nonresponse and home ownership status, but this does not mean that no associations may exist using other categories. For example, it could be that responses by gender or sexual orientation may show associations, or that even survey response type (i.e., mail or online) may reflect significant associations. To conduct these tests, researchers may simply change the indicator of interest and click "Execute Group Comparison" again. This execution will create a new PDF with the Chi-squared result and will deploy a new interactive network representation.

Our reliance on the city of Austin's community survey, and our emphasis on analyzing these data under the most extreme conditions (i.e., not removing nonresponses), aimed to test MDCOR's performance in handling data downloaded directly from the internet. Although we did not show our Twitter data analyses, we compiled a dataset with 2.5 million tweets on the topic "Online Learning" that is being analyzed without the need for further preprocessing or cleaning of the data. This latter point is important because MDCOR does not require formal

English to execute the machine learning classification. As long as words are used by 1% of the responses, they will be retained in the analyses. This attribute is particularly relevant when analyzing social media data, but it is also relevant when including responses from members with divergent levels of formal schooling, for whom their expressions do not adhere to formal English structures, or who prefer to respond in other languages. This heterogeneity of responses highlights once more the relevance of preserving participants' original and unaltered voices.

Integrating MDCOR Outputs into Local Scholarly Brainstorming

One of the distinctive strengths of MDCOR is that its outputs are portable across analytic stages. The interactive word summaries, the most representative response tables, the downloadable fully classified datasets, and the code-to-group association outputs are useful not only for identifying latent thematic structure, but also for supporting subsequent interpretive work. In the broader ecosystem developed throughout this book, these outputs can be brought into ISARI, a local, fully offline, open-source, and multimodal brainstorming environment designed to support scholarly memoing, comparison, synthesis, and evidence-grounded writing.

This possibility matters because machine-driven classification, however rigorous, does not by itself determine what latent codes mean or how they should be narrated in relation to a study's purposes. Researchers still need to compare possible labels, reflect on overlaps and distinctions among codes, consider how code distributions vary across groups, and decide how to translate the resulting structure into qualitative argument. Because ISARI can work with textual notes, screenshots, exported figures, scanned pages, and other image-based or document-based analytic artifacts, researchers can upload MDCOR outputs in multiple forms—for example, screenshots of the interactive quadrant plots, copied text from code-summary panels, PDF exports of code-by-group comparisons, or notes derived from the fully classified datasets—and use them as the basis for local scholarly reasoning.

This workflow should not be mistaken for automation of interpretation. On the contrary, it extends MDCOR's mixed equal-status design. The outputs remain guides, not conclusions. ISARI can help researchers compare latent code interpretations, generate analytic memos, test explanatory contrasts, and begin drafting evidence-grounded prose, but the authority to decide what the codes mean remains with the research team. In that sense, MDCOR and ISARI are methodologically aligned: one classifies and organizes open-ended responses while preserving original voices; the other helps researchers think with those outputs under conditions of privacy, interpretive accountability, and local control. This "original voice preservation" is not only an ethical commitment but also an epistemic one, because it ensures that latent code identification remains accountable to the actual language through which respondents expressed their concerns, reasons, and experiences.

Limitations

Traditionally, the main challenge of latent Dirichlet allocation, or topic modeling, concerns selecting the optimal number of codes. This challenge arises from the unsupervised classification nature of this technique, which implies that the machine learning algorithm does not know the relationship between the texts and latent codes and must instead rely on inductive, data-driven processes that begin with a random allocation of codes (Boyatzis, 1998; Eickhoff and Wieneke, 2018a; Tashakkori and Teddlie, 2008a).

In this respect, MDCOR addresses this challenge by incorporating four metrics designed to identify the most optimal number of codes for latent Dirichlet allocation (Nikita and Chaney, 2020). As further discussed in the appendix, the metrics assessment included in MDCOR employs four different methods, each described in its corresponding methodological paper (Arun et al., 2010a; Cao et al., 2009a; Deveaud et al., 2014b; Griffiths and Steyvers, 2004), to identify the most optimal result. Two of these methods (Arun et al., 2010a; Cao et al., 2009a) aim to find the number of topics that minimize correlation or overlap of terms configuring each code. The other two methods seek to maximize the differences among codes (Deveaud et al., 2014b; Griffiths and M.Steyvers, 2004). In the end, the goal is the same: if we have a three-code solution and the minimization approaches indicate the correlation among these codes is close to zero, their configuring terms may not be overwhelmingly present across codes. The same would be true if the difference among these codes is the maximum compared to other N-code solutions. In our study, we found that, for one minimization method (Cao et al., 2009a) and one maximization method (Deveaud et al., 2014b), the most optimal solution was a four-code solution. As observed in Figs. 6.8 and 6.9, the distance among codes indicated independence, and the distribution of their configuring words within codes corroborated this code independence, which eased the qualitative understanding of participants' responses. From this view, although optimal number of code identification remains challenging, MDCOR was designed and built to provide researchers with statistical and visual tools that may provide clarity and transparency in this process.

Having said this, in other instances, there may be two competing solutions. For example, in the analysis of restricted data (see Figure B1(B), available online at http://dx.doi.org/DOI_of_Chapter6) the metrics assessments indicated that the optimal solutions were between seven and 11 codes. Based on our experience using topic modeling, we recommend selecting the lowest number, given that as the number of codes increases, the chance of them sharing terms also increases, which makes their differentiation less clear and more challenging. For example, even with this seven-code solution in the restricted data example, we observed a code right in the middle of the quadrant (see Figure B2(b), available online at http://dx.doi.org/DOI_of_Chapter6), which clearly indicates that this code shares information with all other codes. In this case, although this seven-code solution resulted in meaningful descriptions (see Appendix), transparency must be the best policy when justifying code number selection, particularly when competing solutions are at play. This selection process is truly important because code number selection guides the ultimate code allocation and all the resulting integrative analyses.

Software Access and Execution

Software Access: Mac users can download MDCOR here: https://cutt.ly/jc7n3OT, PC users here: https://cutt.ly/wc7nNKF. In addition, peer review replication procedures can be found in González Canché (2022a), which executes the software in the Code Ocean server. To execute the application locally with the Mac or Windows versions, you simply need to: (1) extract the zip folder, (2) open the folder starting with "LACOID-App-" and ending with Win or Mac, and (3) double-click on the file called "LACOID-App-(Win or Mac)." The first time MDCOR is executed, you may need to select the menu "View" and "Force Reload" to generate all paths to this newly extracted folder. If, after forcing reload, MDCOR does not display, close the window, execute MDCOR again, and, if needed, force reload once more. The interface will now load. This process is only needed the first time MDCOR is launched on a computer. The software applications have been comprehensively tested for viruses and are housed in a secure server location. Accordingly, all executable files are virus-free but may require administrative permission to be deployed. For the execution of the metrics assessment, Windows requires access to parallelization to speed up the process. This process will also request administrative permission the first time MDCOR is executed on a Windows machine (this process is not required on Mac). Finally, each of these executable files for the Mac and Windows operating systems contains all the source code required to modify and/or expand the application's capabilities in the file called "app.R." Mac users are also recommended (and in some instances they do need) to place the MDCOR software (and all software products discussed in this book) in their applications folder for newer versions can only read and write files from this location.

Future inquiry

We showcased MDCOR with short text survey data, but the framework has also been used with qualitatively coded, machine-learning-coded, or computer-assisted coded textual data drawn from essays, social media posts (i.e., YouTube comments and Facebook posts), open-ended survey responses, and qualitative interview transcripts. In all these instances, our results indicated that the framework and software are versatile enough to handle a variety of inputs and yield meaningful results. MDCOR could also be a valuable tool in questionnaire design. That is, when questionnaires need to have closed-ended questions, researchers could use MDCOR with pilot surveys containing open-ended questions as part of the design process and then apply the procedures depicted here to "close" these responses for use in the final questionnaire.

Finally, regarding areas of improvement, MDCOR incorporates the most recent advancements in machine-driven text classification and code identification in English—but other languages may be added to MDCOR's source code. That is, although non-English responses may be analyzed in the current version, the text mining, including lemmatization, is based on English language rules. Accordingly, the resulting text preparation of non-English text will be less optimal, which may complicate the classification and reconstruction of qualitative meanings. From this perspective, immediate extensions of MDCOR include the need to expand this application to other languages.

MDCOR seeks to democratize access to rigorous, state-of-the-art, and sophisticated cost-free analyses that do not require computer programming experience, ensuring that not only programmers or quantitative experts can realize the benefits that machine learning brings to qualitative, survey, and mixed methods research. Having said this, because this line of research is constantly progressing, we are making all source code available (Mac version available here:

https://cutt.ly/UNyzlxW and Windows version available here: https://cutt.ly/KNyzGUV) so that researchers may modify and expand MDCOR's functionalities (González Canché, 2022b). This openness is methodologically significant. MDCOR's availability as a local, no-code application together with its public source code and Code Ocean peer-review history strengthens reproducibility, inspectability, and long-term usability. These same qualities also make MDCOR especially compatible with the broader local analytic ecosystem advanced in this book, in which outputs can remain under the researcher's control while being revisited, compared, and interpreted across later stages of analysis.

Appendix: Supplementary material

Supplementary material related to this chapter can be found online at: http://dx.doi.org/DOI_of_Chapter6.

References

Albalawi, R., Yeap, T.H., Benyoucef, M., 2020a. Using topic modeling methods for short-text data: a comparative analysis. Front. Artif. Intell. 3, 42.

Alexander, E., Eppler, M.J., Bresciani, S., 2019. Visual replay methodology: a mixed methods approach for group discussion analysis. J. Mixed Methods Res. 13 (1), 33–51.

Arun, R., Suresh, V., Veni Madhavan, C.E., Narasimha Murthy, M.N., 2010a. On Finding the Natural Number of Topics with Latent Dirichlet Allocation: some Observations. In: Zaki, M.J., Yu, J.X., Ravindran, B., Pudi, V. (Eds.), Advances in Knowledge Discovery and Data Mining. Springer, Berlin Heidelberg. https://link.springer.com/chapter/10.1007/978-3-642-13657-3%5C_43, pp. 391–402.

Belotto, M.J., 2018. Data analysis methods for qualitative research: managing the challenges of coding, interrater reliability, and thematic analysis. Qual. Rep. 2160-3715. 23 (11), 2622–2633. Available from: https://nsuworks.nova.edu/cgi/viewcontent.cgi?article=3492&context=tqr.

Boyatzis, R.E., 1998. Transforming Qualitative Information: Thematic Analysis and Code Development, Sage.

Cao, J., Xia, T., Li, J., Zhang, Y., Tang, S., 2009a. A density-based method for adaptive lda model selection. Neurocomputing 72 (7-9), 1775–1781. doi:10.1016/j.neucom.2008.06.011.

Cao, J., Xia, T., Li, J., Zhang, Y., Tang, S., 2009b. A density-based method for adaptive LDA model selection. Neurocomputing. 0925-2312. 72 (7-9), 1775–1781. doi:10.1016/j.neucom.2008.06.011.

Chang, T., DeJonckheere, M., Vydiswaran, V.V., Li, J., Buis, L.R., Guetterman, T.C., 2021a. Accelerating mixed methods research with natural language processing of big text data. J. Mixed Methods Res. 15 (3), 398–412.

Charmaz, K., 2014a. Constructing Grounded Theory: A Practical Guide through Qualitative Analysis. SAGE Publications Sage CA, Los Angeles, CA.

Coffey, A., Atkinson, P., 1996a. Making Sense of Qualitative Data: Complementary Research Strategies, Sage Publications, Inc.

Deveaud, R., SanJuan, E., Bellot, P., 2014b. Accurate and effective latent concept modeling for ad hoc information retrieval. Document Numérique. 1279-5127. 17 (1), 61–84. doi:10.3166/dn.17.1.61-84.

Deveaud, R., SanJuan, E., Belloto, P., 2014a. Accurate and effective latent concept modeling for ad hoc information retrieval. Document Numérique 17 (1), 61–84. doi:10.3166/dn.17.1.61-84.

Eickhoff, M., Wieneke, R. (2018a). Understanding topic models in context: a mixed-methods approach to the meaningful analysis of large document collections. *Proceedings of the 51st Hawaii International Conference on System Sciences*. https://pdfs.semanticscholar.org/9663/20028a68a5dd3d044f2bf1b40239aa2fcf58.pdf

Feinerer, I. (2020). Introduction to the tm package text mining in r. R CRAN. https://cran.r-project.org/web/packages/tm/vignettes/tm.pdf

Fetters, M.D., 2020. The mixed methods research workbook: Activities for designing, implementing, and publishing projects (Vol. 7). Sage Publications, CA, USA. pp. 312.

Fetters, M.D., Curry, L.A., Creswell, J.W., 2013a. Achieving integration in mixed methods designs—principles and practices. Health Serv. Res. 48 (6pt2), 2134–2156.

Fetters, M.D. and Freshwater, D., 2015. The 1 + 1 = 3 Integration Challenge. Journal of Mixed Methods Research 9 (2), 115–117. doi:10.1177/1558689815581222.

Friborg, O., Rosenvinge, J.H., 2013. A comparison of open-ended and closed questions in the prediction of mental health. Qual. Quant. 1573-7845. 47 (3), 1397–1411. Available from: https://link.springer.com/journal/11135. doi:10.1007/s11135-011-9597-8.

Geer, J.G., 1988. What do open-ended questions measure? Public Opin. Q. 0033-362X. 52 (3), 365–367. doi:10.1086/269113.

González Canché, M S., 2026. Network Analysis of Qualitative Data (NAQD): An Analytic Framework and Software for Qualitative Data Science Visualization and Hypothesis Testing. International Journal of Qualitative Methods, 25. doi:10.1177/16094069261439942.

González Canché, M. S., 2025. Ethics and trust in social network analysis: how data science and interactive visualizations impact anonymity, confidentiality, consent, and IRB procedures?. In: Handbook of Social Network Analysis and Education. Edward Elgar Publishing, pp. 114–131. doi:10.4337/9781035312191.00013.

González Canché, M.S. (2022a). Austin survey for mdcor analyses. Mendeley. 10.17632/nb7yvhjvzk.1

González Canché, M.S., 2022b. Mapping, organizing, and visualizing interdependent events (movie): a rigorous analytic framework and cost-free software application designed to model temporal and dynamic complex realist structures in social research settings. Methodol. Innov. 15 (3), 263–288. doi:10.1177/20597991221119012.

González Canché, M.S., 2022c. Network analysis of qualitative data: an integrative software application to visualize and assess similarities in participants' qualitative contributions. J. Mixed Methods Res. 16 (3), 373–377. doi:10.1177/15586898211051584.

Griffiths, T.L., Steyvers, M., 2004. Finding scientific topics. Proc. Natl. Acad. Sci. U. S. A. 0027-8424. 101 (1), 5228–5235. Available from: www.pnas.org. doi:10.1073/pnas.0307752101.

Grün, B., Hornik, K., 2011. Topicmodels: an r package for fitting topic models. J. Stat. Softw. 1548-7660. 40 (13), 1–30. Available from: http://www.jstatsoft.org/v40/i13/paper. doi:10.18637/jss.v040.i13.

Ho, P., Chen, K., Shao, A., Bao, L., Ai, A., Tarfa, A., Brossard, D., Brown, L., Brauer, M., 2021a. A mixed methods study of public perception of social distancing: integrating qualitative and computational analyses for text data. J. Mixed Methods Res. 15 (3), 374–397.

Holsti, O.R., 1969. Content Analysis for the Social Sciences and Humanities. Addison-Wesley (content analysis), Reading. MA.

Humble, Á.M., 2019. Computer-aided Qualitative Analysis Software, SAGE Publications Limited.

Janasik, N., Honkela, T., Bruun, H., 2009a. Text mining in qualitative research: application of an unsupervised learning method. Organ. Res. Methods. 12 (3), 436–460.

Jason, L., Glenwick, D., 2016a. Handbook of Methodological Approaches to Community-based Research: Qualitative, Quantitative, and Mixed Methods, Oxford university press.

Kelley, S., 1983. Interpreting Elections, Princeton University Press.

Kintzer, D.F.C., 1977. Advantages of open-response questions in survey research. Community Jr. Coll. Res. Q. 2 (1), 37–46. doi:10.1080/03616975.1977.10588098.

Kunz, T., Quoß, F., Gummer, T., 2020. Using placeholder text in narrative open-ended questions in web surveys. J. Surv. Stat. Methodol. 9 (5), 992–1012.

Lazarsfeld, P.F., 1944. The controversy over detailed interviews—an offer for negotiation. Public Opin. Q. 8 (1), 38–60.

Lichtman, A.J., Kelley, S., 1984. Interpreting Elections. J. Am. Hist. 0021-8723. 71 (1), 102. doi:10.2307/1899837.

Lynam, T., Damayanti, R., Rialine Titaley, C., Suharno, N., Bradley, M., Krentel, A., 2020a. Reframing integration for mixed methods research. J. Mixed Methods Res. 14 (3), 336–357.

McColl, E., Jacoby, A., Thomas, L., Soutter, J., Bamford, C., Steen, N., Thomas, R., Harvey, E., Garratt, A., Bond, J. (2001a). Design and use of questionnaires: a review of best practice applicable to surveys of health service staff and patients.

Miles, M.B., Huberman, A.M., 1994. Qualitative Data Analysis: An Expanded Sourcebook, Sage.

Nikita, M., Chaney, N. (2020). ldatuning: tuning of the Latent Dirichlet Allocation Models Parameters. (R Project). https://CRAN.R-project.org/package=ldatuning

Parker, E.B., Holsti, O.R., 1970. Content analysis for the social sciences and humanities. Am. Sociol. Rev. 0003-1224. 35 (2), 356. doi:10.2307/2093233.

Popping, R., 2015. Analyzing open-ended questions by means of text analysis procedures. Bulletin of Sociological Methodology/Bulletin de Méthodologie Sociologique 128 (1), 23–39.

Poth, C.N., Bulut, O., Aquilina, A.M., Otto, S.J., 2021a. Using data mining for rapid complex case study descriptions: example of public health briefings during the onset of the covid-19 pandemic. J. Mixed Methods Res. 15 (3), 348–373.

Raftery, A.E., Lewis, S.M., 1992. How many Iterations in the Gibbs Sampler? In: Bemardo, J., Berger, J., Dawid, A., Smith, A. (Eds.), Bayesian Statistics. Oxford University Press (OUP), pp. 763–774. doi:10.1093/oso/9780198522669.003.0053.

Reja, U., Manfreda, K.L., Hlebec, V., Vehovar, V., 2003a. Open-ended vs. close-ended questions in web questionnaires. Dev. Appl. Stat. 19 (1), 159–177.

Saldaña, J., 2021a. The Coding Manual for Qualitative Researchers. SAGE Publications, CA: USA. pp. 1–440.

Schuman, H., Presser, S., 1979. The open and closed question. Am. Sociol. Rev. 44 , 692–712.

Schütze, H., Manning, C.D., Raghavan, P., 2008a. Introduction to Information Retrieval, Vol. 39. Cambridge University Press, Cambridge. https://nlp.stanford.edu/IR-book/.

Shapiro, G., 2020. THE FUTURE OF CODERS: HUMAN JUDGMENTS IN A WORLD OF SOPHISTICATED SOFTWARE. In: Roberts, C.W. (Ed.), Text Analysis for the Social Sciences: Methods for Drawing Statistical Inferences from Texts and Transcripts. Taylor and Francis, United States, pp. 225–238. Available from: http://www.tandfebooks.com/doi/book/9781000106039. doi: 10.4324/9781003064060-16.

Stroll, A., Martinich, A. (2022). *Epistemology*. Encyclopedia Britannica. https://www.britannica.com/topic/epistemology

Tashakkori, A., Teddlie, C., 2008a. Introduction to mixed method and mixed model studies in the social and behavioral sciences. In: Creswell, J.W., Plano-Clark, V.L. (Eds.), The Mixed Methods Reader, pp. 7–26.

Teddlie, C., Tashakkori, A., 2009a. Foundations of Mixed Methods Research: Integrating Quantitative and Qualitative Approaches in the Social and Behavioral Sciences, Sage.

Teddlie, C., Tashakkori, A., 2009b . Foundations of Mixed Methods Research: Integrating Quantitative and Qualitative Approaches in the Social and Behavioral Sciences. SAGE Publications, Inc., Thousand Oaks, CA: USA, pp. 178. doi:10.4135/9781506350288.

Tesch, R., 2013b. Qualitative Research: Analysis Types and Software Tools Qualitative Research: Analysis Types and Software Tools. Taylor and Francis, pp. 1–331. Available from: http://www.tandfebooks.com/doi/book/10.4324/9781315067339. doi:10.4324/9781315067339.

PART IV

Integrative Extensions for Textual, Relational, Spatial, and Affective Analysis

7

Sentiment and Emotion Network Analysis (SENA)

Introduction

Today's data availability and growth are difficult to conceptualize. In 2019, for example, the accumulated amount of data available in the entire digital universe was estimated to be 4.4 zettabytes (ZB)—with one ZB equaling a trillion gigabytes (Joda et al., 2019a). One year later, data availability expanded to a total of 44 ZB, a number that is 40 times larger than the number of stars in the observable universe (Desjardins, 2019a; Marr, 2019a). Despite this tenfold growth in one year, data expansion is expected to continue at even faster rates (Holst, 2021a). Experts have estimated that by 2025, there will be 463 ZB of new data generated *every day* (Andre, 2022a; RACONTEUR, 2020).

There are two interrelated notable attributes of these newly generated data. The first is that the vast majority of these data are in unstructured form, with estimates ranging between 85% (Kumar and Bhatia, 2013) and 90% (Davis, 2019a) of all newly generated data. The main implication of unstructured data is that these data are not ready to be *quantitatively or statistically* analyzed because they do not adhere to the rectangular (or spreadsheet) form, where we have *n* number of rows and *m* number of columns—hence leading to a rectangular shape. The second attribute is that, based on the prevalence of online communication (i.e., email, text messaging) and social media sites, where users may also be content producers (Van Dijck, 2013), the most prevalent form of unstructured data is configured by written information (Kumar and Bhatia, 2013; Marr, 2018a). Bringing these two attributes together, we find that textual data constitute most of the newly generated data, but these data are not ready to be statistically analyzed. This prevalence highlights the need to rely on data science to efficiently manage large amounts of information.

Texts as unstructured data

Although texts are classified as unstructured data (Nandwani and Verma, 2021), texts are not inherently unstructured. Instead, textual or written data are guided by linguistic and grammatical rules that we, as humans, have learned to decode (De Beaugrande and Dressler, 1981). From this view, and as briefly noted above, the reason why texts are considered unstructured from a quantitative analytical perspective is that, in their original format, written information is not stored in matrices or rectangular databases with *n* by *m* dimensions—i.e., rows and columns, respectively. The main implication of this lack of rectangular structure is that, for machine learning, classification, and other statistical analyses of textual data to be possible,

Data Science, Interactive Visualizations, and Generative AI Tools for the Analysis of Qualitative, Mixed-Methods, and Multimodal Evidence. https://doi.org/10.1016/B978-0-443-21961-0.00005-X

specialized preprocessing tools must be applied before we are able to extract the meanings embedded in vast amounts of unstructured written information.

The analytic process followed to structure textual data is referred to as natural language processing (NLP). NLP is "a collection of computational techniques for automatic analysis and representation of human languages, motivated by theory" (Chowdhary, 2020, p. 604). As briefly stated, this use of computer algorithms is, to a great extent, justified precisely by the aforementioned vast amount of data currently available and constantly being generated. Essentially, with NLP techniques, we can normalize and store textual data in matrices or databases so that we can start applying computer- and statistically-based algorithms (Chowdhary, 2020; Kumar and Bhatia, 2013) designed to extract, learn, or detect potentially useful information that highlights the structure embedded in vast amounts of data.

NLP and text analysis may include word-level analysis in the form of term frequencies (word clouds or tag clouds) as well as a variety of machine learning text classification tools (Chowdhary, 2020; González Canché, 2023; González Canché, 2023b). It may also be used to objectively and automatically determine the distribution of sentiments and emotions contained in textual databases (Nandwani and Verma, 2021; Sharma et al., 2024a). When the resulting outputs indicate negative, neutral, or positive sentiments, this latter approach is referred to as sentiment analysis—and this technique has also been referred to as polarity detection (Mäntylä et al., 2018a). When the resulting analyses are presented in terms of emotions that go beyond positive or negative experiences or opinions (i.e., beyond polarity detection), the analytic method is called emotion analysis. From this view, emotion analysis represents efforts to move beyond polarity detection to gain more nuanced understandings of written information (Mäntylä et al., 2018a).

Considering that polarity detection is a useful technique to aggregate positive and negative sentiments, whereas emotion analysis yields potentially more nuanced understandings, in our study, we offer both a sentiment and emotion analytic framework for the resulting outcomes, integrating both procedures. However, as further explained below, we also deviate from traditional sentiment and emotion analysis (SEA) approaches by minimizing the risk of aggregation bias (James, 1982), which may lead to losing nuance in our understandings by failing to discover potential differences across groups of participants. We also depart from the traditional use of SEA by illustrating how our proposed framework may be applied to mixed methods and qualitative academic research—as opposed to being used mostly as a marketing analysis tool, as we further elaborate below.

Study contributions and purpose

In this study, we focus on the intersection of linguistics, NLP, statistical and relational modeling, and computer science (Chang et al., 2021a; Kumar and Bhatia, 2013; Taboada, 2016) to offer an analytic framework designed to (1) detect sentiments and emotions embedded in our participants' speeches, discourses, or textual contributions across seventeen languages (these languages are: Basque, Catalan, Danish, Dutch, English, Esperanto, Finnish, French, German, Irish, Italian, Latin, Portuguese, Romanian, Spanish, Swedish, Welsh), (2) assess emotional inconsistencies (i.e., emotional entropy) in the content of these texts both individually and by groups, and (3) formally test (i.e., hypothesis testing) whether the prevalence of emotions is similarly distributed across groups of participants. In other words, our proposed framework is saliently different from the traditional use of SEA because our framework goes beyond aggregate analyses of the distributions of sentiments and emotions in a collection of texts (i.e., corpus) and integrates measures of speech or content inconsistency at both individual and group levels.

The application of NLP to structure textual data, as well as the subsequent data processing, sentiment and emotion identification, visualization, and hypothesis testing, requires advanced technical expertise in data engineering. This expertise continues to serve as an important barrier to the mainstream use of data science in general, and NLP and text classification in particular (Chang et al., 2021a; González Canché, 2023; González Canché, 2023b). From this perspective, another contribution of this study is our goal to expand access to data engineering and data science tools by providing a standalone, free-to-use, and free-to-distribute software that automatically implements all analytic steps discussed below (see Fig. 7.1) without requiring any computer or statistical programming expertise. In this sense, SENA contributes not only

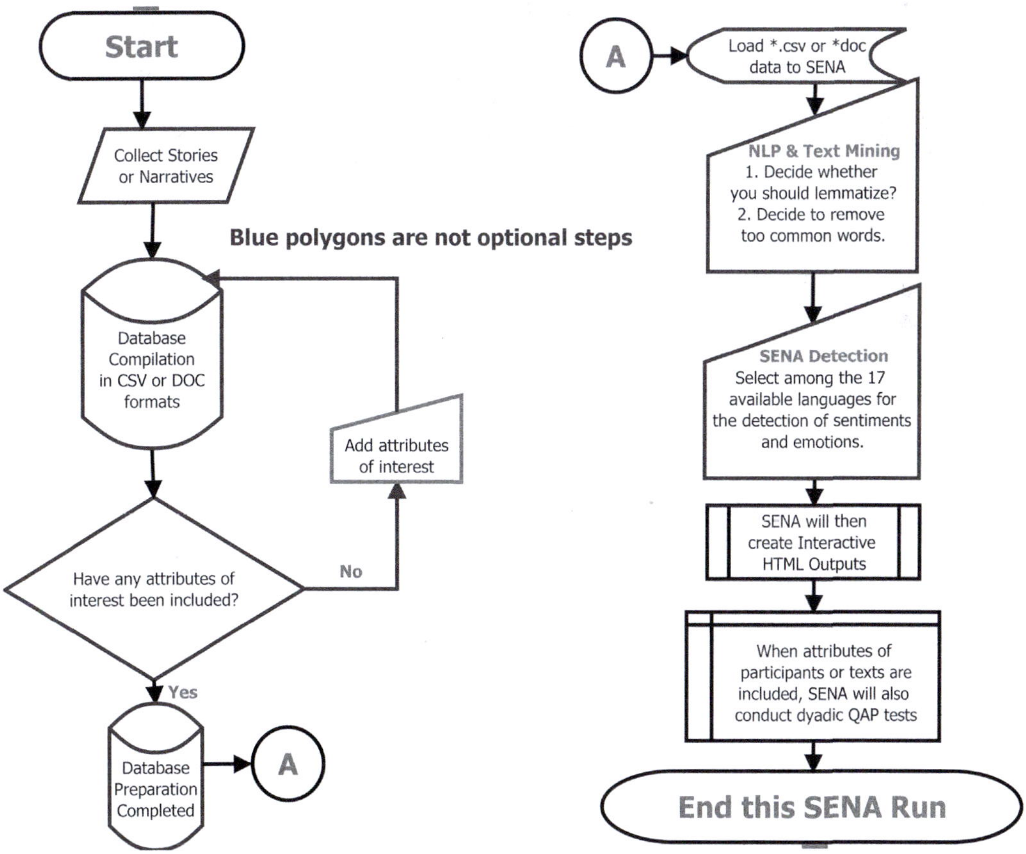

FIG. 7.1 Flow Diagram for the Entire Methodological Process, Including Database Construction.

to efficiency and accessibility, but also to interpretive transparency by making the path from unstructured text to sentiment and emotion distributions more inspectable, revisitable, and documentable. That is, all the steps, which include the structuring and normalization of textual data, the identification of sentiments and emotions, and the application of relational or network modeling for the visualization of findings and hypothesis testing, may be executed by simply uploading the collection of texts to be analyzed.

As part of the versatility of this *no-code software*, the analytic framework is implemented by either uploading a collection of Microsoft Word documents or uploading a collection of texts stored in a spreadsheet file. An important attribute of this software is that it was programmed to conduct SEAs in seventeen different languages without the need to upload datasets to any server—hence running all analyses locally. This local architecture is not merely a technical convenience; it is also an ethical and methodological choice. Because many of the texts that qualitative and mixed-methods researchers analyze may contain sensitive, identifiable, institutionally restricted, or emotionally charged content, keeping sentiment and emotion analyses on the researcher's own machine can be preferable to relying on cloud-based infrastructures that require external data transfer. Based on our reliance on relational modeling to identify potential group variation in sentiment and emotion distributions, the resulting analytic framework and its accompanying software are referred to as Sentiment and Emotion Network Analysis (SENA). Finally, with the goal of providing qualitative and mixed-methods researchers with the opportunity to become fully familiarized with these analytic tools, all databases used to showcase this analytic framework and software tool are being made publicly available so that researchers may interact firsthand with the SENA software to implement the analytic framework discussed in the remainder of this document.

Purpose

With this brief introduction in mind, the purpose of this study is to offer a free-to-use, free-to-distribute, and modifiable analytic framework and software tool designed to conduct SENAs in seventeen different languages. The goal of offering this no-code tool is to expand access to data science and feature engineering outputs such as interactive visualizations and statistical modeling comparative results without any software programming or statistical coding requirements. The relevance of these outputs extends beyond sentiment and emotion detection alone. Once distributions of sentiments and emotions have been rendered, emotional entropy has been estimated, and group differences have been tested, researchers still need to determine what those patterns mean, which cases deserve closer memoing, how contradictory or mixed messages should be interpreted, and how these findings should be translated into qualitative argument. In the broader architecture of this book, SENA outputs can therefore be understood not as endpoints, but as structured guides that may also be brought into ISARI, a local, fully offline, open-source, and multimodal brainstorming environment designed to support scholarly comparison, memoing, synthesis, and evidence-grounded writing. In this way, SENA and ISARI are methodologically aligned: SENA identifies and structures distributions of sentiments and emotions while preserving local control; ISARI helps researchers think with those outputs under conditions of privacy, interpretive accountability, and data sovereignty. With our data science democratization goal in mind, all the data used in this paper (see Fig. 7.2 for all data access) are provided with this submission so that researchers may interact firsthand with the SENA software interface. This software tool is available for Mac (access here: https://cutt.ly/QwhYruBr) and Windows (access here: https://cutt.ly/YwhJJKvO) operating systems.

To address this purpose, the rest of this paper is structured as follows. In the next section, we present the conceptual and methodological underpinnings and current applications of NLP, as well as SEAs. In this section, we also showcase how the incorporation of participants' attributes and network modeling leads to deeper and more nuanced understandings. Subsequently, we discuss the flow diagram of the entire SENA process (see Fig. 7.1), including SENA's software user interface (UI). In the following section, we showcase the analytic

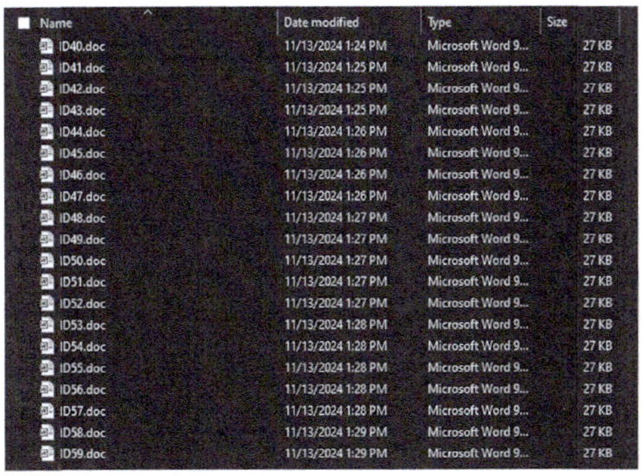

(a) Essays without gender categories, access available at https://cutt.ly/leJfF4RW

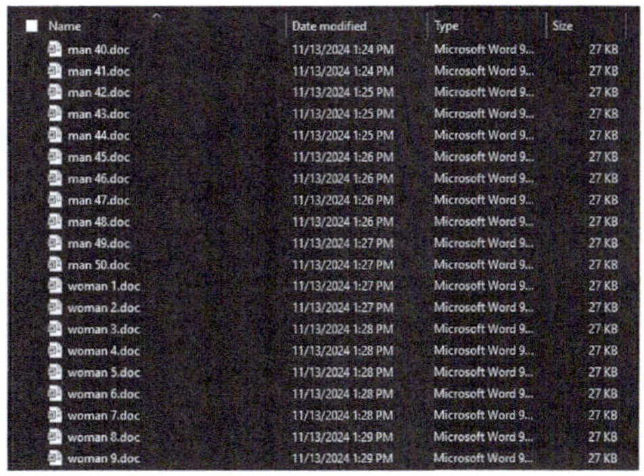

(b) Essays with gender categories, access available at https://cutt.ly/aeJfFbZt

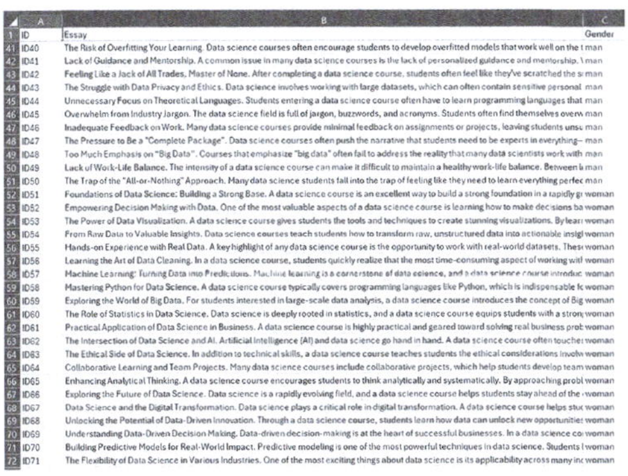

(c) Database with all 99 essays, access available at https://cutt.ly/zeJfHr0M

(d) Database of classified YouTube comments, access available at https://cutt.ly/ReLN5yH8

FIG. 7.2 Replication data. All links provide full access to these textual datasets. (A) Essays without gender categories, access available at https://cutt.ly/leJfF4RW, (B) Essays with gender categories, access available at https://cutt.ly/aeJfFbZt, (C) Database with all 99 essays, access available at https://cutt.ly/zeJfHr0M, and (D) Database of classified YouTube comments access available at https://cutt.ly/ReLN5yH8.

capabilities of our proposed framework and software. As part of this presentation, we discuss the types of narratives or texts that may be analyzed, their data format, and the rationale followed for testing differences in the distribution of emotions across groups in both Word document and database formats. Specifically, we illustrate how Microsoft Word metadata (i.e., the name of each file) may be used to capture attributes of the text content. Similarly, we demonstrate how these same attributes may be captured when texts are stored in a spreadsheet format. Our presentation of the SENA findings discusses the multiple interactive HTML outputs automatically generated by our software, including the quadratic assignment procedures (QAPs) correlations that test for differences in emotions across groups. The discussion section presents limitations, elaborates on the current strengths of SENA, and highlights potential future areas of improvement.

Current applications

Text analytics is the process of examining unstructured data in the form of text to gather insights on patterns and topics of interest (Chang et al., 2021a; Chowdhary, 2020; Kumar and Bhatia, 2013). With text analytics, we can extract meaningful information from social media posts, emails, text messages, advertisements, interview transcripts, essays, abstracts, and much more. Using these tools, we may also understand sentiments and emotions expressed by users regarding the services they received or experienced. Indeed, most applications of SEAs relate to reviews of food and services, movies, and airlines (Taboada, 2016), which is why companies and marketers have placed primary interest in SEA as a tool that can be used for marketing, understanding consumer behaviors and preferences, and driving for-profit expansion (Nandwani and Verma, 2021; Sharma et al., 2024a). From this perspective, the availability and constant expansion of textual data that can be used to better understand services and customer satisfaction have fueled the relevance of sentiment and emotion modeling for marketing and market research (Kennedy and Inkpen, 2006; Sharma et al., 2024a).

Another marketing-related application of SEA focuses on political campaigns, including political movements or initiatives (Chowdhary, 2020; Hitesh et al., 2019; Taboada, 2016; Wang et al., 2012a). With respect to this latter point, well-established for-profit companies, like Determ and socialays.com, offer sentiment analysis services for political campaigns as a strategy to gain valuable insights into the overall political landscape. For example, an important selling point that socialays.com features on its website is the use of sentiment analysis in two successful political campaigns: President Obama's 2012 campaign and the 2016 Brexit referendum.

Overall, much less represented in the NLP and sentiment analysis literature is the analysis of sentiments and emotions as an analytic tool to be employed in academic *research settings* (Pooja and Bhalla, 2022; Rambocas and Pacheco, 2018; Sharma et al., 2024a; Taboada, 2016; Yu et al., 2018b). For example, Yu et al. (2018b) relied on extracting the sentiments in text-based, self-evaluated comments written by students and used those sentiments as early predictors of academic failure. This study employed a method similar to that used by Luo et al. (2015a) who asked students to write comments after each lesson, expecting to understand students' learning attitudes toward the methods, content complexity, and perceived difficulties and challenges.

In both instances, the results indicated that SEAs, when used in students' comments regarding their own evaluation of the class content (Yu et al., 2018b) or evaluations regarding the class content (Luo et al., 2015a) show promise in the early detection of students who may need extra support. In this line of research, authors (Calvo and D'Mello, 2010; Yu et al., 2016b; Dolianiti et al., 2018) have consistently emphasized the relevance of providing teachers with easy-to-use tools that would enable the implementation of these real-time and objective SEA analyses.

Even less represented in the academic literature is the use of SEAs in qualitative and mixed methods research (see Mäntylä et al., 2018a, for a review of the literature on sentiment analysis that describes the absence of its use in qualitative and mixed methods research since the first publication on the topic in 1945). Considering that qualitative "research involves finding out what people think, and how they feel" (Rambocas and Gama, 2013a, p. 14), "[w]hen integrated with qualitative research, sentiment analysis can be used as a tool that promotes rigor and structure to an otherwise flexible and subjective data collection and data analysis process" (Rambocas and Gama, 2013a, p. 15). Although informative, note that this latter statement focuses on the use of SEA in marketing, and when the authors mention qualitative research, they are referring to the use of SEA on written information rather than qualitative data obtained from interviews or focus groups, for example, which are typical of qualitative studies.

What is common across the statements provided by Rambocas and Gama (2013a) and our conceptualization of qualitative and mixed methods applications is that qualitative evidence is overwhelmingly composed of texts. This is why we argue that an application of SEA may enable qualitative and mixed methods researchers to comprehensively take advantage of the insights that can be gained from the analysis of *how our research participants perceived their experiences about a given topic of interest, or about what sentiment and emotions were reflected in our participants' episodic memories* (Byrne, 2017; Ramachandran et al., 2002).

An important point of departure of SENA from current SEAs is our emphasis on providing analyses that enable the identification of individual- and group-based sentiment and emotion distributions. This implementation, which enables individual-level analyses, is particularly important when using this analytic framework in self-evaluation and students' comments (Calvo and D'Mello, 2010; Luo et al., 2015a; Yu et al., 2018b, 2016b; Dolianiti et al., 2018). Group analyses, on the other hand, also enable us to understand whether the content analyzed was seen more favorably by participants based on personal attributes and characteristics. That is, when these attributes are based on personal factors like gender or ethnicity, we can conduct group analyses using these indicators. However, the analyses may also form groups based on the date when their comments or documents were collected, thereby forming temporal groups. Naturally, the analyses can then even include the interaction of individual and temporal-based attributes, as well as the integration of more than one attribute, such as both gender and ethnicity, as we discuss in Fig. 7.3.

In sum, our literature review on the current use of SEA indicates that SEA is primarily used in marketing and political analyses. Notably, an interesting and important, yet less frequently employed analysis consists of using SEA for students' self-evaluations and comments on class content and delivery. From this perspective, our integration of analytic techniques that enable both individual- and group-level analyses represents an asset for the use of SENA in school or

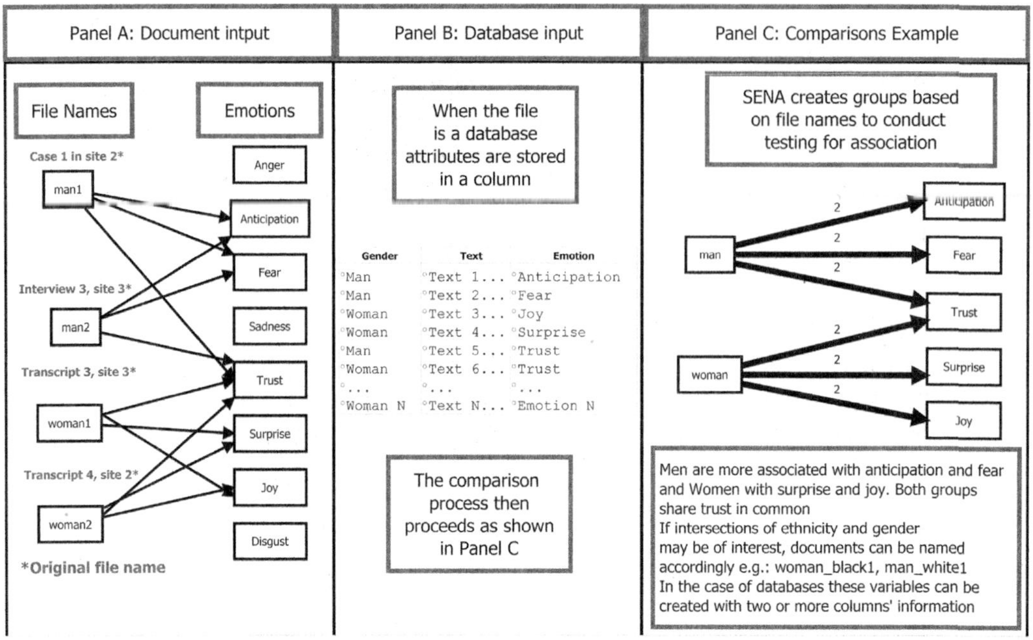

FIG. 7.3 Document-based attribute comparison.

educational settings. Finally, absent from the applications of SEA is its use in more traditionally focused qualitative and mixed-methods academic research. From this perspective, let us note that, although the SENA software tool described in this paper may be applied to gain marketing and political insights, which aligns with the current use of these analytic tools, our examples and discussions focus on how to use SENA to better understand our research participants' experiences and opinions, including examples of its applications to students' self-reports, social media posts, and more traditional qualitative interviews or focus group transcripts. That is, we offer a multifaceted application of SENA as a tool that teachers and qualitative and mixed-methods researchers may easily use for the analysis of written qualitative evidence in a variety of settings and data formats and with a multiplicity of purposes. This academic orientation is not incidental. It reflects a deliberate methodological choice to reposition sentiment and emotion analysis away from its historical confinement to market research and toward the analysis of lived experience, evaluative reflection, and context-rich qualitative evidence in educational and social research settings.

Conceptual and methodological underpinnings

The main premise of SENA is that SEAs should consider both individual and group understandings, in addition to aggregate outputs. This premise also implies that the resulting outputs should be treated as guides to interpretation rather than as interpretations in themselves.

Sentiment and emotion distributions can reveal structure, intensity, divergence, and contradiction, but researchers must still determine what those patterns mean in light of the original qualitative evidence, the context in which the texts were produced, and the purposes of the study. For example, returning to analyses that may predict class performance, if analyses are left at the group level, no targeted interventions can be conducted, but if the resulting analyses show that a group of students or individual students reported worse experiences, then early interventions can be designed and implemented.

Although individual and group analyses lead to more nuanced understandings, their implementation currently requires either computer and statistical programming skills or expensive proprietary software. Unlike these features, SENA is free to use and distribute and does not require any programming proficiency to render interactive visualizations. This section is designed to detail SENA's back-end processes that are implemented in its UI.

Flow diagram for the entire SENA process

To illustrate the conceptual and analytical process followed by SENA—from database creation to software execution—we present Fig. 7.1. With the purpose of providing as much transparency and clarity of the entire process as possible, every step (or polygon) in the flow diagram presented in Fig. 7.1 can be conceptualized as follows:

1. Collect stories, essays, narratives, or any written information.
   a. SENA allows loading a collection of Word documents or a spreadsheet in which one column contains the text to be analyzed.
2. If there are attributes of interest that enable the analysis of groups (these attributes can be at the individual or temporal levels—like the time of posting in social media data, or a combination of them, as discussed below and illustrated in Figure 7.3), you can add them as follows.
   a. In the case of Word documents, the name of the document should contain a reference to the group. For example, if you have men and women and want an analysis classified in these categories, you can name the files "woman1.doc," "woman2.doc,", "man1.doc," and "man2.doc" (see Fig. 7.3) and SENA will automatically create analyses of men and women categories. For temporal attributes, you can start naming the documents by the month of data gathering, like "month_one.1.doc," "month_two.1.doc," "month_one.2.doc," and "month_two.2.doc." In addition to group analyses, individual-level outputs are also conducted using network analyses—more on this below.
   b. In the case of a spreadsheet, SENA requires a unique ID column, a text column, and optionally a group column (i.e., men and women attributes). If group information is provided, SENA produces group and individual-level analyses.
3. Once data are loaded, SENA conducts NLP and text-mining procedures.
   a. As part of these analyses, users need to decide whether lemmatization (or the normalization of words to retain their full functional form) is to be implemented. Note that, as further explained below, although lemmatization could lead to small changes in

sentiment and emotion meanings, it also allows more words to be found in the seventeen sentiment and emotion lexicons applied by SENA.

4. If attributes were added (in point 2), after clicking on "Executing SENA," the analyses will report a set of
   a. Word clouds to summarize text content (one per group—see Fig. 7.4).
   b. Sentiment and emotions distributions per group—see Fig. 7.5B.
   c. Network visualizations enable individual-level analyses (sociograms, one per group—see Fig. 7.6).
   d. QAP correlations compare all possible group combinations (i.e., dyadic analyses) to assess whether the distribution of emotions varies by group—see Fig. 7.7.
5. If no group information is added (i.e., temporal or individual attributes), the results show one aggregated depiction of the distribution of sentiments and emotions, an aggregate word cloud, and an aggregate sociogram or network visualization. In this case, since no groups are formed based on individual (or temporal) attributes, no QAP tests are conducted.

In the following pages, we expand on the details of all back end processes needed to implement the flow diagram described here, which are conducted by the SENA software.

Natural language processing and text preparation

Once the collection of texts (in *.doc format or a *.csv file) is uploaded, the NLP begins with the creation of the corpus or collection of texts. It is during this process that group identification, if it exists, is displayed. This grouping process is achieved if Word documents are named

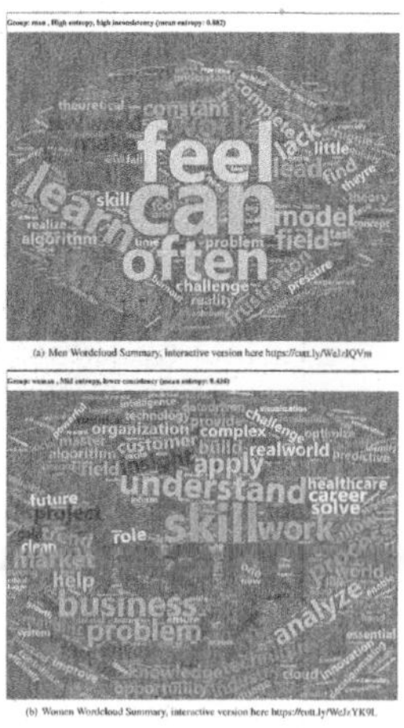

FIG. 7.4 Example of Wordcloud summary representation of students' evaluations/comments about a data science course. (a) Men's Wordcloud Summary, interactive version available here: https://cutt.ly/WeJzIQVm (b) Women's Wordcloud Summary, interactive version available here: https://cutt.ly/WeJzYK9L.

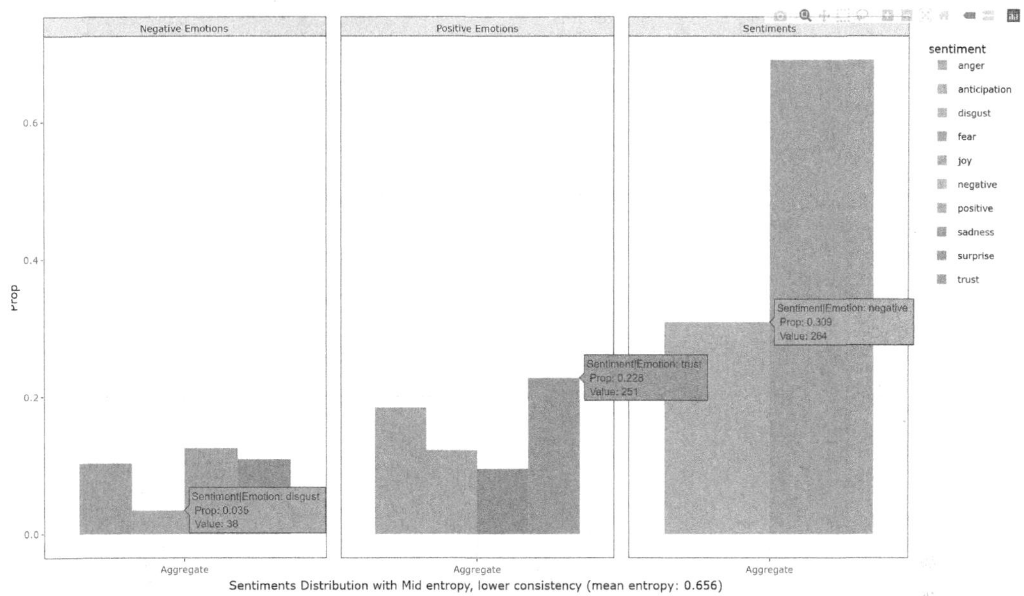

(a) Aggregate Analysis of SEA, interactive version here https://cutt.ly/FeJdsGo0

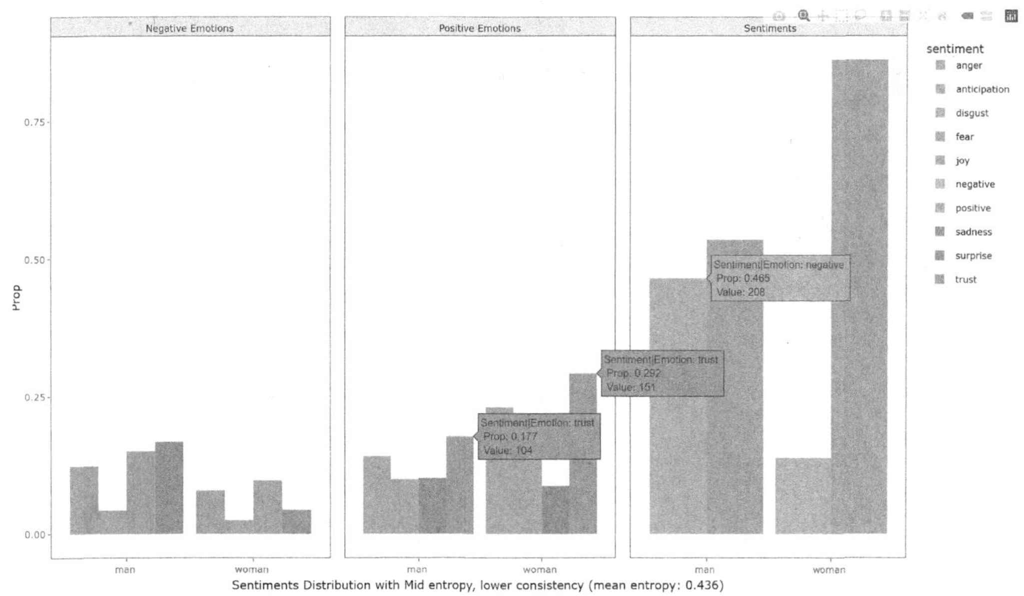

(b) Gender Disaggregated Analysis of SEA, interactive version here https://cutt.ly/leJddzfw

FIG. 7.5 Sentiment and Emotion Analyses of Students' Evaluations/Comments About a Data Science Course. (A) Aggregate Analysis of SEA, interactive version available here: https://cutt.ly/FeJdsGo0 (B) Gender-Disaggregated Analysis of SEA, interactive version available here: https://cutt.ly/leJddzfw.

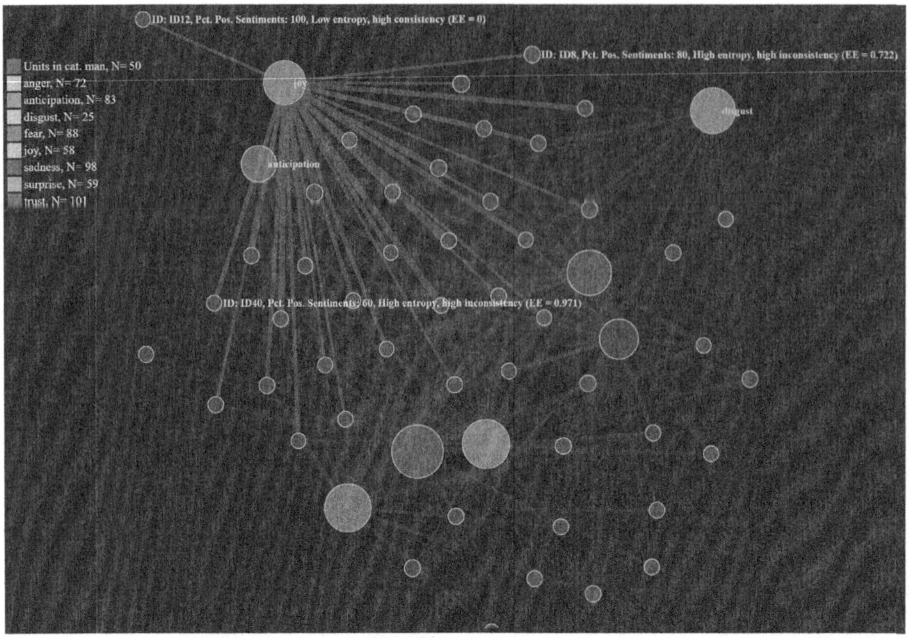

(a) Men Individual Level SENA, interactive version here https://cutt.ly/QeJfNb9t

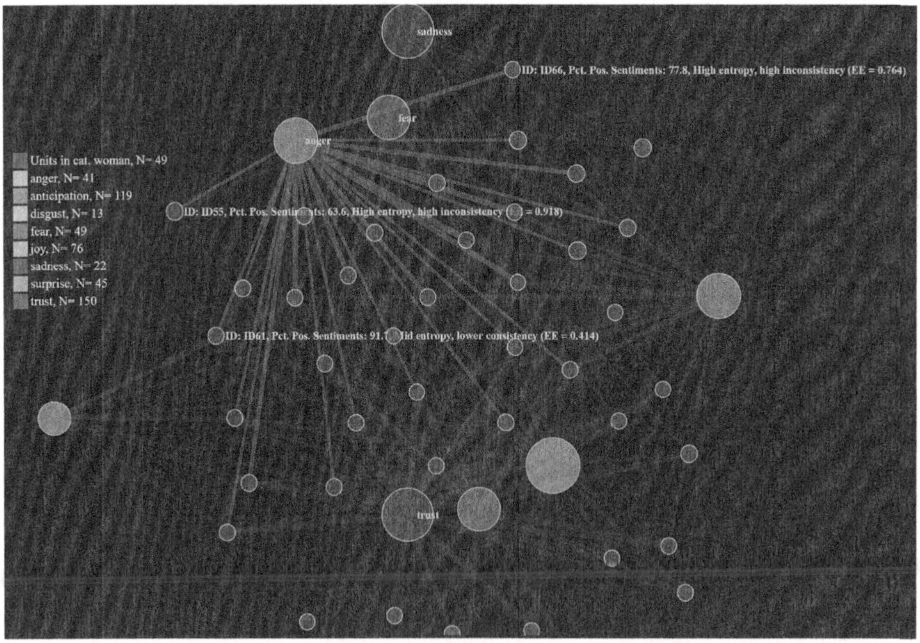

(b) Women Individual Level SENA, interactive version here https://cutt.ly/DeJfNILc

FIG. 7.6 Sentiment and Emotion Network Analyses of Students' Evaluations/Comments About a Data Science Course. (A) Men's Individual-Level SENA, interactive version available here: https://cutt.ly/QeJfNb9t (B) Women's Individual-Level SENA, interactive version available here: https://cutt.ly/DeJfNILc.

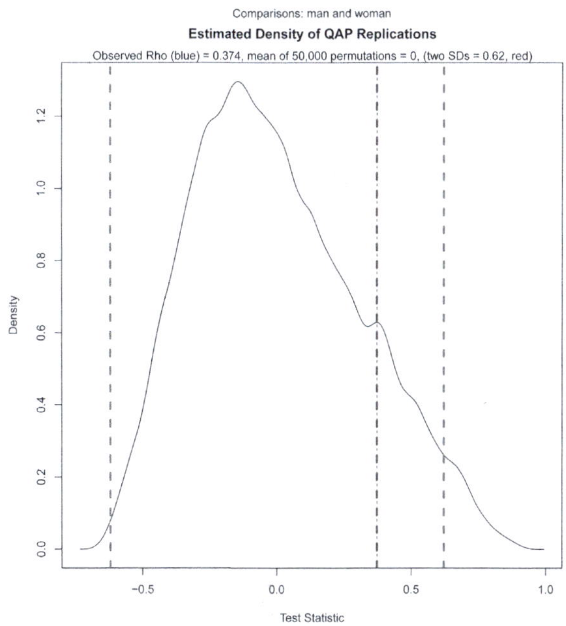

(a) Example a QAP testing whether women's and men's text had different emotions in their essays regarding a data science course

(b) Trumps classified topics about assasination attempt

FIG. 7.7 QAP testing across groups for small figure (a) and big data figure (b). (A) Example of QAP testing to determine whether women's and men's texts expressed different emotions in their essays about a data science course. (B) Trump's classified topics about assasination attempt.

based on individual attributes, or if a column in the database contains these attributes. If no attributes are added, the dialog displayed in SENA simply describes the number of Word documents (*.doc) uploaded or the number of rows with valid text read in the Excel (*.csv) file.

After this initial step is completed, standard text mining procedures are implemented

(Feinerer, 2020a; Silge and Robinson, 2017a). First, all nonstandard characters such as "&," "Ÿ," "#," and "\r\n" are removed. In addition, all characters are transformed to lower case to avoid creating duplicate words (e.g., "Learn," "learn," and "LEARN"). SENA also removes terms classified as "stop words" across the 17 languages included in SENA. Stop words, like "at," "by," "for," "with," "on," "off," "over," "under", are words that, while needed to form thoughts and sentences, add no meaning alone.

To exemplify the process of removing stop words, let us consider the following sentence: "**the** mere thought **of** losing **you** gives **me the** horrors." In this sentence, we have highlighted, using bold font, all stop words (the, of, you, me, the). Therefore, a sentence that does not contain stop words reads as: "mere thought losing gives horrors." This implies that once stop word removal is applied, the resulting vocabulary would contain information that could be classified into emotions and sentiments.

In continuing with text mining and normalization, SENA allows users to choose whether they want to implement word lemmatization, a text normalization technique typically employed in NLP (Silge and Robinson, 2017a). Lemmatization applies a morphological analysis to all words with the goal of reducing variants while preserving the full meaning of a word based on its form (i.e., pronoun, verb, noun, adjective). We selected lemmatization instead of stemming, another normalization technique, because the latter preserves just the root or base of a word, resulting in a loss of meaning and usually incomplete words that will not be found in emotion and sentiment dictionaries. For example, in the string "educational," "educated," "educate," "education," "educative," applying stemming results in "educ," "educ," "educ," "educ," "educ." Applying lemmatization to the same original string renders "educational," "educate," "educate," "education," "educative." From this example, we can infer that lemmatization is more computationally demanding in both the initial transformation and the length/complexity of the resulting vocabulary. That is, in stemming, this example renders one word (i.e., one column in the word index vocabulary length), whereas lemmatization yields four words, which translates into four columns of the resulting document-term matrix (see González Canché, 2023b; González Canché, 2023).

Like any other normalization technique, lemmatization may potentially change some of our participants' meanings, which is a particularly salient issue for SEAs. For clarity, let us build from our previous sentence example, which reads: "the mere **thought** of **losing** you gives me the **horrors**." In this case, we highlighted with bold font all the words that can be located in a dictionary of sentiments and emotions (as we describe below, this dictionary is also referred to as a lexicon of emotions and sentiments). In this case, if we decide to apply lemmatization, the resulting sentence becomes "the mere **think** of **lose** you give me the **horror.**" (Note that "gives" changed to "give", but neither gives nor give are included in the sentiment and emotions dictionary.) The main implication of these changes is that the lemmatized words altered the assumed meaning as follows:

- Thought was changed to think, leading to two issues. The first is that think is not part of a lexicon or dictionary of sentiments and emotions. The second is that thought is part of such a lexicon is associated with the emotion of anticipation. This means that lemmatization led to the loss of this emotion.
- Lose was changed to losing. In this case, losing does exist in the lexicon, and this word is associated with anger, disgust, fear, sadness, and surprise. In the case of lose, lose is also found in the lexicon, but its meaning is only associated with sadness. From this view, by lemmatizing, we added more emotions than those associated with the original text.
- Horror was changed to horrors. In this case, both forms are also located in the lexicon. Similar to the previous case, horror is associated with more emotions than horrors (i.e., anger, disgust, fear, sadness, and surprise, versus only fear in the case of horrors).

Even though we risk altering sentiments and emotions when we lemmatize, by not lemmatizing we face a greater risk of being unable to locate our participants' words in the lexicon or dictionary. For example, assume we have the following sentence: "Manuals are no panaceas." If we do not lemmatize, we would lose the words "manuals" and "panaceas" because most lexicons only contain the singular forms "manual" and "panacea," not their plural forms. On a positive note, however, our analyses of hundreds of thousands of texts suggest that the issues we discussed in our "the mere **thought** of **losing** you gives me the **horrors**" example are quite unusual. Contrary to this issue, in most cases, lemmatization is recommended to locate more words in the lexicon. With the pros and cons of lemmatization in mind, researchers should conduct sensitivity checks by running models where lemmatization is applied and models where it is not, and observe whether those decisions yield different results or whether the results remain consistent. The good news is that these checks are as easy as executing SENA twice: once with lemmatization and once without.

Having decided whether to lemmatize, we need to execute the text mining button in SENA's UI. This process will result in a summary of the top 20 most common words.

This frequency of word occurrence within texts is captured in a document-term matrix. In this matrix, the documents (word documents or cells of a column in a database) constitute the rows of the matrix, and the terms (i.e., words or vocabulary) constitute the columns. All words that are not used in a given document appear with a frequency of zero in the respective intersection of that row and column in this matrix. Word frequencies are used to calculate sparsity. Sparsity indicates that some words, while they may frequently appear in one document, may not necessarily be part of the vast majority of the remaining documents, thus not being relevant for the overall vocabulary (Silge and Robinson, 2017a). Our UI was programmed to detect texts containing very uncommon words (appearing in less than three percent of all documents) and remove them from the analyses.

Finally, following Schütze et al. (2008a) we provide researchers with the option of removing the most "extremely" common words across all texts. This is an important step because, in some instances, these extremely common words may add little value to the analyses (Schütze et al., 2008a). This potential issue may be particularly relevant when these common words are synonyms of the main research goal or a central topic in an interview protocol, for example. That is, if researchers are interested in students' opinions about a *data science* course, the transcriptions or texts may include common words such as "data," "science," "student," and "course." If researchers believe that keeping these words in the corpus would add little value for the detection of sentiments and emotions, then they should remove these words. Similarly, when dealing with social media data, we may decide to transform emojis into their official descriptions. In cases like this, "face" may appear frequently because many emojis use this word.

It is with the goal of identifying common words that SENA's UI shows the top 20 most frequent words, along with their frequencies across all documents. It is at this point that we may decide to remove *x* number of them. This optional removal step completes the text preparation and mining procedures—we illustrate in detail how to remove these words in our applications section.

Summarizing texts via word clouds

Once the collection of texts has been cleaned and normalized, it can be presented as a summary indicating word relevance. A convenient and visually appealing method to achieve this is via a word cloud (or tag cloud). A word cloud is a visual or graphical representation of text data (DePaolo and Wilkinson, 2014; Xie and Lin, 2019). In these representations, attributes of words like size, weight, and colors can be used to highlight the relevance of those words as a function of their presence in the corpora. These graphical summaries may allow researchers to get a quick or "big picture," yet accurate, representation or assessment of the patterns that may emerge from the text data (DePaolo and Wilkinson, 2014).

Although in SENA we propose the use of word clouds as a descriptive or summarizing tool, authors like (DePaolo and Wilkinson, 2014) and (Xie and Lin, 2019) have proposed their use as a big-picture guideline in assessment to gain insights into students' understanding of the material covered. Specifically, (DePaolo and Wilkinson, 2014) suggest that since word clouds can be used with any type of data, we can use them to assess understanding in short-answer responses, compare changes in pretest and posttest responses, course evaluations, and, overall, students' learning and attitudes toward the course content. In the case of Xie and Lin (2019) they recommended using word clouds to support students' knowledge integration from online inquiry as demonstrated by blog posts, tags, and concept maps. In this latter case, the authors investigated whether word clouds may be useful in students' retention and comprehension. In their study, Xie and Lin (2019) concluded that "the word cloud was more effective in orchestrating students' attention to the prominent concepts than [a] list [of concepts]" (Xie and Lin, 2019) p. 489).

Building on the visual and analytical benefits that word clouds provide, SENA offers these outputs in both aggregate and group analyses. That is, when comparisons are selected, word clouds are generated by group. As discussed in this section, this exercise provides important insights. For example, Fig. 7.4 contains the word clouds of men's (Fig. 7.4A) and women's (Fig. 7.4B) reflections or essays regarding a data science course. A closer observation of these word clouds clearly suggests that women had more positive views or experiences than men. As with all outputs, these representations are interactive, so that when placing the cursor on top of each word, we can observe its frequency in the texts. When no comparisons are requested, the word cloud will contain all texts as opposed to only including the text provided by each of the groups included in the analyses.

Sentiment and emotion lexicons or dictionaries

SENA's flexibility to analyze sentiments and emotions in 17 languages locally—without having to upload data to any server—is based on its use of sentiment and emotion dictionaries provided by the National Research Council Canada (NRC). The NRC Emotion Lexicon is a list of words and their associations with eight basic emotions (anger, fear, anticipation, trust, surprise, sadness, joy, and disgust) and two sentiments (negative and positive). As described by Mohammad and Turney (2013a), the classification of sentiments and emotions was manually performed through crowdsourcing. Similar to Wikipedia, NRC was created with the collaboration of a large number of people or coders, as a form of peer-production. Specifically, NRC relied on Amazon's Mechanical Turk, "an online crowdsourcing platform that is especially suited for well-defined tasks that can be completed over the internet through a computer or a mobile device" (Mohammad and Turney, 2013a, p. 2).

It is important to note that coders were given tasks to carefully annotate term-emotion associations based not only on individual or isolated words but also by considering their context in sentences and paragraphs. From this perspective, the training or identification of term-emotion associations using crowdsourcing is similar to the training that generative artificial intelligence (AI) follows to assess whether the text produced by ChatGPT (and similar generative AI platforms) is "human-like" enough. That is, via crowdsourcing, NRC has been able to consider context when creating the term-emotion associations. These associations included nouns, verbs, adjectives, and adverbs.

Based on crowdsourcing, the NRC lexicon was built through random checks to ensure that no erroneous or misleading annotations were created (Mohammad and Turney, 2013b). SENA's reliance on NRC is strategic, as this lexicon includes both polarity, which is a classification of terms as negative or positive, as well as an eight-emotion classification of those terms. When words or terms were associated with more than one emotion based on context, NRC respected those annotations. To ensure coder annotation quality, each coder's tasks involved two steps. In the first step, there was a correct response that a coder needed to address, and only if this response was correct was the annotation of that coder considered a good candidate for further quality assessment. In their 2013 paper, Mohammad and Turney (2013b) mention the following example.

First, they ask the coder "what word is closest in meaning to **shark**?" where "shark" is the target term. To answer this question, the team offered the following options: car, tree, fish, and olive. If the coder selected anything other than "fish," the answer is not considered. However, if the coder correctly selected "fish," they were then asked how they would associate the word in terms of positive and negative sentiments (polarity) or in one or more of the following emotions: anger, fear, joy, sadness, disgust, surprise, trust, and anticipation.

Each of the terms was annotated, classified, or coded by five people. Mohammad and Turney (2013b) reported that in 74.4% of the annotations, all five coders agreed on the emotions, and that in 16.9%, four out of five annotators agreed with one another. These levels of intercoder agreement are a notable attribute of NRC. However, an arguably more salient attribute of NRC is its availability in 16 additional languages, in addition to English. This is important because, as Mohammad et al. (2016a) mentioned, sentiment analysis has predominantly been English-centric.

From this perspective, the fact that NRC's lexicon has been applied to the Basque, Catalan, Danish, Dutch, English, Esperanto, Finnish, French, German, Irish, Italian, Latin, Portuguese, Romanian, Spanish, Swedish, and Welsh languages is truly remarkable and adds significant value to SENA's engine and its contribution to this line of research.

Minimizing aggregation bias

Aggregation bias, as applied to NLP, consists of classifying large amounts of text under a single element, even when such text may be composed of a myriad of classes or elements (González Canché, 2023). When applied to SEA, aggregation bias involves the assumption that an aggregate representation applied to all participants in the sample is sufficient to capture the accurate distribution of emotions and sentiments, whereas, in reality, significant variations in the underlying distribution of SEA may be observed at both group and individual levels.

To illustrate this issue, let us present Fig. 7.5. This figure is divided into (a) and (b), with Fig. 7.5A showing the aggregate distribution ignoring participants' attributes and Fig. 7.5B considering participants' gender. The data analyzed consist of 99 brief essays or reflections on a data science course, with 50 of these essays being provided by men and 49 by women. In the aggregate representation, we can see that most sentiments and emotions were positive; however, Fig. 7.5B shows that these positive distributions of sentiments and emotions were led by women, whereas men had more mixed reflections with close to parity in positive and negative sentiments.

We will discuss these data and findings in more detail. For now, please note that, by far, the most common application of sentiment analysis has relied on aggregated analyses of the content of document-term matrices. In some fewer instances, these analyses may have included changes over time. For an example of how time is traditionally incorporated in SEA, let us look at an output presented by Determ, one of the for-profit software tools we mentioned earlier in the "Current Applications" section. This output, published in January 2023, can be accessed at https://cutt.ly/fwlZUgdA. There, analysts at Determ showed how sentiments (classified as negative, neutral, and positive) changed from late October to late January (note that no year was specified in Determ's example).

Although a temporal analysis is useful, as it allows us to observe positive and negative peaks over this time frame, an even more nuanced understanding of these patterns may be attained by incorporating electors' individual attributes, such as those shown in Fig. 7.5B. That is, it may be useful to know if the negative peaks observed on December fifth in that figure were driven by electors from specific demographics—women voters, for example.

From this perspective, we argue that with a more nuanced level of analysis, better decisions or plans of action may be implemented. For example, if spikes in negative sentiments were driven by women, a campaign could be designed to address their concerns, or, with this information, we might decide to gather more data (e.g., focus groups or interviews) to better understand the concerns of this group. Accordingly, although disaggregated and group SEA analyses have remained underused, they may deepen our understanding not only in marketing settings (Kennedy and Inkpen, 2006; Taboada, 2016; Turney, 2002) but also in academic and education research (Calvo and D'Mello, 2010; Luo et al., 2015a; Yu et al., 2018b, 2016b; Dolianiti et al., 2018). These analyses, however, while important, do require specialized statistical and computer programming knowledge that has prevented their use among qualitative and mixed-methods researchers. Accordingly, a notable contribution of our proposed analytic framework and software is that they were designed to purposefully account for participants' attributes when conducting SEA procedures without requiring any programming experience.

Attributes in documents using meta data

Each of the documents uploaded to SENA will be analyzed to detect the distribution of sentiments and emotions in its content. If we decide to add attributes at the document name level, SENA will automatically detect these attributes and conduct analyses to test whether the distribution of emotions varies across each of the groups. Following the example shown in Fig. 7.5, let us assume we have two groups: one contains the texts of women, and the second contains the texts of men. SENA will compare the distribution of their emotions via QAPs (more on this below). If we had another group indicating "other," for example, then the distributions would be compared as women versus men, women versus other, and men versus other, so that we complete all possible pairs of comparisons, which is also referred to as dyadic comparisons.

Dyadic comparisons consist of comparing each of the resulting groups, two at the time. That is, if we have two groups of interest, there will be one statistical QAP test correlation to be conducted (i.e., group 1 vs group 2); with three groups, there will be three QAP tests (i.e., group 1 versus group 2, group 1 versus group 3, and group 2 versus group 3); with four groups there will be six QAP tests, and so on. In other words, the SENA software has been programmed to apply a combination equation where the statistical tests will be comparing two groups at a time (i.e., dyadic comparisons). This combination equation take the following form, $C(n,2)=\frac{n!}{(2!(n-2)!)}$, where 2 captures dyadic comparisons, and n is the total number of groups included in our analytic sample (i.e., the number of text groups included in our corpora).

Panel A in Fig. 7.3 shows the rationale followed to make these comparisons in SENA. In this figure, we only have two categories: men and women. To achieve the dyadic analyses, we only need to name the documents based on the attributes of interest and add a number at the end of the file name to differentiate these names. However, if we do not want to conduct comparative analyses, each document name may be different or may only contain one category, as we further illustrate below.

In Panel A in Fig. 7.3, we are showing that each document could have had a different name during data collection. For example, we present a file called "Case 1 in site 2," which is the original file name, but if we intended to add categories, we could add a gender attribute by renaming it to "man1." The second file, called "Interview 3, site 3," should then be named "man2" to account for the category "man." In the case of women, we also present the original and modified names. The naming based on attributes is case-sensitive. That is, if we have "woman3" and "Woman4," SENA will consider the latter as another attribute, thus resulting in a different category for analysis. We strongly recommend keeping track of these name changes. Note that in Panel C in Fig. 7.3, we also mention that we could add more than one attribute. For example, if we aimed to add ethnicity to this example, in addition to gender, we could then name files: "hispanic_man1," "black_woman1," "black_woman2," "hispanic_man2," for example, and SENA will conduct all corresponding group analyses.

In the datasets we are providing for replication, document names that do not include attributes are all called ID*n*, where *n* simply indicates that it is a different file or case. In this case, SENA will assess how many groups are present in the analyses, and the resulting outputs will simply be called "ID." Finally, if each document has completely different names, like the case shown in Panel A in Fig. 7.3, the outputs will be called "Aggregate" rather than ID, but the results will be exactly the same as if only one group (i.e., ID) was included in the analyses.

From Fig. 7.3, we can also see that individual-level analyses of distributions of sentiments and emotions are possible as part of this analytic process. That is, one of the outputs presented by SENA, which involves the use of network visualizations, resembles Panel A in Fig. 7.3. In this output, the strength of participant-emotion relationships is also captured. These results are particularly relevant when analyses involve participants' class evaluations, as described in our literature review (Calvo and D'Mello, 2010; Luo et al., 2015a; Yu et al., 2018b, 2016b; Dolianiti et al., 2018). Based on the potential usefulness of SENA for this education-based task, the datasets we are providing include students' perspectives, in the form of short essays, regarding a data science course.

Attributes as a column in a database

Panel B in Fig. 7.3 represents how attributes of participants may also be included in databases. This representation is more conventional for texts in these data formats, which are stored in cells (i.e., the intersection of a row and a column). In this format, rows may represent different individuals or the same individuals with different text contributions—over time, for example. In Panel B, we show, for example, that Man in the first row provided Text 1 and illustrate that, as part of the analyses, SENA identified "Anticipation" as one of the emotions present in that text. This is important because our databases do not include emotions; instead, it is the goal of SENA to apply the lexicons to the text columns and identify those emotions and sentiments.

Following the example shown in Panel A in Fig. 7.3, the attribute of interest in our example shown in Panel B is gender, wherein, for consistency (and simplicity) purposes, it only includes two categories. Note that this decision is entirely up to the research team, as long as those categories are consistently written (i.e., these categories are case-sensitive as well, so if we write "Woman and woman", these would be two distinct groups rather than only one). SENA will identify groups as indicated in Panel C in the same figure.

The datasets we are providing cover these three case scenarios described here (i.e., Word documents with and without categories, and a dataset). In the first case, we provide access to one corpus configured by 99 Word documents with gender categories—available here: https://cutt.ly/aeJfFbZt. The second case is a corpus that also contains the same 99 Word documents but without these categories—available here: https://cutt.ly/leJfF4RW. Finally, the third corpus is stored in an Excel spreadsheet with three columns and 99 rows—each of these rows contains one of the essays stored in their Microsoft Word format versions. In this spreadsheet, the first column is the ID, which matches the order of the Word documents without categories. The second column contains the essays, and the third column contains the gender attributes. For clarity, note that this database, available here: https://cutt.ly/zeJfHr0M, completely matches the Word document essays. As explained in our application section, if SENA users want to account for attributes, they simply need to select the gender column, and SENA will conduct group analyses.

Emotional entropy or mixed messages

Although the identification of sentiments and emotions is reliable based on the lexicon used, there remains the possibility that some texts contain mixed or even contradictory messages. Accordingly, to offer a sense of how prevalent these mixed messaging practices are, all SENA outputs (i.e., word cloud summaries—see Fig. 7.4—, SEA histograms—see Fig. 7.5—, and network analyses—see Fig. 7.6) include a measure of emotional entropy or contradictory messaging, as described next.

Emotional entropy "can be thought of as a measure of unpredictability and surprise based on the consistency or inconsistency of the emotional language contained in a given text. [*In other words, a*] text with conflicting emotional language may be said to express or contain a mixed message" (Jockers, 2023a, para. 37). For example, if a text reads, "I hate loving you," this would be a mixed message with high entropy. As explained next, SENA normalized this value to range between 0 and 1, with the latter representing mixed or confusing messages and zero indicating that the content of the text was configured by emotions that were all positive or all negative. For example, "I love loving you" and "I hate hurting you" both have emotional entropy values of zero.

Procedurally, emotional entropy is measured as the sum of positive and negative words included in a sentence or text. For example, let us assume we have the following text: "I love hating you, and I would be happy if you die." This sentence has two positive words ("love" and "happy") and two negative words ("hating" and "die"). In texts like this example, ties in the number of positive and negative words lead to high emotional entropy values. However, as the number of words increases and the evenness in the distribution of positive and negative sentiment words decreases, entropy distributions become less extreme. For example, if instead of 4 words we have 11 and the sentiment distribution is 3 negative and 8 positive, the entropy value would be 0.84. But with the same 11 sentiments, if only one is negative, the entropy is 0.43. When no negative or positive sentiments are mixed with each other, the entropy is zero. These values are obtained from the following two equations

$$frequencies = \frac{sentiment\iota}{\Sigma sentiment} \tag{1}$$

where sentiment is a row vector with two elements, the first counts the number of negative words in the text, and the second counts the number of positive words in the same text (i.e., a word document or a cell in a database). The denominator in this equation is the sum of sentiments observed in that same text. The result of this equation is the proportion of negative and positive sentiments with respect to the total detected. With this information, we apply the following equation:

$$entropy = -\Sigma(frequencies * log2(frequencies)) \tag{2}$$

Following Jockers (2015a), entropy is standardized to range between 0 and 1 using a log base 2, or binary logarithm log2 transformation. This transformation is useful because values of 0 mean no change, while negative values indicate reductions and positive values indicate increases. The negative symbol ensures that entropy is always positive. In addition, it works well with both small and large databases. In the emotional entropy value implemented in SENA, every word in the text is counted in the final "sentiment" vector, even in the presence of repetitions. That is, if one person used the word "love" three times, this count is represented in "sentiment," as opposed to counting it only once, as implemented by Jockers (2015a). This change is relevant to our goal of preserving the accuracy of sentiment representation as observed in our participants' words.

SENA computes entropy values at the text level so that each participant's text is evaluated following equations (1) and (2). To display individual or text-level results, SENA relies on its network interactive visualization. As shown in Fig. 7.6, when we click on an individual (i.e., blue node), a message box displays the estimated emotional entropy of the text that the individual provided. This same dialog box displays the proportion of words per participant that were associated with positive sentiments as another tool to understand the mechanisms behind each participant's text contributions.

In addition, SENA also offers the average emotional entropy by group when presenting the distribution of sentiments and emotions, as well as the summaries of the most frequently used words. Specifically, each title of Fig. 7.4 contains the average entropy of the groups represented in each word cloud. Finally, the average entropy values across all texts, irrespective of groups, are shown on the X-axis of the sentiment and emotion histograms, as illustrated in Fig. 7.5.

For convenience, each of these values, at both individual and group levels, indicates whether the entropy is considered low, mid, or high. Since the entropy ranges between 0 and 1, we relied on tertiles to capture these values. Specifically, the messages displayed are: "High entropy, high inconsistency" with values greater than 0.66, "Mid entropy, lower consistency" with values over 0.33 and below 0.67, and "Low entropy, high consistency" with values below 0.34.

Pragmatically, we offer emotional entropy estimates with the goal of identifying texts that deserve more attention. For that reason, emotional entropy should be understood as an invitation to deeper qualitative analysis rather than as a self-sufficient indicator: high-entropy texts signal the need for interpretive return to the original language, context, and possible complexity of participants' meaning-making. For example, if NLP misses sarcasm or the content is more complex by nature, involving a variety of sentiments, such texts should be more carefully analyzed. Accordingly, by identifying these texts, we can qualitatively assess whether the sentiment and emotion analyses were able to capture that complexity or sarcasm, or whether we should be concerned with the resulting classification. In addition to this "manual qualitative analysis" texts with high levels of entropy could be analyzed with intelligent systems, like ISARI, as we discuss in chapter 9. This AI fueled analysis, is not a replacement for researchers' qualiatitve understandings, instead it should be seen as complimentary during the scholarly process.

Network modeling and data transformations

As discussed above, SENA relies on network analysis methods or relational thinking to identify each unit's connection strength with the sentiments and emotions embedded in their textual contributions. These relationships are conceptually shown in Fig. 7.3 and are operationalized in Fig. 7.6. Procedurally, recall that each text is analyzed to detect sentiments and emotions, and this process may be represented as a matrix, wherein the rows contain units (i.e., text providers) and the columns represent the sentiments and emotions identified across those texts. Returning to our working example, since we have 99 essays, after the sentiment and emotion identification, we would have a matrix of 99 rows and 10 columns—with eight of these columns accounting for the emotions discussed above and two columns containing positive or negative sentiments (or polarity).

In SEA, the number of columns will always be 10. This is based on our NRC lexicon, which contains eight emotions and two sentiments. From this perspective, if our total number of texts to be analyzed is 99 and we ignore participants' attributes, the resulting relational matrix will have 99 rows and 10 columns as its dimensions. From now on, we will represent these dimensions as [99,10]. Nonetheless, when we construct these matrices considering participants' attributes, the row dimensions will account for the number of elements in each group. That is, since we have 50 essays written by men and 49 essays written by women, the resulting dimensions for each group are [50,10] and [49,10] for men and women, respectively. In sum, regardless of whether we consider these attributes, all the matrices handled by SENA are configured by two different elements (i.e., texts in the rows and emotions in the columns of the resulting matrices).

In network modeling, these matrices are referred to as bipartite or two-mode matrices (Breiger, 1974; González Canché, 2023; Kolaczyk and Csárdi, 2014) because they allow us to capture relationships taking place among two types of elements—i.e., the sentiments and emotions embedded in those texts. With these matrices, we can add the relational or network attribute to SEA, making it SENA. This is how SENA enables us to build sociograms or network representations like those shown in Fig. 7.6. In these networks, the intersection of a row and a column captures the strength of that relationship with true values. That is, if row 88 and column 8 have a value of 24, this means that for this individual's speech or text, the emotion contained in column 8 appeared 24 times in her essay or any type of written information. However, if this value is zero, it means that such an emotion was not present. The resulting interactive visualizations (see https://cutt.ly/QeJfNb9t and https://cutt.ly/DeJfNILc for Fig. 7.6A and B, respectively) contain this information so that when we place the cursor on top of a connecting line or link, the message box displays the intensity or the number of times that individual mentioned words classified as that emotion. In this network representation, the absence of a line indicates that such an individual did not mention words classified under that particular emotion.

As described earlier, the matrices obtained from the NRC lexicon contain both sentiment and emotions. Note further that sentiments reflect polarity with negative and positive categories. That is, in the NRC lexicon, sentiments are not measured on the same scale as emotions (Jockers, 2015a, 2023; Mohammad and Turney, 2013b). Based on this discrepancy in measurement, our network representations, along with the QAPs discussed in the following section, do not include sentiments but focus on emotion distributions. Having noted this, we are not discarding the distribution of sentiments at the individual level. Instead, we are displaying positive sentiment prevalence in the network visualizations as an attribute of the participants. Specifically, we show an estimate of the proportion of sentiments per actor that were classified as positive—which, as discussed in Equation (1), is one of the steps followed to compute the emotional entropy estimates. For example, in Fig. 7.6A, we can see that a dialog box shows that for ID8, 80% of his words located in the sentiments lexicon were classified as positive. In Fig. 7.6B, for participant ID61, 91.7% of her words located in the NRC lexicon were also positive. In addition to this individual-level depiction, we also include sentiment distribution in the histograms of sentiments and emotions, as shown in Fig. 7.5.

In terms of data transformations, the main implications of focusing on emotions for the visualization and the hypothesis testing via QAPs are that the two-mode matrices will be $[n,8]$, where n represents the number of texts and 8 is the number of emotions. Moreover, also note that the QAP analyses require matrices that are square, meaning they contain the same elements in the rows and the columns. As explained in the QAP section discussed below, we needed to project or transform these two-mode matrices to their one-mode form, retaining the information contained in the emotions (or the columns) without altering the underlying structure of these two-mode matrices. We achieved these transformations following (Breiger, 1974) projections, as we detail in the QAP section.

Before moving to the next section, let us note that our network visualizations weight the size of the emotions via betweenness centrality. Centrality in network modeling serves to highlight the relevance of the actors or units in a network based on their role (Borgatti, 2006; González Canché, 2023; Kolaczyk and Csárdi, 2014). Our selection of betweenness centrality is based on our interest in highlighting the emotions that connected more actors in the network or served as bridges. In this centrality, a unit is relevant to the extent that such a unit falls in between two other units (Borgatti, 2006; González Canché, 2019; Kolaczyk and Csárdi, 2014). That is, emotions that are connecting or bringing together actors or texts have a larger size. In the case of men, this emotion was fear (in green), and for women, it was trust (in purple) (see Fig. 7.6A and B, respectively). From this view, our emphasis on these visualizations is to highlight emotions' relevance while at the same time identifying clusters of cases that are more similar by being connected to the same emotions. That is, units that are located close to each other (blue dots in the figures) tended to mention the same emotions and with a similar intensity as units that are located further away, as we further discuss in our findings section.

Hypothesis testing via quadratic assignment procedures

When group information is added to texts, SENA automatically tests the hypothesis that some groups' emotions are more similar than others as a function of the participants' attributes. For example, in Panel 3 of Fig. 7.3, women seemed to have more positive emotions than their male counterparts, being ascribed more frequently to surprise and joy, whereas men's texts were the only ones associated with fear. Although qualitative descriptions like this are important, such descriptions do not allow us to test whether these results are more robust than chance alone.

To better understand whether the SENA outputs we observed may present differences among groups, SENA relies on QAPs, as discussed next.

QAP is a nonparametric analytic procedure that does not rely on normality assumptions and does not assume independence (Krackhardt, 1988). Indeed, QAP "builds into the test statistic the kind of row/column interdependence that is assumed in network data" (Krackhardt, 1988, p. 363). Accordingly, given that QAP enables testing for statistical significance using network data in matrix form (Whitbred, 2011a), it is particularly well-suited to help us test "the null hypothesis that two network variables are uncorrelated" (Krackhardt, 1988, p. 362).

As mentioned in the "Network Modeling and Data Transformations" subsection above, this matrix form is obtained by identifying the emotions associated with each of the texts configuring the corpus. We further elaborate on network projections via matrix algebra below. For now, let us mention that QAP takes the form of correlation and regression analyses. In correlation analyses, one can measure the extent to which two networks are statistically associated without any assumption of directionality in this relationship. In regression models, directionality exists, and an outcome matrix is explained by a set of network attributes, also measured in matrix form, to account for all dyadic relationships (Krackhardt, 1988). In both cases, i.e., correlation and regression, QAP translates into measuring whether every corresponding dyad or pair of connections in two (or more) matrices tends to vary in the same direction (positive correlation or association), the opposite direction (negative correlation or association), or whether they are independent.

Procedurally speaking, QAP can be classified under the random permutation test family, which is sometimes referred to as a randomization test (Phipson and Smyth, 2010). The random permutation tests begin by recording a given statistic of interest across two mutually exclusive and exhaustive groups. As discussed above, these statistics can include mean differences, correlation coefficients, and regression coefficients, for example. These observed estimates are recorded, and then the actual values across these groups are shuffled randomly X number of times. Each time this random shuffle or permutation occurs, the statistics obtained from this process are recorded. This procedure enables the recreation of a sampling distribution wherein the coefficients obtained through a random process should have a mean of zero, given the lack of inherent structure upon which the data-generating process rendering these coefficients was founded. Once this sampling distribution is obtained, analysts can then compare the number of times the randomly generated estimates were greater than the estimates obtained with the actual observed data.

In using a correlation coefficient, for example, we could estimate the proportion of times that the randomly generated correlation coefficients between two networks were larger than the correlation obtained with the observed data. Once more, the fundamental rationale is that randomly obtained coefficients should be distributed around zero, on average. Hence, if this randomly generated coefficient exceeds the actual observed coefficient, one would conclude that our observed correlation was not better than what one would expect to see by chance alone. If the proportion of times that randomly generated correlation coefficients are greater in magnitude than our coefficient obtained with actual data approximates zero, then one can conclude that our observed correlations are better than chance by 1 minus the proportion of times randomly generated coefficients were greater. Specifically, assume that 0.005 is the proportion of times out of 50,000 permutations that the random coefficients surpassed the magnitude of our observed coefficients. Then we can conclude that 99.5% of the time, we would expect our observed results to hold true if we had access to other networks of the same size and originating from the same population of interest.

The QAP correlation begins by estimating a Pearson correlation coefficient across the corresponding cells of two matrices (Phipson and Smyth, 2010). This requires that the units in the networks are the same and follow the same order across matrices. For example, the following representation has three units. Note that these matrices contain the same actors, and they follow the same order (A1, A2, and A3). The process takes the dyadic correlation among A*i* and A*j* in matrices 1 and 2 to assess whether a given dyadic intersection in one matrix, classified under trust, for example, as shown in Fig. 7.3, is realized with a similar magnitude in the other network. In our example, these networks are the projection of emotions obtained from men and women, following Fig. 7.3 as our working example. The second step is depicted in the shuffled section of the matrix below, where all the rows and cells of that matrix are randomly permuted (Krackhardt, 1988). Each time this permutation takes place, its corresponding Pearson coefficient is recorded.

	Observed Matrix men			Observed Matrix women		
	A1	**A2**	**A1**	**A1**	**A2**	**A1**
A1	2	1	5	4	2	5
A2	1	4	0	0	5	0
A3	1	0	6	2	0	7

	Observed Matrix men			Shuffled Matrix women	
A1	2	1	5?	?	?
A2	1	4	0?	?	?
A3	1	0	6?	?	?

Matrix projections

The matrices built from qualitative data analyzed by SENA are the one-mode representation of the actors and their ascription to emotions, as represented in Fig. 7.3. This one-mode transformation is achieved by network projections in the form $A^{T} * A_{binarized}$ (Breiger, 1974). As described above, the dimensions of the SENA matrices are rows accounting for the number of participants or texts (i.e., 50 and 49 men and women, respectively), and the columns are the emotions to which their texts were classified.

In the case of men, the matrix is [50,8], and in the case of women, the matrix is [49,8]. Following (Breiger, 1974), when the goal is to preserve the information contained in the columns, which renders an emotions matrix, we can transpose the matrix and multiply it by the original (but binarized) matrix—more on this below. Assuming we call this matrix A, we have that

$$A^{T} * A_{binarized} \tag{3}$$

This will yield a one-mode projected network or matrix of emotions. Note that the dimensions of this multiplication, in the case of men, are [8,50] $*$ [50,8], which, following matrix algebra, results in a [8,8] matrix. In the case of women, the dimensions are [8,49] $*$ [49,8], also resulting in a [8,8] matrix. These matrices have the attribute that the order of the row and column information will be the same across men's and women's projected matrices. Moreover, these projected matrices are square and contain the eight emotions described above.

The element A binarized in equation (3) indicates that the matrix will only contain 0s and 1 s when a participant is linked to an emotion. This binarization avoids double-counting relationships when transforming weighted two-mode matrices (González Canché, 2023).

When the presence of attributes is included in the texts uploaded to the UI, SENA automatically produces a PDF containing all dyadic comparisons. For example, Fig. 7.7 shows the results of comparing the distribution of emotions contained in the responses provided by women and men at the end-of-course evaluation of a data science course. The output produced by SENA shows the comparisons conducted; in this case, it reads "Comparisons: man and woman." Then, it shows the statistics of interest. Specifically, SENA prints the observed correlation coefficient, Rho, as a blue dotted line and a density distribution based on 50,000 permutations. The value of zero represents the average correlation across these permutations. The red lines indicate two standard deviations from the mean obtained from 50,000 permutations. Notably, when the blue line falls between the red lines, this indicates that the distribution of emotions across these groups is different—that is, the matrices are not statistically correlated. Conversely, when the blue line falls outside the red lines, the conclusion is that both groups have emotions that are more similar than chance alone, meaning there are no significant discrepancies in these distributions (Fig. 7.8).

SENA's user interface

SENA's UI is available for Mac (access here: https://cutt.ly/QwhYruBr) and Windows (access here: https://cutt.ly/YwhJJKvO) operating systems. As shown in Fig. 7.9, this UI has a maximum of seven steps. The eighth step is optional in case users need to reset the application or change the data format.

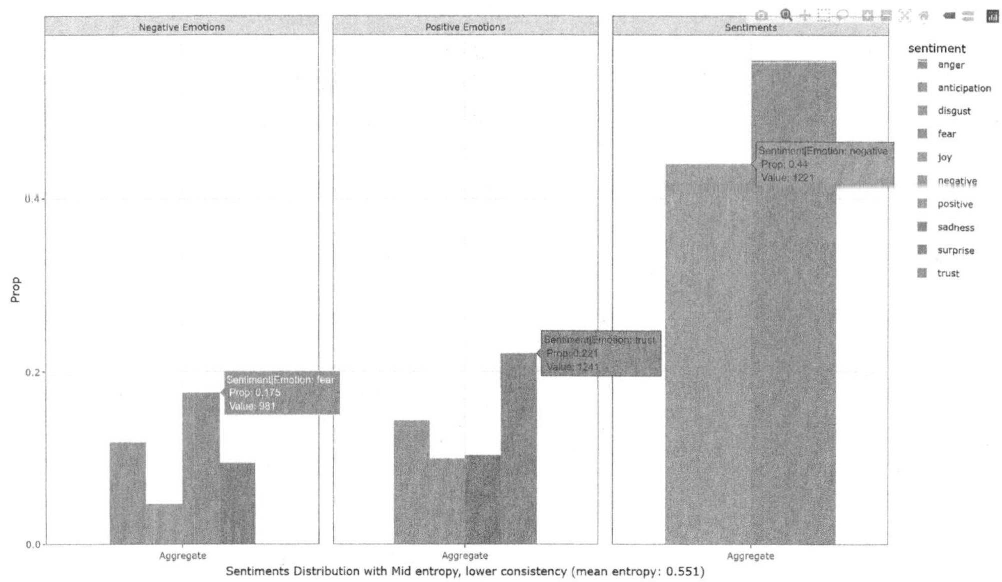

(a) Aggregate Analysis of SEA, interactive version here https://cutt.ly/lrx53Pej

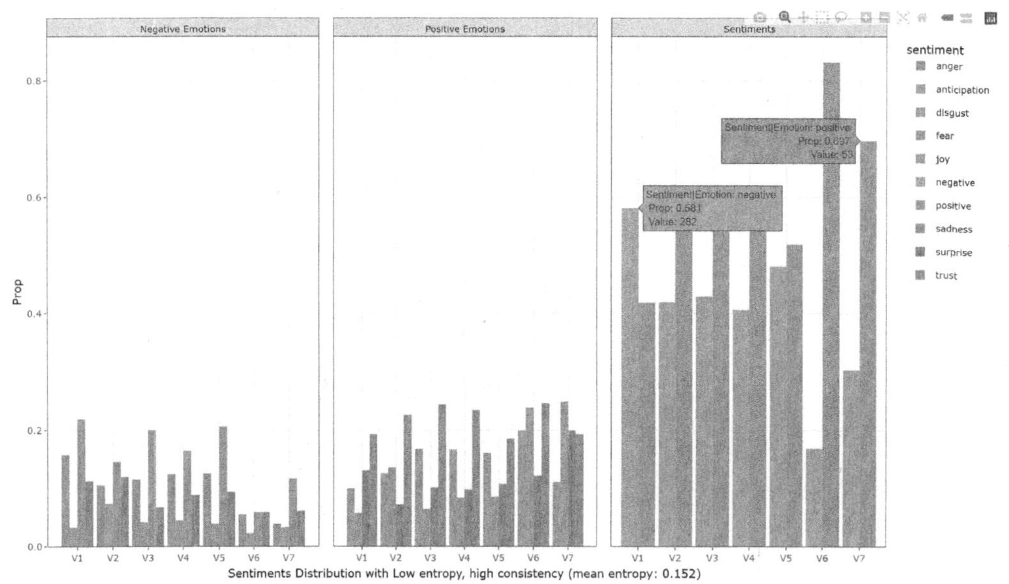

(b) Topic Disaggregated Analysis of SEA, interactive version here https://cutt.ly/Mrx51Osc

FIG. 7.8 Sentiment and Emotion Analyses of Classified Comments About Trump's Assassination Attempt. (A) Aggregate Analysis of SEA; interactive version available here: https://cutt.ly/FeJdsGo0. (B) Gender-disaggregated Analysis of SEA; interactive version available here: https://cutt.ly/leJddzfw.

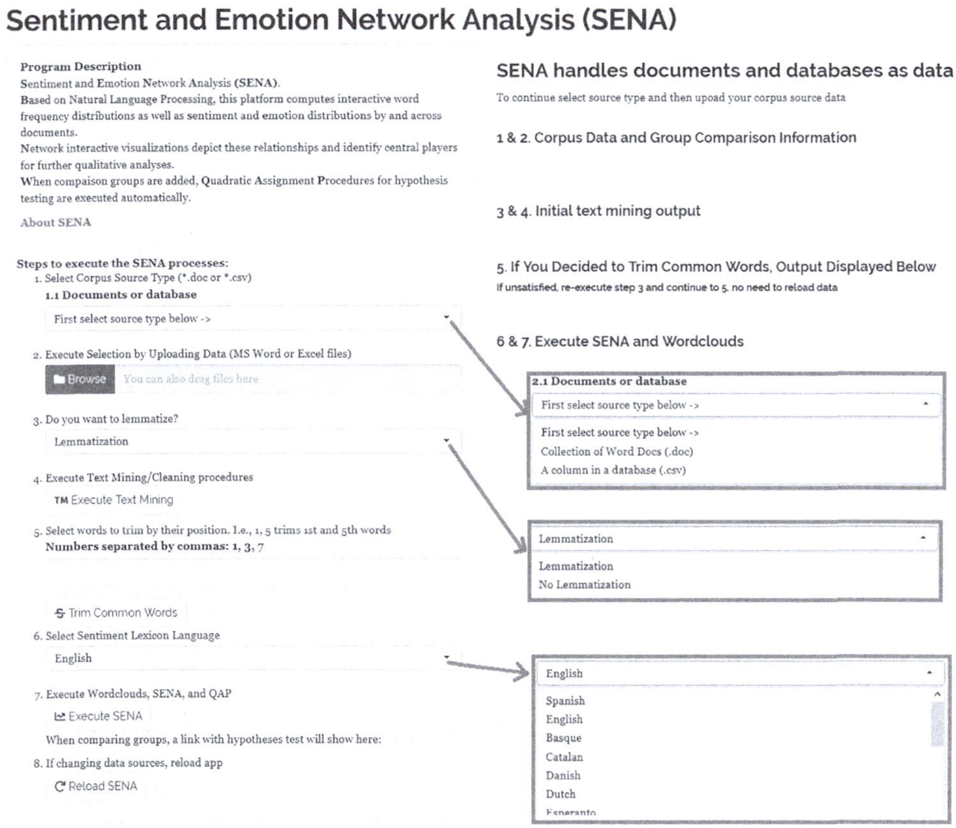

FIG. 7.9 SENA's User Interface. Available in Mac (access here:https://cutt.ly/QwhYruBr) and Windows (access here: https://cutt.ly/YwhJJKvO) operating systems.

First step: Data format selection

In the first step, users need to select Word documents (*.doc) or a comma-separated value (*.csv) spreadsheet. After this selection, step two will ask users to locate their databases according to the selection made in step one. If the Word documents option was selected, users can go to the folder where those documents are stored and select all. SENA is programmed to detect only *.doc documents. That is, if the folder selected has multiple document formats and we select all, only *.doc documents will be uploaded to SENA. Note that multiple *.doc files may be uploaded at once, but all these files need to be stored in the same containing folder. If attributes in the names are included, the results will mirror Panel B in Fig. 7.10. If not, they will resemble Panel A in the same figure.

If the option *.csv file was selected, users also need to locate this file. In this case, only one file can be uploaded at a time. Note that, as shown in Panel C in Fig. 7.10, once a dataset is loaded, the resulting database is displayed in the UI. After this uploading process is completed, users need to select the columns of interest. After all data are loaded, and in the case of the

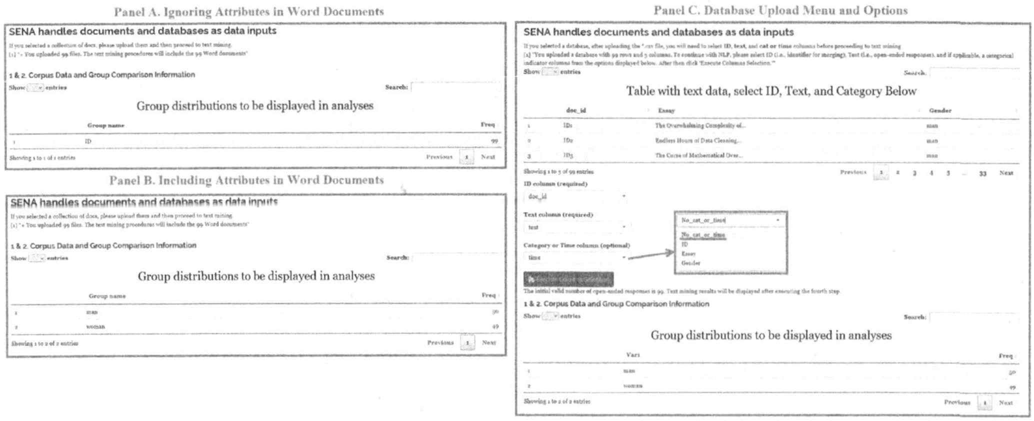

FIG. 7.10 Options and results of those options in SENA.

*.csv files, columns are selected, SENA displays the comparison groups or, if no attributes were selected, the aggregate frequencies. Due to differences in Word document and spreadsheet files, the following subsection elaborates further on these steps, paying close attention to the inclusion of attributes.

Second step: Aggregate or disaggregate analysis decisions

Fig. 7.10 contains three panels that illustrate how different analytic decisions will impact whether the outputs will present aggregated or disaggregated outputs. Specifically, panel A shows the output displayed when uploading documents without attributes. As can be seen in this panel, the results display the number of files uploaded, and then a table of group distributions indicates that all analyses will be aggregated under the Category ID. As briefly noted above, if every file had a different name, instead of displaying "ID," SENA would have displayed "Aggregate" in the table of group distributions.

Panel B in Fig. 7.10 shows the same result, but in cases where the names of the Word documents included gender information. Here, we can interpret this table as indicating that the analyses will contain two groups, and the output includes the frequency of each group.

Finally, Panel C in the same Fig. 7.10 shows the results of loading a database. When a database is loaded, SENA displays the content of this table. Moreover, in this case, users need to select whether they want to include a category for comparison or not. By default, SENA selects the option "No_cat_or_time." This option implies that SENA analyses can also involve temporal elements across participants or may include a combination of attributes across time in the form: "woman_time_1," "woman_time_2", for example. If the option "No_cat_or_time" is retained, then the output group distribution table will mirror the word output, including only the "Aggregate" categorization. In the case of the essay analyses, we selected the column "Gender," which translated into the "man" and "woman" categories displayed in Panel C. In the case of Trump's comments (more on this dataset below), we selected the option "code," which resulted in analyzing texts classified in one of the seven classes learned via machine-driven text classification—as we elaborate further in the datasets section below. Finally, note that to formalize the column selection, SENA requests users to "Execute Columns Selection." After this execution, a table indicating the number of texts per class or category (i.e., gender and topics in our examples) is displayed.

Third, fourth, and fifth steps: NLP

In the third step, users are asked whether they want to lemmatize (default option) or continue without lemmatization. Once this selection is made, we can then proceed to conduct text mining or NLP. As part of this text preparation, and as discussed in the "natural language processing and text preparation" subsection, we have included an optional fifth step. In this step, users may decide to remove words that are too common and therefore add little nuance to our analyses, as we elaborate further next (Fig. 7.2).

In our essay data example, we decided to remove the first four words of the resulting vocabulary. This output is shown in Panel A. There, we can see that these words were "datum" (freq = 287), "science" (freq = 176), "student" (freq = 155), and "course" (freq = 94). In our case, we deleted these words because they essentially resembled the prompt we used to ask students to write their essays—this prompt is shown in the "Short Essays Examples" subsection below. To remove these words in SENA, we just need to type their positions in the fifth step. Specifically, since these words were the first four most frequent words in our vocabulary, to exclude them we typed "1, 2, 3, 4" without quotation marks. Let us assume that, in other cases, the 12th word would also need to be removed. In this case, we achieve this by typing "1, 2, 3, 4, 12" also without quotation marks. After typing these numbers, we click on "Trim Common Words," and the result is shown in Panel B, also in Fig. 7.11.

Sixth and seventh steps: Lexicon and SENA execution

In the sixth step, we need to select the lexicon language. As stated above, SENA currently includes lexicons in 17 languages. Following this multilingual capability, SENA is programmed to remove stopwords in all these 17 languages. However, the SEA analyses per se need to be conducted for one language at a time.

Once this language selection is completed, we can execute SENA as the seventh step. After the "Execute SENA" button is clicked, the word clouds, network visualizations, and sentiment emotion analyses will be displayed as HTML pages. In addition, after the processes conclude, a message is displayed below the sixth and seventh steps on SENA's main panel. An example of this message is shown in Panel C of Fig. 7.11. Specifically, SENA displays that since we included two groups (men and women, as shown in our applications section below), we have two word clouds, two network visualizations, and one interactive plot including the SEA analyses. Finally, when group information is included, SENA also renders a PDF with all applicable dyadic QAP correlations. To open and save this file, you can find the link below the seventh step at the bottom of the left panel.

A. Full Vocabulary

```
Weighting            : term frequency (tf)
        words freq
1       datum  287
2     science  176
3     student  155
4      course   94
5         can   79
6       learn   77
7        feel   49
8        many   42
9       often   39
10       work   37
11      skill   36
12       like   33
13       make   30
14      model   26
15  realworld   25
16      field   25
17 understand   24
18    problem   21
19       tool   20
20      leave   18
```

B. Trimmed Vocabulary

```
Weighting            : term frequency (tf)
        words freq
1         can   79
2       learn   77
3        feel   49
4        many   42
5       often   39
6        work   37
7       skill   36
8        like   33
9        make   30
10      model   26
11      field   25
12  realworld   25
13 understand   24
14    problem   21
15       tool   20
16   analysis   18
17      leave   18
18    process   18
19  algorithm   17
20      build   17
```

C. Summary of Outputs Produced by SENA

6 & 7. Execute SENA and Wordclouds

```
[1] "Based on the comparison groups in your data, you should see"
[1] "2 wordclouds and 2 SENA interactive outputs. You will also obtain one"
[1] "interactive output of sentiment comparison and one PDF with all"
[1] "dyadic QUAP correlations across all groups."
```

FIG. 7.11 Examples of SENA outputs displayed in its user interface.

These outputs are also portable across analytic stages. The interactive word clouds, sentiment and emotion distributions, individual-level sociograms, and downloadable QAP comparisons are useful not only for identifying patterns, but also for supporting subsequent interpretive work. In the broader ecosystem developed in this book, researchers may bring these outputs into ISARI by uploading screenshots of the interactive maps, copied text from the summary panels, exported figures, or notes derived from the HTML outputs. Because ISARI can operate locally and work with both textual and image-based artifacts, these SENA outputs can become the basis for scholarly brainstorming, memoing, comparison of alternative interpretations, and evidence-grounded writing without moving sensitive materials outside the researcher's control.

SENA applications to small and big data

In this section, we present two applications of SENA. The first consists of 99 short student evaluations/reactions about an undergraduate data science course offered at the Guangzhou Institute of Science and Technology in Baiyun District, Guangzhou, China, during the summer of 2024. The second dataset consists of 22,995 comments scraped from YouTube using the R package "vosonSML" by Gertzel et al. (2022a). These comments correspond to five videos analyzing the "second-by-second" evolution of an assassination attempt on Donald J. Trump in Butler, Pennsylvania, on July 13, 2024 (To watch these videos, you can visit this playlist). This second dataset analyzed serves to demonstrate how SENA may handle big data and may be integrated with existing software to incorporate machine learning text classification—more on this below.

Short essays examples

As described above, the 99 essays included in this section consist of 50 essays in which students responded to the prompt: After having spent the last eight weeks learning about data science applications, what are your reactions or your assessment regarding data science uses, including the pros and cons you have observed?

As part of this data-gathering exercise, we obtained participants' gender attributes, which resulted in 50 men and 49 women. To illustrate SENA's performance and flexibility, in the following analyses, we present all data formats that SENA can handle, separating the following subsections into Microsoft Word and spreadsheet formats, including and not including participants' attributes.

Word documents format

Fig. 7.2A and B show the data formats that we represented in Panel A of Fig. 7.3. Recall that when no participants' attributes are added to the Word document metadata, the results will only show one aggregate estimate of SEA (see Fig. 7.5A), one aggregate word cloud, and one network visualization. This network visualization allows for obtaining individual estimates of emotions, positive sentiment prevalence, and emotional entropy estimates. This Word document data format is shown in Fig. 7.2A, and access to this database is available at https://cutt.ly/leJfF4RW.

When attributes are included in the Word document, as shown in Fig. 7.2B, the outcomes match Fig. 7.5B with analyses by each of the groups of interest (i.e., men and women in this case) and two word clouds, as shown in Fig. 7.4A and B. In addition, this process will render a network visualization per group, with all individual-level estimates discussed earlier (i.e., emotional entropy, prevalence of positive sentiments, and intensity with all applicable emotions). Finally, note that the QAP analyses will be automatically computed, allowing users to download all dyadic analyses in a PDF. Access to this database is available at https://cutt.ly/aeJfFbZt.

Database analyses of short essays

Fig. 7.2C shows an example of the data structure in a spreadsheet format. This dataset contains the 99 essays described above, but this information is stored in a total of three columns. The first column contains the ID information, the second column is the text content, and the third column contains participants' attributes that can be used to conduct comparative analyses. Access to this database is available at https://cutt.ly/zeJfHr0M.

Note that the SEA procedures do not allow for duplicate IDs in the database. In case your dataset has duplicate IDs, an ID differentiation needs to be executed, such as including row numbers. Having noted this, analyses that ignore participants' attributes need to simply account for the ID and text columns. As in the case of Word documents, these analyses will yield aggregate outputs, including a SEA analysis as depicted in Fig. 7.5A, an aggregate word cloud, and one network visualization.

To incorporate group analyses, users will need to select the column in the dataset that includes group (or temporal) information. In our dataset shown in Fig. 7.2C, this information is called "Gender" and is located in the third column. As with the case of Word documents that include participants' attributes, the outputs include SEA analyses as shown in Fig. 7.5B, two word clouds as shown in Fig. 7.4A and B, network visualizations per group shown in Fig. 7.6, and the QAP results compiled in a PDF shown in Fig. 7.7A.

Application to machine learning classes of big data

As briefly described above, the YouTube comments database can be analyzed by video source, that is, by using each posting account as an attribute and then analyzing whether there are sentiment variations across these accounts (i.e., the New York Times, The Wall Street Journal, Channel 4 News in the UK, as well as two individual accounts). Alternatively, this database can also be analyzed in SENA following a different approach that involves machine learning.

Our analytic strategy consisted of merging SENA with machine learning text classification and then analyzing whether the resulting machine-learned topics or classes vary in their distribution of emotions and sentiments. Specifically, we first determined machine-driven text classes via topic modeling as implemented by Machine driven classification of open-ended responses (MDCOR) (González Canché, 2023) and then analyzed whether variations exist in the distribution of these sentiments and emotions as a function of the machine-learned classes. This approach enabled us to ask whether machine-learned classes are configured by different distributions of sentiments and emotions, as well as the level of emotional entropy those classes contain. From this perspective, SENA may be easily integrated with existing software and machine learning text classification applications.

Note that a description of topic modeling via latent Dirichlet allocation is beyond the scope of this paper. For now, let us simply state that the result of implementing MDCOR with 5000 MCMC/Gibbs resamplings and 500 burn-in samples yielded the following seven topics as the optimal solution for this text classification learning process: V1. Criticism of the video accuracy, V2. Middle East comments/support, V3. Surprise at the failure to stop this attempt, V4.

Multiple conspiracy theories, V5. Belief that it was staged, V6. Love and support for Trump, and V7. Laughing in disbelief—data access here: https://cutt.ly/ReLN5yH8. For more details about text classification using MDCOR, please read the methodological paper that provides access to the software on Windows and Mac operating systems (see González Canché, 2023).

Findings

This section is divided into two subsections. The first presents the analyses of short essays, both aggregated and by groups. The second shows the outputs of the classified comments retrieved from YouTube.

Small(er) data analyses

In this section, we illustrate how aggregated results may be misleading. Specifically, note that Fig. 7.5 contains the aggregate (Fig. 7.5A) and the group analyses (Fig. 7.5B) of students' comments. In this case, it is clear that only relying on aggregate analyses would have led us to believe that the vast majority of sentiments (about 70% of them) were positive and that this result may apply to all participants in the sample. This latter issue is referred to as aggregation bias (James, 1982), which, as depicted earlier, may be minimized by conducting analyses across groups (González Canché, 2023b; González Canché, 2023).

Before moving forward, let us further describe the content of Fig. 7.5A. Each SEA figure presents the outputs separated by negative emotions (left panel), positive emotions (center panel), and sentiment distribution (right panel). This output is normalized so that sentiment distributions (i.e., positive and negative) add up to 1. That is, these distributions are represented in proportions. Similarly, the left and center panels together add up to 1, as both account for emotions. In other words, the distribution of the eight emotions represented in this figure adds up to 1. The results of the sentiment distribution indicate that there were 264 words across all 99 essays (or 30.9%) that were associated with negative sentiments. In terms of emotions, the most prevalent emotion was trust, with a frequency of 251 representing 22.8%, and the least prevalent was disgust, with a frequency of 38 (or 3.5%). The interactive visualization displays these values when hovering the cursor over each bar. In this example, we show the proportions described in this paragraph for the trust and disgust emotions.

Following our example, Fig. 7.5B shows the resulting group analysis output. In that figure, we can clearly see that women's responses were more positive than their male counterparts. The interactive visualization (available here: https://cutt.ly/leJddzfw) shows that 86.2% of women's words identified in the lexicon were classified as positive sentiments. Subgroup emotion analyses reflected that men tended to have more affiliation with anger and disgust emotions, whereas women's words were associated with trust, joy, and anticipation. As displayed in this figure, we also see that women's trust almost doubled the trust emotion found in men's texts (29.2% versus 17.7%).

Although these differences are notable, we still do not know if these distributions are statistically significantly different. This hypothesis was tested automatically by SENA, and the results may be found in Fig. 7.7A. In this QAP output, we had enough evidence to conclude that the emotional distributions of women are different from their male counterparts and that, considering the histogram shown in Fig. 7.5B, the former tended to be more positive than men's emotions about this data science course. Specifically, the observed matrix correlation had a magnitude of 0.374, which was not far from the simulated mean of around 0. That is, the distribution of these matrices is uncorrelated. These findings mean that we can describe the distributions shown in Fig. 7.5B as being statistically significantly different.

Fig. 7.6A and B show individual analyses. These figures are particularly useful in situations where the identification of cases with negative or positive emotions may be relevant, as described in research where students' evaluations were used as predictors of subsequent success (Calvo and D'Mello, 2010; Luo et al., 2015a; Yu et al., 2018b, 2016b) (see; Dolianiti et al., 2018). As mentioned above, each of these cases also contains emotional entropy estimates that indicate clarity in text content. Texts with a mix of positive and negative emotions are classified as confusing, whereas others are clearer, indicating either negative or positive emotions. Once cases of interest are identified, we could read the texts of these cases to gain a deeper understanding of their concerns or praises regarding their experiences. Within the broader analytic architecture of this book, those same case-level outputs may also function as prompts for local AI-supported scholarly reasoning in ISARI, where researchers may compare alternative readings of sentiment and emotion patterns, memo possible explanations, and begin translating those patterns into evidence-grounded prose while preserving the primacy of participants' original voices This AI-fueled analytic help may be particularly relevant in capturing nuance or complexity of thought in texts with high levels of emotional entropy but should not be seen as a replacement for researchers understandings and expertise. Moreover, following network analysis principles, we can observe how subgroups are formed. For instance, in Fig. 7.6A and B, those participants located in the bottom-right sections are more related to a specific set of emotions than participants located in the middle of the graph or in the top-right section of this sociogram. For women, bottom-right location indicated their affiliation with trust, surprise, and joy. For men, participants located in the bottom-right were associated with anger, fear, and surprise.

Finally, note that a more descriptive analysis may be gained with word clouds. Fig. 7.4A and B, in addition to summarizing content relevance, also show average emotional entropy estimates. These estimates suggest that women have half the emotional entropy values on average compared to men. That is, men's texts were less clear regarding their inclusion of positive and negative emotions.

Big data analyses of machine-learned topics

As mentioned above, the analysis of YouTube comments involved a classification of these comments and then analyzing whether the emotion distribution of those comments was different or similar. Just as in the short essay case, Fig. 7.8A shows that an aggregate analysis of the seven topics identified via machine learning could be understood as having more positive sentiments. However, the disaggregate analysis shown in Fig. 7.8B suggests that topics 6 and 7 ("Love and support for Trump" and "Laughing in disbelief or nervously", respectively) were associated with more positive sentiments and emotions. Of the other five topics, only topic 1 (Criticism of the video's accuracy) was associated with more negative sentiments and emotions than all the other remaining topics.

Access to all word clouds and SENA outputs per topic can be gained by following this link: https://cutt.ly/ieZlGeOI. For this example, let us address the question of whether emotion distributions are similarly distributed across topics. To this end, we can use the QAP output. However, given that seven groups or topics exist, the resulting dyadic combinations rendered 21 comparisons. These comparisons are shown in Fig. 7.7B.

Overall, these results confirmed that the discrepancies observed in Fig. 7.8B are important. Across all comparisons that included topics 6 and 7 ("Love and support for Trump" and "Laughing in disbelief or nervously", respectively), the findings consistently indicated that the positive emotions captured in these topics were different from those captured in topics 1 to 5. Of these latter topics, topic 1 captured the most negative emotions. However, topic 1 (Criticism of the video accuracy) was only significantly different from topic 2, which included Middle East comments/support. Note that when an extremely potentially viral topic exists on social media, groups tend to "profit" from that topic by posting their support for issues that they care about. In this case, the issues were current conflicts in the Middle East. Finally, topics 6 and 7 were highly correlated, thereby allowing us to conclude that they are configured by similar distributions of emotions, which, as shown in Fig. 7.8B, are overwhelmingly positive.

In closing this section, let us note that a study focused on the analysis of these datasets would proceed to qualitatively analyze the most relevant texts. This identification can be done using cases shown in the SENA outputs, either individually or by group location, like our discussion of the bottom-right location in Fig. 7.6A and B. After this identification is done, researchers then proceed to qualitatively explain why group differences were found, preferably using participants' own voices, as captured in their original essays or comments (see González Canché, 2023b; González Canché, 2023), for examples of the relevance of preserving these voices).

Discussion, limitations, and future steps

As discussed above, our motivation for integrating SEAs with qualitative research is that the latter "involves finding out what people think, and how they feel" (Rambocas and Gama, 2013a, p. 14). From this perspective, we believe that "[w]hen integrated with qualitative research, sentiment analysis can be used as a tool that promotes structure to an otherwise flexible and [perhaps more] subjective data collection and data analysis process" (Rambocas and Gama, 2013a, p. 15). Accordingly, our study presented an analytic framework and multiplatform software application developed to conduct SENAs. This is a contribution to qualitative and mixed-methods literature because most current applications of sentiment and emotion modeling focus on providing aggregated analyses rather than allowing for the testing of group variation and individual experiences.

From this brief discussion, a salient contribution of the analytic framework we are proposing consists of offering a statistical test of sentiment changes across groups. This is another related but distinct departure from traditional SEAbecause current applications of SEAhave remained purely descriptive in nature, even in cases where temporal elements have been incorporated (as we discussed in the determ's example on page 20). This lack of testing for statistical variations is precisely due to the fact that, other than time-based variations, SEA frameworks are yet to include participants (or texts' attributes) that may be used for statistical comparisons. In our analytic framework and software, these statistical tests are conducted via QAPs correlations, as described above. In sum, we believe that with SENA, researchers can comprehensively take advantage of the insights that can be gained from analyses of *how our research participants perceived their experiences about a given topic of interest, or about what sentiment and emotions were reflected in our participants' episodic memories* (Byrne, 2017; Ramachandran et al., 2002).

Based on this discussion, let us then summarize that the SENA framework and software contribute to this literature in the following aspects:

1. SENA focuses on the use of SEA for academic research, even though the software does not prevent its use for marketing purposes.
2. SENA allows for the incorporation of participants' (i.e., texts') attributes into the analytic process in order to gain more nuanced understandings of potential group variation.
   a. Moreover, the resulting interactive visualizations will also allow the identification of individual cases (i.e., texts) that show particularly strong sentiments and emotions, as well asprovide deeper insights into the underlying patterns and trends.
   b. Discrepancies or inconsistencies in the presence of positive and negative sentiments and emotions (i.e., emotional entropy) in speeches, based on the use of positive and negative emotions (Nandwani and Verma, 2021) occur at both the individual and group levels.
3. SENA also allows for the statistical dyadic comparison of similarities and differences in emotion distributions across these groups of interest via QAPs (Hubert and Arabie, 1989; Krackardt, 1987).

SENA aligns with an equal-status integrative mixed-method design (González Canché, 2023b, 2024; González Canché, 2023). That is, once the outputs are produced, a qualitative analysis must follow to gain deeper and more nuanced understandings of our participants' sentiments and emotions on the topic of interest.

Integrating SENA outputs into local scholarly brainstorming

One of the strengths of SENA is that its outputs are portable across analytic stages. The interactive word clouds, sentiment and emotion histograms, individual-level sociograms, and dyadic QAP comparisons are useful not only for identifying affective structure in large corpora, but also for supporting subsequent interpretive work. In the broader methodological ecosystem developed throughout this book, these outputs can be brought into ISARI, a local, fully offline, open-source, and multimodal brainstorming environment designed to support scholarly memoing, comparison, synthesis, and evidence-grounded writing.

This possibility is especially important because sentiment and emotion distributions, however informative, do not determine their own meaning. Researchers still need to ask why certain groups expressed more trust than fear, why some texts showed especially high emotional entropy, how contradictory emotional signals should be interpreted, or why machine-learned classes differed so sharply in their emotional composition. Because ISARI can work with textual notes, screenshots, exported figures, scanned pages, and other image-based or document-based artifacts, researchers can upload SENA outputs in multiple forms—for example, screenshots of word clouds or sociograms, copied text from HTML information panels, QAP output PDFs, or notes derived from individual-level cases—and use them as the basis for local scholarly reasoning.

This workflow should not be mistaken for automation of interpretation. On the contrary, it extends SENA's equal-status mixed-method design. The outputs remain guides, not conclusions. ISARI can help researchers compare sentiment and emotion interpretations, generate analytic memos, test explanatory contrasts, and begin drafting evidence-grounded prose, but the responsibility for deciding what these affective distributions mean remains with the research team. In that sense, SENA and ISARI are methodologically aligned: one structures and visualizes affective patterns in textual data while preserving local control; the other helps researchers think with those outputs under conditions of privacy, interpretive accountability, and data sovereignty. Accordingly, the distinction is methodological rather than oppositional: SENA remains the local engine for sentiment and emotion detection, whereas ISARI becomes a local environment for downstream interpretation of those outputs.

As discussed above, SEA provides us with time-efficient insights into the distribution of positive or negative experiences or opinions (i.e., sentiment analysis) and the range of feelings or emotions (i.e., emotion analysis) expressed by our informants, customers, users, or participants (Nandwani and Verma, 2021). In this respect, and although typically SEAs are used for profit purposes (i.e., marketing, consumer behaviors, and preferences), our discussion focused on the use of SENA in academic settings (i.e., to better understand our research participants' experiences and opinions). We were purposeful in this use, for we hope SENA may become a tool that qualitative and mixed-methods researchers may use for the analysis of qualitative evidence.

Limitations

SENA currently does not rely on server-based generative AI for the identification of sentiments and emotions. Instead, it uses locally executable lexicon-based procedures in order to preserve confidentiality, avoid external data transfer, and maintain analytic control within the researcher's own computing. This analytic strategy, however, does not preclude the later use of local AI in the interpretive phase. As argued throughout this book, SENA outputs may still be brought into ISARI for offline scholarly reasoning, memoing, comparison of alternative interpretations, and evidence-grounded writing, without requiring that the original sentiment and emotion detection itself depend on external AI services. From this view, local natural language processing and local generative or reasoning AI analyses are paramount in our pursuit to protect participants' confidentiality and anonymity in academic research (González Canché, 2025).

More to ths integrative strategy point, we also showcased how SENA may be easily incorporated into existing machine learning procedures, such as those implemented with no-code software like MDCOR and "Latent Code Identification." In these latter applications, we first need to learn or classify the raw text into codes or topics and then apply SENA to these learned classes to test whether the distribution of emotions is similar or statistically divergent. This analytic flexibility is notable because we can include machine learning and AI in texts that will then be analyzed via SENA without compromising the integrity, anonymity, and confidentiality of the analyses.

Software access and architecture

Mac users can download SENA here: https://cutt.ly/QwhYruBr, and Windows users can download it here: https://cutt.ly/YwhJJKvO. To execute the application for the first time on a computer, the following steps need to be performed:

1. Extract the zip folder
2. Open the folder starting with "SENA-." It contains two subfolders: SENA-App-Win and SENA-App-Mac, for Windows and Mac operating systems, respectively
3. Double-click on the file called "SENA-App-Mac" or "SENA-App-Win"

The first time SENA is executed, you may be required to select the menu "View" and "Force Reload" to generate all paths to this newly extracted folder. If, after forcing a reload, SENA does not display as shown in Fig. 7.9, close the window and execute SENA again; if needed, force reload once more when the software is launched.

SENA is housed in a secure server location, and all executable files are virus-free but may require administrative permission to be deployed. This is particularly true for Mac users, as they need to go to systems security and select to execute SENA, even if this software was not developed by Apple. Mac users are also recommended (and in some instances they do need) to place the SENA software (and all software products discussed in this book) in their applications folder for newer versions can only read and write files from this location. Finally, each of these executable files for Mac and Windows operating systems contains all the source code required to modify and/or expand the application capabilities in the file called "app.R."

SENA is also supported through a public GitHub repository that centralizes software files, tutorials, example data, and backend materials (access here: https://cutt.ly/atLaMqGw). This strengthens the framework's reproducibility and transparency by allowing researchers not only to replicate the procedures described in this chapter, but also to inspect, preserve, and extend the software environment over time.

Note that SENA has also been peer-reviewed by Code-Ocean, a multisoftware platform that seeks to ensure quality, reproducibility, and transparency in scientific research. To view SENA in Code-Ocean, follow this link: https://cutt.ly/BrhEnyfY.

This openness is methodologically significant. Because SENA runs locally, stores outputs in reusable formats, and has undergone external software review, the framework supports not only accessibility but also reproducibility, inspectability, and long-term usability. These same qualities make SENA especially compatible with the broader local analytic ecosystem advanced in this book, in which outputs can remain under the researcher's control while being revisited, compared, and interpreted in subsequent stages of analysis.

Final thoughts and next steps

The present document focuses on the analysis of textual or unstructured data, with an emphasis on academic research and how researchers may benefit from this tool by considering the temporal evolution of emotions or the changes in the distributions of such emotions across groups. Although future iterations of the framework and software we are presenting may more formally include machine learning or AI applications, until personal servers can be easily and affordably implemented, for now, and due to privacy concerns, we are presenting a NLP implementation that runs completely locally and does not require the uploading of any output or data to any server. Having said this, the level of anonymity varies depending on the source. For example, while YouTube comments do not require any type of confidentiality protection, as they are made public when users share them online, essay data or interview transcripts, for example, need to be handled with more caution.

In terms of confidentiality protection, note that for the essay database, although we are presenting our interactive outputs online, before doing so we made sure to (1) anonymize all texts, (2) ensure such texts did not include any potentially harmful information, (3) request written permission from participants, and (4) clarify that the online posting of these outputs is completely optional. Because interactive sentiment and emotion outputs may include original wording, emotionally charged expressions, and contextual clues that could weaken confidentiality in small, distinctive, or otherwise sensitive samples, the public sharing of HTML outputs should always be treated as optional and governed by the same principles of confidentiality, data minimization, and participant protection that guided the original qualitative or mixed methods design (González Canché, 2025). In many cases, local storage, restricted sharing, or the use of selected screenshots and excerpts may be methodologically and ethically preferable to full public release (González Canché, 2025). That is, the use of SENA does not, in any way, imply that the outputs will automatically be made available online; instead, this is an extra step that is completely based on the research team's goals, interests, and motivation.

Finally, we emphasize that sentiment and emotion analysis' historical focus on marketing or market research (Nandwani and Verma, 2021; Sharma et al., 2024a) can still be implemented with SENA. However, we developed SENA with a particular interest in applying it as a tool to better inform mixed-methods and qualitative research. As such, our interest in providing multiplatform software, along with all data used in this study, aimed to facilitate qualitative and mixed-methods researchers' firsthand interaction with SENA. We hope this strategy may ease access to data science applications without the need to learn computer and statistical programming, which, aside from monetary costs, arguably remains the most significant barrier to benefiting from these tools in qualitative and mixed-methods research. This is why we are offering these tools completely free of charge, so that no financial barriers exist to their implementation. Within the broader infrastructure developed in this book, SENA's affective and relational outputs may also serve as structured prompts for local scholarly reasoning in ISARI, where those patterns can be compared, memoed, and translated into evidence-grounded qualitative and mixed-methods explanation.

References

Andre, L. (2022a). Important statistics about how much data is created every day. *Finances Online.Reviews for Business*. https://financesonline.com/how-much-data-is-created-every-day/

Borgatti, S.P., 2006. Identifying sets of key players in a social network. Computational & Mathematical Organization Theory 12 (1), 21–34. doi:10.1007/s10588-006-7084-x.

Breiger, R.L., 1974. The duality of persons and groups. Soc. Forces 53 (2), 181–190 doi:10.1093/sf/53.2.181.

Byrne, J.H., 2017. Learning and Memory: A Comprehensive Reference 2nd. Academic Press, Elsevier Inc., United States, p. 658. http://www.sciencedirect.com/science/book/9780128052914.

Calvo, R.A., D'Mello, S., 2010. Affect detection: an interdisciplinary review of models, methods, and their applications. IEEE Trans. Affect. Comput. 1 (1), 18–37 doi:10.1109/T-AFFC.2010.1.

Chang, T., DeJonckheere, M., Vydiswaran, V.V., Li, J., Buis, L.R., Guetterman, T.C., 2021a. Accelerating mixed methods research with natural language processing of big text data. J. Mixed Methods Res. 15 (3), 398–412. doi:10.1177/15586898211021196.

Chowdhary, K.R., 2020. Natural Language Processing. Springer Science and Business Media LLC, pp. 603–649. doi:10.1007/978-81-322-3972-7_19.

Davis, D. (2019a). Ai unleashes the power of unstructured data. Cio. WWW-dokumentti. Saatavilla:https://www.cio. com/article/3406806/ai-unleashes-the-power-of-unstructured-data. html [viitattu 20.5. 2021].

De Beaugrande, R.-A., Dressler, W.U., 1981. *Introduction to Text Linguistics* (Vol. 1)., longman London.

DePaolo, C.A., Wilkinson, K. (2014). Get your head into the clouds: using word clouds for analyzing qualitative assessment data. Tech Trends 58 (3), 38–44. doi:https://doi.org/10.1007/s11528-014-0750-9 .

Desjardins, J., 2019a. How much data is generated each day. World Econ. Forum, 17. https://www.weforum.org/stories/2019/04/how-much-data-is-generated-each-day-cf4bddf29f/.

Dolianiti, F.S., Iakovakis, D., Dias, S.B., Hadjileontiadou, S., Diniz, J.A., Hadjileontiadis, L. (2018). Sentiment analysis techniques and applications in education: a survey. *International conference on technology and innovation in learning, teaching and education*, 412–427.

Feinerer, I. (2020a). Introduction to the tm package text mining in r. *R CRAN*. https://cran.r-project.org/web/packages/tm/vignettes/tm.pdf

Gertzel, B., Borquez, F., Graham, T., Ackland, R., 2022a. Vosonsml: Collecting Social Media Data and Generating Networks for Analysis. R package version 0.32.7. VOSON Lab, Virtual Observatory for the Study of Online Networks. https://github.com/vosonlab/vosonSML.

González Canché, M.S., 2019. Geographical, statistical, and qualitative network analysis: a multifaceted method-bridging tool to reveal and model meaningful structures in education research. High. Educ. Handb. Theory Res. 34, 535–634. doi:10.1007/978-3-030-03457-3_12.

González Canché, M.S., 2023a. Latent code identification (LACOID): a machine learning-based integrative framework [and open-source software] to classify big textual data, rebuild contextualized/unaltered meanings, and avoid aggregation bias. Int. J. Qual. Methods 22, 1–29. 16094069221144940.

González Canché, M.S., 2023b. Machine driven classification of open-ended responses (mdcor): an analytic framework and no-code, free software application to classify longitudinal and cross-sectional text responses in survey and social media research. Expert Syst. Appl. 215, 1–21, 119265.

González Canché, M.S., 2023. Spatial Socio-econometric Modeling (SSEM): A Low-code Toolkit for Spatial Data Science and Interactive Visualizations Using R, vol. 1, Springer Texts in Social Sciences, Springer Nature, p. 520. https://link.springer.com/book/10.1007/978-3-031-24857-3. doi:10.1007/978-3-031-24857-3.

González Canché, M.S., 2024. Graphical retrieval and analysis of temporal information systems (gratis): an integrative mixed methodology and open-access software to analyze the (non-)linear chronological evolution of information embedded in textual/qualitative data. J. Mixed Methods Res. 18 (1), 71–103. doi:https://doi.org/10.1177/15586898231166968.

González Canché, M.S., 2025. Ethics and trust in social network analysis: how data science and interactive visualizations impact anonymity, confidentiality, consent, and IRB procedures?. In: Handbook of Social Network Analysis and Education. Edward Elgar Publishing, pp. 114–131. doi:10.4337/978103531 2191.00013.

Hitesh, M., Vaibhav, V., Kalki, Y.A., Kamtam, S.H., Kumari, S. (2019). Real-time sentiment analysis of 2019 election tweets using word2vec and random forest model. *2019 2nd international conference on intelligent communication and computational techniques (ICCT)*, 146–151.

Holst, A., 2021a. Amount of data created, consumed, and stored 2010-2025. Technol. Telecommun. Retr. 1, 06–29. https://www.statista.com/statistics/871513/worldwide-data-created/.

Hubert, L., Arabie, P., 1989. Combinatorial data analysis: confirmatory comparisons between sets of matrices. Appl. Stoch. Models Data Anal. 5 (3), 273–325. doi:10.1002/asm.3150050309.

James, L.R., 1982. Aggregation bias in estimates of perceptual agreement. J. Appl. Psychol. 67 (2), 219. doi:10.1037/0021-9010.67.2.219.

Jockers, M. (2015a). *Syuzhet: Extract sentiment and plot arcs from text.* https://github.com/mjockers/syuzhet

Jockers, M. (2023a). Introduction to the syuzhet package. *The Comprehensive R Archive Network.* https://cran.r-project.org/web/packages/syuzhet/vignettes/syuzhet-vignette.html

Joda, T., Waltimo, T., Probst-Hensch, N., Pauli-Magnus, C., Zitzmann, N.U., 2019a. Health data in dentistry: an attempt to master the digital challenge. Public Health Genomics 22 (1-2), 1–7. http://content.karger.com/ProdukteDB/produkte.asp?Aktion=JournalHome&ProduktNr=224224. doi:10.1159/000501643.

Kennedy, A., Inkpen, D., 2006. Sentiment classification of movie reviews using contextual valence shifters. Comput. Intell. 22 (2), 110–125. doi:10.1111/j.1467-8640.2006.00277.x.

Kolaczyk, E.D., Csárdi, G., 2014. *Statistical Analysis of Network Data with R* (Vol. 65), Springer. doi:10.1007/978-1-4939-0983-4.

Krackardt, D., 1987. Qap partialling as a test of spuriousness. Soc. Netw. 9 (2), 171–186 doi:10.1016/0378-8733(87)90012-8.

Krackhardt, D., 1988. Predicting with networks: nonparametric multiple regression analysis of dyadic data. Social Networks 10 (4), 359–381. doi:10.1016/0378-8733(88)90004-4.

Kumar, L., Bhatia, P.K., 2013. Text mining: concepts, process and applications. J. Glob. Res. Comput. Sci. 4 (3), 36–39.

Luo, J., Sorour, S.E., Goda, K., Mine, T., 2015a. Predicting Student Grade Based on Free-style Comments Using Word2vec and Ann by considering Prediction Results Obtained in Consecutive Lessons, International Educational Data Mining Society

Mäntylä, M.V., Graziotin, D., Kuutila, M., 2018a. The evolution of sentiment analysis—a review of research topics, venues, and top cited papers. Comput. Sci. Rev. 27, 16–32 doi:10.1016/j.cosrev.2017.10.002.

Marr, B., 2018a. How much data do we create every day? the mind-blowing stats everyone should read. Forbes 21, 1–5. https://www.forbes.com/sites/bernardmarr/2018/05/21/how-much-data-do-we-createevery-day-the-mind-blowing-stats-everyone-should-read/?sh=3b3f208d60ba.

Marr, B., 2019a. What Is Unstructured Data and Why Is It so Important to Businesses? an Easy Explanation for Anyone. Forbes. https://www.forbes.com/sites/bernardmarr/2019/10/16/what-is-unstructured-data-andwhy-is-it-so-important-to-businesses-an-easy-explanation-for-anyone/?sh=76e4255a15f6.

Mohammad, S.M., Salameh, M., Kiritchenko, S., 2016a. How translation alters sentiment. J. Artif. Intell. Res. 55, 95–130.

Mohammad, S.M., Turney, P.D., 2013a. Crowdsourcing a word-emotion association lexicon. Comput. Intell. 29 (3), 436–465.

Mohammad, S.M., Turney, P.D., 2013b. Nrc emotion lexicon. Natl. Res. Counc. Can. 2, 234.

Nandwani, P., Verma, R., 2021. A review on sentiment analysis and emotion detection from text. Soc. Netw. Anal. Min. 11 (1), 1869–5450. doi:10.1007/s13278-021-00776-6.

Phipson, B., Smyth, G.K., 2010. Permutation p-values should never be zero: calculating exact p-values when permutations are randomly drawn. Stat. Appl. Genet. Mol. Biol. 9 (1), 1–12. https://arxiv.org/pdf/1603.05766. doi:10.2202/1544-6115.1585.

Pooja,, Bhalla, R., 2022. A review paper on the role of sentiment analysis in quality education. SN Comput. Sci. 3 (6), 469. doi:10.1007/s42979-022-01366-9.

RACONTEUR. (2020). A day in data. *RACONTEUR*. https://www.raconteur.net/infographics/a-day-in-data

Ramachandran, V.S., et al. 2002. Encyclopedia of the Human Brain tech. rep. Academic Press.

Rambocas, M., Gama, J., 2013a. The role of sentiment analysis. FEP Work. Pap. 489, 17–18. https://wps.fep.up.pt/wps/wp489.pdf.

Rambocas, M., Pacheco, B.G., 2018. Online sentiment analysis in marketing research: a review. J. Res. Interact. Mark. 12 (2), 146–163. http://www.emeraldinsight.com/info/journals/jrim/jrim.jsp. doi:10.1108/JRIM-05-2017-0030 .

Schütze, H., Manning, C.D., Raghavan, P., 2008a. *Introduction to Information Retrieval* (Vol. 39). Cambridge University Press Cambridge. https://nlp.stanford.edu/IR-book/.

Sharma, N.A., Ali, A.S., Kabir, M.A., 2024a. A review of sentiment analysis: tasks, applications, and deep learning techniques. Int. J. Data Sci. Anal. 23644168, 1–38. https://www.springer.com/journal/41060. doi:10.1007/s41060-024-00594-x.

Silge, J., Robinson, D., 2017a. Text Mining with R: A Tidy Approach. "O'Reilly Media, Inc." https://www.tidytextmining.com/.

Taboada, M., 2016. Sentiment analysis: an overview from linguistics. Annu. Rev. Linguist. 2, 325–347.

Taboada, M., 2016. Sentiment Analysis: An Overview from Linguistics. Annu. Rev. Linguist. 2 (1), 325–347. doi:10.1146/annurev-linguistics-011415-040518.

Turney, P.D. (2002). Thumbs up or thumbs down? semantic orientation applied to unsupervised classification of reviews. arXiv preprint cs/0212032 .

Van Dijck, J., 2013. Social media platforms as producers. Producing the Internet. Crit. Perspect. Soc. Media 1, 45–62. https://www.diva-portal.org/smash/get/diva2:1534758/FULLTEXT01.pdf#page=47.

Wang, H., Can, D., Kazemzadeh, A., Bar, F., Narayanan, S. (2012a). A system for real-time twitter sentiment analysis of 2012 us presidential election cycle. *Proceedings of the ACL 2012 system demonstrations*, 115–120. Association for Computational Linguistics (ACL) United States. https://aclweb.org/.

Whitbred, R., 2011a. Quadratic assignment procedure (qap). In: A, B.G. (Ed.) Encyclopedia of Social Networks. SAGE Publications, Inc, pp. 733–734.

Xie, Y., Lin, S.-Y., 2019. Using word clouds to support students' knowledge integration from online inquiry: an investigation of the process and outcome. Interact. Learn. Environ. 27 (4), 478–496. doi: Available from: www.tandf.co.uk/journals/titles/10494820.asp. doi:10.1080/10494820.2018.1484774.

Yu, L.-C., Lee, C.-W., Pan, H., Chou, C.-Y., Chao, P.-Y., Chen, Z., Tseng, S., Chan, C., Lai, K.R., 2018b. Improving early prediction of academic failure using sentiment analysis on self-evaluated comments. J. Comput. Assist. Learn. 34 (4), 358–365.

Yu, L.-C., Lee, L.-H., Hao, S., Wang, J., He, Y., Hu, J., Lai, K.R., Zhang, X. (2016b). Building Chinese affective resources in valence-arousal dimensions. *Proceedings of the 2016 conference of the North American chapter of the association for computational linguistics: Human language technologies*, 540–545.

8

GeoStoryTelling

Introduction

Our life stories evolve in specific and contextualized places. Although our homes may be our primary shaping environment (via our parents or guardians), our homes are themselves situated in neighborhoods that expose us to the immediate "real world," or at least the version of the world where our homes are located (i.e., low-income neighborhoods look quite different from high-income neighborhoods, as we elaborate further below). This exposure to the places where we have experienced or are currently experiencing life plays a fundamental role in shaping our beliefs, fears, perceptions of the world, and even our prospects for social mobility (R. Chetty and N.Hendren, 2018). In fact, this influence is so strong that researchers have shown that most of us resemble a weighted average of our childhood neighborhoods' socioeconomic and demographic attributes (R. Chetty and N.Hendren, 2018; Chetty et al., 2014a). That is, if the norm in the neighborhood where we grew up was that its inhabitants held bachelor's degrees, we were expected to attain such a degree. On the other hand, if our childhood neighbors were prone to serving time in jail, this would also have been a likely possibility for ourselves or some members of our family and friends. In both instances (i.e., positive or negative neighborhood effects), the longer our exposure times to these neighborhoods were, the stronger these impacts have been estimated to be (R. Chetty and N. HendHendren, 2018; Chetty et al., 2016a; Chetty et al., 2014a). From this perspective, then, a methodical effort to understand the geo-contextualized information of the places where life happens, or stories evolve, may serve to better understand our overall perceptions of the world and our outcomes. More to the point, geo-contextualization may help us reach a more nuanced understanding of our life stories, health and life outcomes, and decision-making processes.

Aside from serving as an introspective tool to better understand ourselves based on our life experiences, when applied to research, the incorporation of information that considers the attributes of the places where our research participants' stories are happening, or have happened, may be instrumental in gaining a deeper and more nuanced understanding of their perceptions of the world and may help better inform the types of conclusions and implications we offer in our studies (Ajayakumar et al., 2019a; Corbett and Keller, 2005; Jung and Elwood, 2010; M.-P. Kwan and G.Ding, 2008; Matthews et al., 2005a). More formally, the relevance of geo-contextualizing the stories of our research participants to better understand: (1) their patterns and processes and (2) the overall influence of their environments on their behaviors and decisions (Mennis et al., 2013a) explains the academic interest that health scientists have

Data Science, Interactive Visualizations, and Generative AI Tools for the Analysis of Qualitative, Mixed-Methods, and Multimodal Evidence. https://doi.org/10.1016/B978-0-443-21961-0.00006-1

placed on the overt incorporation of geospatial techniques and methodologies as part of the research process (Ajayakumar et al., 2019a; Mennis et al., 2013a). Indeed, "[h]ealth researchers incorporating spatial data have argued that we need to consider-the-local because of its importance in designing effective interventions" (Ajayakumar et al., 2019a, p. 2). Here, the effectiveness of the intervention is assumed to be strengthened based on the nuances gained when including geo-contextualized details. Note, however, that details in this context do not simply refer to place-based indicators, as in quantitative spatial modeling where it "suffices" to include poverty or income inequality in spatial regression models to "control" for spatial context. Instead, Ajayakumar et al. (2019a) refer to the inclusion of *context* in the form of narratives or *geo*stories, thus going well beyond merely adding layers with quantitative estimates of sociodemographic and socioeconomic place-based indicators. In their words, the interest is not only in indicating that something occurred but, more importantly, in explaining why it did or how it did (Ajayakumar et al., 2019a).

The benefits of *geo*contextualization, however, are not limited to health research. In fact, most, if not all, qualitative and mixed-methods studies, regardless of specific disciplines, may benefit from incorporating this *geo*contextualization as part of the research methodology implemented (M.-P. Kwan and G.Ding, 2008). More to the point, we argue that *geo*contextualization does not need to be the primary analytic framework or methodology in a given study to strengthen our understanding. Even when used as an ancillary technique (i.e., playing a supportive role in our research endeavors), the inclusion of *geo*contextualized information may ultimately also lead to more nuanced conclusions and recommendations.

Despite the immediate impact of the places where we experience life in shaping our perceptions of the world, our outcomes, and, more importantly, in better understanding our life stories and decisions, place-based narratives, stories, or information have typically remained untold and unmapped (Palis, 2022a). More to the point, although it is evident that technologies like social media applications, which can seemingly merge stories with location (i.e., geotagged pictures, videos, or posts), have been widely appreciated for over a decade (M.-P. Kwan and G. Ding, 2008; Matthews et al., 2005a; Mooney et al., 2013a), to date, most qualitative, ethnographic, and mixed-methods researchers continue to forego the analytic power that *geo*contextualizing our and our participants' narratives brings to our analyses, research findings, conclusions, and plans of action (Ajayakumar et al., 2019a).

The main reason for the underuse of geo-contextualization in qualitative and mixed-methods research in the social sciences is that, so far, its application has required extensive specialized training in geographical information systems (GIS), geo-referencing and mapping, web-development, computer programming, and data science and visualization techniques. A notable recent effort to lower some of these barriers is QualNotes, a collaborative and cross-disciplinary mobile mapping approach that supports the geo-tagging of qualitative data such as participant observations, mobile mapping notes, and audio or video interviews (Drozdzewski and Berengueres, 2024). Although such developments are important, they still leave open the need for a fully no-code, locally executable, and multimedia-rich storytelling environment that allows researchers to compile, visualize, and interpret geocontextualized narratives without requiring specialized GIS workflows, external server uploads, or costly proprietary platforms. It is precisely this remaining gap that GeoStoryTelling seeks to address.

Moreover, as Ajayakumar et al., 2019a) mentioned, it remains unclear and even daunting how to (1) transform narratives into geo-referenced or geo-tagged data, (2) further include multi-media attributes that may take the form of texts, stories or narratives, images, audio, or videos, and (3) ultimately obtain fully interactive HTML outputs that capture these now georeferenced stories.

Relatedly, another remarkable challenge is the current absence of free-to-use, user-friendly, and multi-platform software [i.e., available on Windows and Mac operating systems (OS)] capable of transforming these multi-media data into interactive HTML-mapped outputs with absolutely no computer programming experience and/or GIS training. That is, currently available software like ArcGIS StoryMaps (https://storymaps.arcgis.com/) starts at a cost of $550 per year and still requires training in GIS. Mapbox's Storytelling (https://www.mapbox.com/solutions/interactive-storytelling) allows users to "pay-as-they-go" but can become expensive quite quickly (see https://www.mapbox.com/pricing), with starting prices as low as $16 that may reach $3200 per month. Moreover, Mapbox's Storytelling requires both training in GIS as well as statistical and computer programming or coding skills. A recent open-source software tool that integrates natural language processing, in addition to adding location pins to these written narratives, is called WordMapper (Ajayakumar et al., 2019a) (Access to the software available at https://cwru.app.box.com/s/89zlahn4drrmgqfr29xz7cfn3977vn6n). WordMapper, by itself, does not require programming skills nor specialized data formats (i.e., shapefiles storing points, lines, and polygons that are geo-located) and is free to use. However, although the use of WordMapper does not require specialized training, the data preparation may pose challenges. For example, in the case study presented by Ajayakumar et al., 2019a), specialized hardware equipment for video and geo-tagging, as well as specialized software (Contour Storyteller https://contour.com/pages/storyteller) for data preparation, were required. It is this use of specialized software that necessitates specialized training, which may be a barrier to the broad adoption of WordMapper among the qualitative, health, and mixed methods communities. Finally, in terms of its scope regarding OS and its multimedia handling capabilities, WordMapper is only available for the Windows OS, and in its current version, it does not accommodate multi-media formats such as images, videos, audio, or external links.

Study purpose and contribution

As noted above, despite the relevance that *geo*contextualized information brings to qualitative and mixed methods research, the current and prevalent lack of access to multi-platform, free-to-use, and user-friendly software that requires no specialized hardware for data collection or preparation remains a significant barrier to the widespread use of multi-media, geo-coded narratives, or stories in social and health sciences research. Accordingly, the purpose of this study is to fill these methodological and software-related gaps by introducing GeoStoryTelling, a completely cost-free, multiOS platform, and user-friendly geo-visualization methodology, along with the Code Ocean (https://codeocean.com/) peer-reviewed software tool (https://doi.org/10.24433/CO.7385885.v4), which allows non-GIS experts to benefit from the resulting fully interactive *geo*visualizations and multi-media HTML *geo*narratives. The relevance of these outputs extends beyond map production alone. Once geolocated narratives have been rendered, multimedia context has been attached, and spatial relationships or routes have been made visible, researchers still need to determine what those patterns mean, which stories deserve closer memoing, how place-based and temporal context should be interpreted, and how these geocontextualized findings should be translated into qualitative or mixed-methods argument. In the broader architecture of this book, GeoStoryTelling outputs can therefore be understood not as endpoints, but as structured guides that may also be brought into ISARI, a local, fully offline, open-source, and multimodal brainstorming environment designed to support scholarly comparison, memoing, synthesis, and evidence-grounded writing. In this way, GeoStoryTelling and ISARI are methodologically aligned: GeoStoryTelling makes splace-based narrative structure visible; ISARI helps researchers think with those outputs under conditions of privacy, interpretive accountability, and local control.

Specifically, GeoStoryTelling is both an analytic framework or methodology (i.e., a well-specified set of analytic steps to reach interactive outputs, as shown in Fig. 8.1) and a computer programming-free and cost-free software that renders state-of-the-art, esthetic and visually appealing interactive HTML maps. These maps support and integrate multi-media inputs (i.e., text, images, videos, sound recordings or music, and links to other websites) to provide further context to the geo- or place-based stories being shared. From this perspective, the overarching purpose of this study is to provide a detailed step-by-step description of every component of the methodology and to offer full and unrestricted access to the software tool, which is available for both Mac and Windows OS (see access and execution instructions in the subsection called "Software Access and Execution" at the end of this chapter).

Another vitally important goal of this study is to provide a detailed account of every step of the data-building process, demonstrating that no specialized software or hardware is needed to create our databases. Nonetheless, in addition to guiding researchers interested in using this methodology and software tool to build their own datasets, we are also providing a toy database so that researchers may interact first-hand with GeoStoryTelling and, while familiarizing themselves with this tool, fully replicate the outputs discussed in this study.

The multimedia, place-contextualized narratives that can be shared via GeoStoryTelling accommodate maps with a worldwide scope based on our reliance on OpenStreetMap.org servers, which contain free, open geographic databases available in 96 languages and variants for map data (see https://www.openstreetmap.org). In this respect, although the production of the HTML output requires an active internet connection to access these maps free-of-charge, GeoStoryTelling does not upload our databases to any server; all back-end processes are computed locally. In other words, we download maps from OpenStreetMap.org, but we do not upload our data to any server to conduct the operations, which naturally protects the privacy of our data inputs. This access to global maps, as implemented with GeoStoryTelling, strengthens its potential reach to the mixed methods, ethnographic, health, and qualitative scientific communities worldwide. Moreover, the cost-free and programming-free (or no-code) distribution of GeoStoryTelling aims to democratize access to data science tools by removing any form

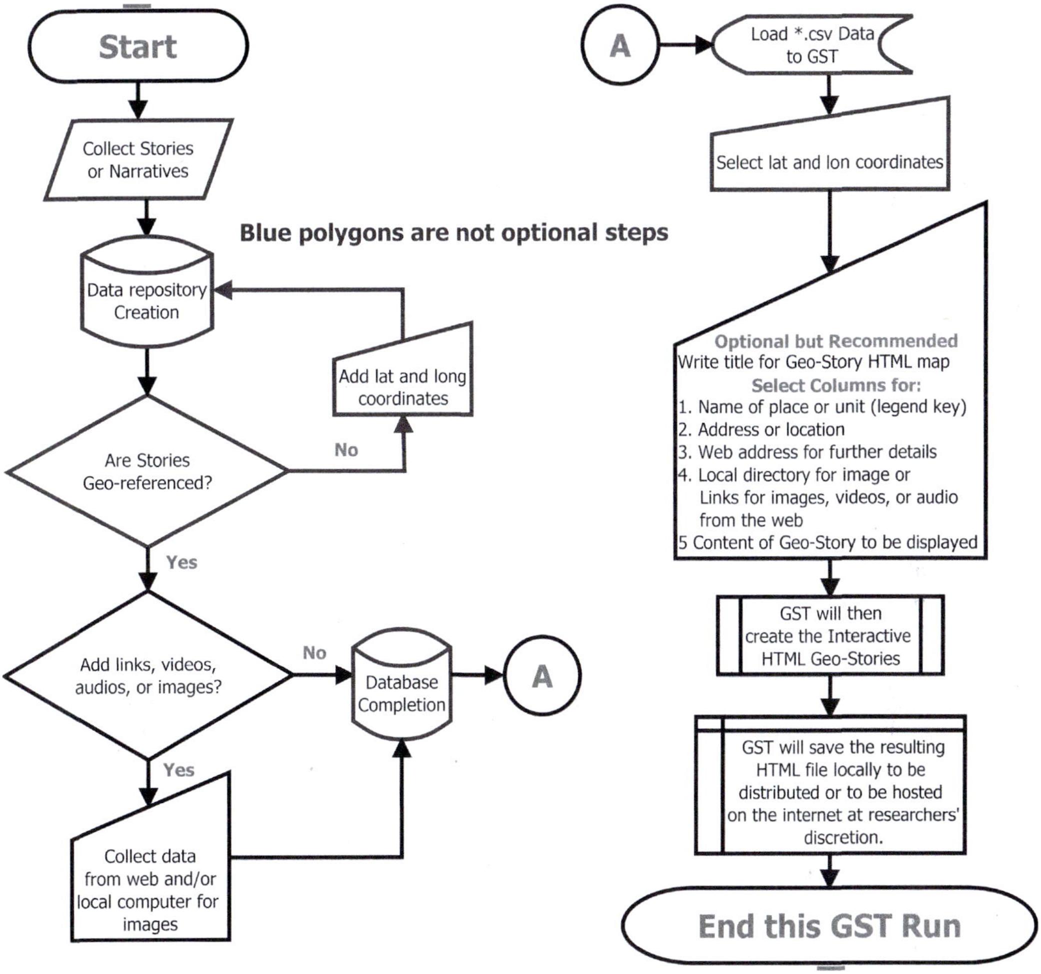

FIG. 8.1 Flow Diagram for the Entire Methodological Process, Including Database Construction.

of financial and computer coding requirements to produce high-quality and fully interactive HTML outputs that integrate a variety of multi-media inputs. In this sense, GeoStoryTelling contributes not only to accessibility and geo-visualization, but also to interpretive transparency by making the spatial basis of narrative claims more inspectable, revisitable, and documentable.

The remainder of this study is organized as follows: we next present a brief summary of the conceptual underpinnings of spaces and places and their relevance for GeoStoryTelling. Subsequently, we provide a summary of previous literature on geo-narratives and geo-stories, with an emphasis on illustrating how GeoStoryTelling, as a methodology, contributes to this movement.

In the next section, we describe the flow diagram guiding the complete execution of GeoStoryTelling, from database creation to the execution of the back-end processes. Accordingly, our description of the flow diagram illustrates the data gathering and preparation processes and discusses the types of decisions researchers may need to make when preparing their databases. In this respect, we have paid particular attention to detailing the steps that researchers need to follow to create their own databases, which can then be fed into GeoStoryTelling's user interface (UI). Notably, the creation of these databases does not require any form of specialized software and can be compiled using free-to-use tools, such as Google Maps (https://www.google.com/maps), Bing Maps (https://www.bing.com/maps), or global positioning system (GPS) Visualizer (https://www.gpsvisualizer.com), for example. After this discussion, we present the GeoStoryTelling UI and execute GeoStoryTelling with a preloaded toy dataset we are providing so that users may replicate all the outputs presented in this study. Finally, we describe the outputs and close with conclusions, limitations, recommendations, and future steps.

Conceptual underpinnings and prior/current applications

Where we grew up shaped—or at least impacted—our experiences. Take a moment and think about your childhood neighborhood. What do you see? How clean does it look? How does it smell? What sounds do you hear? Are there trees? Are there parks? Is it safe to walk around? What about public libraries? How far was the closest college? As discussed in our introductory section, the answers to these questions, to a great extent, shaped how you experienced life growing up in those neighborhoods and the types of memories you created, over and above the specific experiences and memories you had and created inside your home. Even as an adult, these constant exposures to our neighborhoods continue to shape our experiences and behaviors. If we live in a dangerous or unsafe neighborhood, we would not feel comfortable walking late at night and would probably be extra careful with home security. On the other hand, some neighborhoods feel safe at any time of the day (González Canché, 2023d).

Our quest to geo-contextualize our storytelling requires the blending of spaces and places, which, although eventually merged in our geo-stories, may also be decomposed for further clarity, particularly during the data construction process. Spaces, and more specifically *geographical spaces*, are inherently physical and therefore objectively measurable, identifiable on, and delimitable around Earth's surface (Cresswell, 2008; Tuan, 1977). Spaces are abstract (Tuan, 1977) for they may be represented with grids or systems of mathematical coordinates (Agnew, 2005; Agnew and Livingstone, 2011) that are not necessarily earth bounded (i.e., we can map our brain activities or measure distances across galaxies following these same mathematical principles). However, when this grid or coordinate system is applied to Earth's surface, we refer to spaces as being *geo*located or *geo*referenced, for the *geo* root or prefix comes from the Greek meaning of Earth or *gê*.

In GeoStoryTelling, *geo*referenced spaces are relevant because they enable us to *geo*locate the narratives or stories we seek to contextualize. However, mathematical concepts (i.e., spaces) without context limit our ability to understand social experiences (González Canché, 2023d). It is this goal or challenge of contextualizing abstract concepts where places become central in our meaning-building process. In other words, spaces themselves have no meaning because they are more abstract than real (Agnew, 2005; Tuan, 1977). It is only when these objects located within spaces become meaningful (i.e., schools, parks, blocks) because we have experienced them (Cresswell, 2008; Tuan, 1977) that the true analytic power of GeoStoryTelling becomes more feasible to attain. After all, GeoStoryTelling seeks to more deeply understand how space-based social structures may affect our narratives and experiences. Building from (Cresswell, 2008), GeoStoryTelling may be expressed as location + meaning, and together they form "splaces" (González Canché, 2023d).

This simple equation that renders splaces, captures the mental process we follow when asked to remember our childhood neighborhoods. That is, as part of the exercise of retrieving our memories associated with those childhood places, we may also remember measurable attributes such as distances we traveled from home to school, along with other physical attributes that may have affected our experiences associated with those places (e.g., temperature, elevation). These attributes may also go beyond physical ones and include socially based indicators like poverty, crime, and educational attainment levels. From this perspective, places differ from yet complement spaces in that places have meanings that transcend their mere physical properties (Cresswell, 2008; Tuan, 1977) but are experienced in those spaces. That is, although places are inherently linked to spaces (Agnew, 2005; Cresswell, 2008; Tuan, 1977) because they are physically located on or linked to Earth (e.g., geolocated or georeferenced), places also go beyond physical measurable attributes and notions.

This brief discussion aimed to elaborate on how places and spaces intermingle on a daily basis, forming "splaces" (González Canché, 2023d) and how their joint yet differentiable contributions to GeoStoryTelling may enrich our understanding of processes, mechanisms, and the explanations of the outcomes and decisions observed. From this perspective, to the extent that the stories we aim to geo-locate add place-based contexts (i.e., such as the answers to the questions we posed at the beginning of this section), our resulting narratives or stories would be richer in our efforts to understand our participants' experiences while "consider[ing]-the-local because of its importance in designing effective interventions" (Ajayakumar et al., 2019a, p. 2) (Fig. 8.2).

Geostorytelling through the space-time continuum

Note, however, that our experiences are always evolving through the space-time continuum. Accordingly, our GeoStories must not be constrained to specific times; instead, our analyses should freely consider our participants' experiences in a time and space that are meaningful to them and their experiences. For example, assume we want to understand the life path or trajectory followed by a scholar who, despite having grown up in poverty, achieved becoming a professor at a worldclass university. In this case, the GeoStory may begin by depicting the context *where* and *when* this scholar grew up, describing her childhood neighborhood, and then moving to her experiences during college, situating this description in the time and area where she attended college. It would then progress to graduate school and her academic employment trajectory, documenting the multiplicity of positions she has held since becoming a faculty member. In this example, the resulting visualization will locate these stories across multiple spaces (i.e., Earth's surface via latitude and longitude coordinates) while covering several

(a) Coordinates Retrieved From Bing Maps (left) and Google Maps (right)

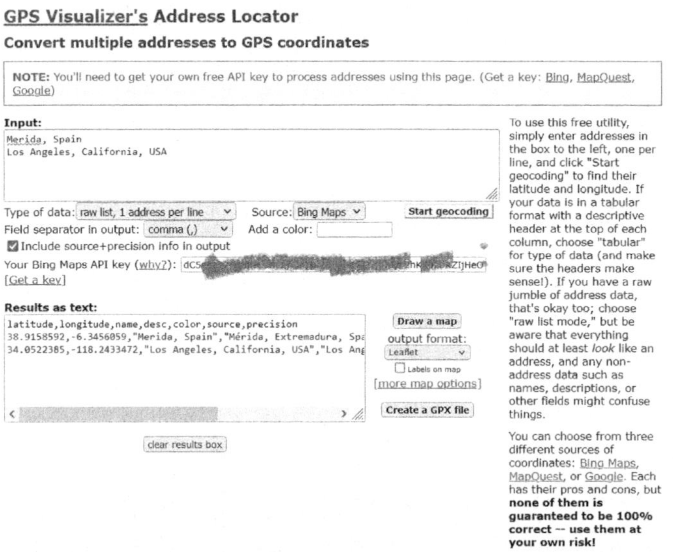

(b) Example of Batch Georeferencing from GPS Visualizer https://www.gpsvisualizer.com/geocoder/

FIG. 8.2 Geocoding or georeferencing processes. (a) Coordinates Retrieved From Bing Maps (left) and Google Maps (right) (b) Example of Batch Georeferencing from GPS Visualizer (https://www.gpsvisualizer.com/geocoder/).

decades in time. In this case, it is our responsibility as GeoStoryTellers to provide clarity and sufficient space and time context as part of our depiction of our participant's trajectory. That is, we must start by adding an ID or name that clearly conveys the time when the event took place, along with some meaningful yet succinct description that will be added as a legend to the resulting GeoStory, accompanied by a particular color for the pin on the map (see, for example).

In addition, the GeoStoryTelling software allows us to add video, audio, or hyperlinks to other websites that illustrate her academic trajectory, such as the types of colleges she attended or the types of neighborhoods where she grew up. This information may be added in addition to the narratives we selected to share with each point in her trajectory map—we will illustrate all these mapping and visualizing possibilities below.

By a similar token, our maps may present the GeoStories of different people and events. For example, in the HTML map shown in Fig. 8.3, which can be replicated using the toy database and the software tool we are providing (see https://cutt.ly/9r2wkXav), we include six points that illustrate different and disconnected stories. This disconnection is strategic, as it serves to showcase the versatility of GeoStoryTelling in capturing multiple types of narratives to be shared. For instance, one of these stories is a *behind-the-scenes* look at the film called "Kensington Blues" (https://kensingtonblues.com/), which documents a drug epidemic in Philadelphia. Another story is a presentation that took place in late 2016 on spatial and network analysis applications at the University of Pennsylvania. Yet another story showcases an audio recording of a 2022 concert, "Avalon Emerson-Live." In addition, three locations highlight examples of places where traditional and nontraditional cultural and literacy-oriented activities take place. These cases include a home, a cultural center for (mostly) undocumented families, and a well-established museum. What is common across these stories is that we demonstrate

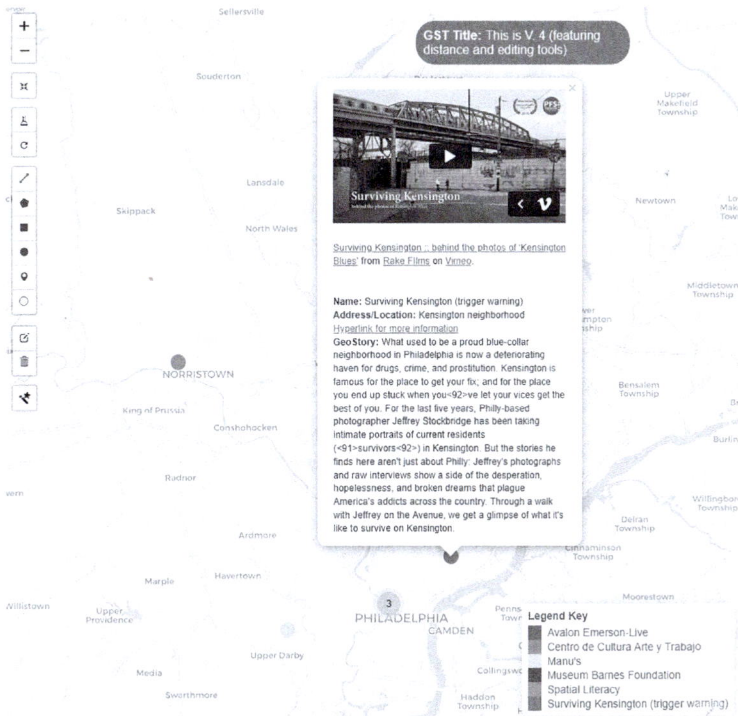

FIG. 8.3 GEOSTORYTELLING MAP

the inclusion of hyperlinks to find more information about each case. For example, in the case of Kensington, we provide a link to a Time Magazine report on the opioid crisis (https://time.com/jeffrey-stockbridge-kensington-philadelphia-photographs/), in addition to the option of playing the documentary directly from the GeoStoryTelling interactive map. As can be seen, the resulting HTML map includes a legend key that enables us to easily locate the GeoStory of interest and display more information, including multimedia inputs. That is, the legend key reveals more information about each point based on attributes in the name column. Users may decide to add categories instead of unique names, also under this column. For visualization purposes, GeoStories located in close proximity are grouped together using a cluster algorithm based on a radius of 80 pixels, or roughly 1.3152E-5 miles. For example, the number "3" indicates that three events were clustered together. Clicking on each cluster will automatically zoom in, revealing the points located within that cluster. Users may refresh the browser to reset the HTML map or hit the reset button located on the top left side of the map. Note that one of the stories shared deals with the opioid epidemic, which may be triggering; accordingly, viewer discretion is advised (as indicated in the legend).

Although we describe this output in more detail when discussing the implementation of GeoStoryTelling, in the following section, we present a summary of how similar methodologies have been discussed and implemented in the existing literature. This discussion also allows us to highlight how GeoStoryTelling is contributing to this line of inquiry and methodology.

How have geostories or geonarratives been implemented?

Two of the first methodological discussions on how to integrate GIS with narratives that we were able to find were published by Matthews et al. (2005a) and Corbett and Keller (2005). In their study, Matthews et al. (2005a) sought to merge or integrate GIS with ethnographic data, relying on a multi-site study of low-income families and their children. With this integration, referred to as *geo-ethnography*, the authors aimed to better situate the actions and experiences of their participants in time and space to enhance data analysis and interpretation. Similar to GeoStoryTelling, Matthews et al. (2005a) also elaborated on the relevance of further incorporating other data inputs, such as photographs, audio, and video files. An important cautionary note raised by these authors, which we also share with our GeoStoryTelling approach, is that the resulting visualizations, although useful, are not the final analytic product but merely a means toward gaining a deeper understanding. In their words, "[w]e still need ethnography to help us piece together how many of the contextual data sets that can be spatially joined are relevant to low-income families and low-income communities" (Matthews et al., 2005a, p. 70). Similar to the StoryMaps (https://storymaps.arcgis.com/) description (see also Carroll, 2020) in the introduction, Matthews et al. (2005a) wrote an extension to be executed in the proprietary software ArcGIS (https://www.esri.com/), which has potentially high costs and requires specialized training in GIS.

Like (Matthews et al., 2005a; Corbett and Keller, 2005), the integration of GIS with qualitative data was also demonstrated in a study involving disadvantaged groups. With the goal of discussing notions of empowerment versus marginalization, Corbett and Keller referred to their integration as participatory GIS (PGIS). The authors argue that although the discussion of integrating GIS with social and community research has been traced back to at least 1995, contemporary applications were referred to as PGIS (Corbett and Keller, 2005). The term "PGIS" was adopted with the goal of combating the idea that GIS by itself was becoming "a positivistic and technocratic tool that supported the more powerful sectors of society, often at the expense of weaker groups" (Corbett and Keller, 2005, p. 92). From this perspective, GeoStoryTelling aligns with this PGIS notion by providing a methodology and software tools that require absolutely no monetary cost or specialized training. That is, like PGIS, GeoStoryTelling seeks to empower qualitative and mixed-methods researchers—and the general public—who, despite not having had the opportunity to receive specialized training, may benefit from these open-source data science and visualization technologies.

One of the most cited papers on this topic (M.-P. Kwan and G.Ding, 2008) also aimed to extend GIS capabilities for the analysis and interpretation of narrative materials, including oral histories and biographies. In their discussion, M.-P. Kwan and G.Ding (2008) argued that the use of GIS methodologies may "complement or triangulate (i.e., verify results using multiple data sources) the knowledge acquired through the qualitative component of the research" (M.-P. Kwan and G.Ding, 2008) p. 444). As in the cases of Richardson et al. (2020) (with StoryMaps), Matthews et al. (2005a), and M.-P. Kwan and G.Ding (2008) also developed an analytic component in ArcGIS (https://www.esri.com/). Similarly, it is likely that the Wordmapper software developed by Ajayakumar et al., 2019a) was inspired by Kwan and Ding's description of the need to incorporate text processing capabilities into GIS. Not only do Ajayakumar et al., 2019a) cite M.-P. Kwan and G.Ding (2008), but Ajayakumar et al. also included natural language processing capabilities in the form of word clouds in their use of geonarratives.

More recently, Mennis et al. (2013a) introduced a prototype implementation of an interactive, cartographic visualization software application for exploring qualitative activity space data that was also based on geonarratives. They referred to this tool as the Qualitative Activity Space Data Viewer (QAV). Although the description of this tool highlights important features incorporated in GeoStoryTelling, such as the integration of multi-media data, there was no information on how to access the QAV prototype. Similarly, an extensive web search yielded no results regarding the existence of this software. From this perspective, the open-source software provided by Ajayakumar et al., 2019a) is perhaps the closest cost-free and no-code option to GeoStoryTelling that we were able to locate. Unfortunately, as of April 2023, we were not able to get this software to work.

Two alternatives to Wordmapper (see Ajayakumar et al., 2019a) that have become popular options due to their support for multi-media inputs for narratives are Story Maps (Richardson et al., 2020) and Mapbox's Storytelling (https://www.mapbox.com/solutions/interactive-storytelling).

Despite their popularity, these software tools require specialized knowledge (e.g., GIS, coding, or both) and can become quite expensive.

The most recent development in this movement was presented by Li et al. (2023a) and featured at the 2023 Conference on Human Factors in Computing Systems. This tool, called GeoCamera, aligns with the goal of GeoStoryTelling to reduce the barriers to incorporating geographical information system data. However, instead of using textual narratives, audio, and images, GeoCamera was developed to tell stories with geographic data videos. Specifically, GeoCamera seeks to produce compelling camera movements for storytelling. In short, as Li et al. (2023a) stated, GeoCamera aims to lower the barrier to crafting diverse camera movements for geographic data videos, and its UI was designed to empower users to tell appealing geographic stories with tailored camera movements. GeoCamera renders visually appealing outputs; however, we were unable to access the UI.

Aligned with the empowering notions depicted by Corbett and Keller (2005) and Li et al. (2023a), GeoStoryTelling offers access to user-friendly, no-code, and no-cost software that, although built on state-of-the-art mapping and visualization options, requires no training in GIS to link texts or narratives to their abstract spaces—in the form of longitude and latitude coordinates. Moreover, GeoStoryTelling was designed with the purpose of adding "hot-links," as described by Matthews et al. (2005a). These "hot-links" are *live hyperlinks* (i.e., active text or graphic links on a web page) that allow users interacting with GeoStoryTelling to explore other websites where more information about the narratives of interest may be gained. In addition, these "hot-links" may also display audio, images or photographs, and videos directly in the interactive map rendered by GeoStoryTelling—all without any need to code or pay for specialized equipment or software.

In sum, this subsection summarized applications related to GeoStoryTelling. Notably, as depicted here, our proposed methodology and software tool stand as the only no-code, nontechnical, multiplatform solution that is completely free to use and distribute. Given the potential salience and impact of GeoStoryTelling, the following section details the set of specific steps that users may follow to build the databases that GeoStoryTelling will then transform into interactive HTML outputs. With this goal in mind, we now proceed to discuss the methodological steps as described in the flow diagram operationalized in GeoStoryTelling.

Flow diagram for the entire geostorytelling process

Fig. 8.1 represents the complete set of steps required to execute GeoStoryTelling—from database creation to software execution. For the purpose of providing as much transparency and clarity as possible regarding the entire process, every step (or polygon) in the flow diagram presented in Fig. 8.1 can be conceptualized as follows.

1. Collect stories or narratives and select texts to be featured in the GeoStoryTelling interactive visualization.
2. Begin the database creation by placing the selected texts in a comma-separated value file in Google Sheets or Microsoft Excel, for example. As we further explain below, this database will require a maximum of seven columns.

(a) The column containing the ID or name of the story will be used to add a legend key to the interactive map. If repeated names are used, all those stories will share the same color, but if different IDs are typed, each resulting pin in the map will have a unique color, thus easing the identification of GeoStories and the navigation in the HTML map.

1. Situate these texts on Earth's surface (i.e., *geo*code or *geo*reference them) by adding latitude and longitude coordinates.
2. Although optional, we recommend adding the so-called "hot-links" to other websites with further information about the selected texts, as well as links to videos, audio, or images. Note that photographs can also be uploaded from your local computer or accessed via a live link from the web.
3. Load the database into the GeoStoryTelling UI, which is available on Mac and Windows OS. It runs all processes locally, so there is no need to upload your data to any server.
4. Select your columns with a few clicks. The GeoStoryTelling UI also allows you to type a customizable title for your interactive map. Adding this title is recommended but optional. The only required columns for GeoStoryTelling to work are the latitude and longitude coordinates. Accordingly, for flexibility, all the remaining columns in the database (shown in Table 8.1, as discussed below) are optional.

Table 8.1 Illustration of the Steps for Database Creation.

Panel A: Empty State						
Name or ID	Location or Address	Link to Webpage	Video/Image/Audio Link*	Longitude	Latitude	Geo-Story
Panel B: Adding Texts or Narratives						
$ID_{a,t=1}$						1st story text example…
$ID_{a,t=2}$						2nd story text example…
…						…
$IDz_{,t=t+N}$						Nth story text example…
Panel C: Geolocating those Stories						
$ID_{a,t=1}$	Mérida, Spain			-6.3456	38.9159	1st story text example…
$ID_{a,t=2}$	Los Angeles, CA			-118.2433	34.0522	2nd story text example…
Panel D: Embedding Videos, Audios, or Links						
$ID_{a,t=1}$	Mérida, Spain	https://www.…	https://www.…	-6.3456	38.9159	1st story text example…
$ID_{a,t=2}$	Los Angeles, CA	https://www.…	C:/Users/Desktop…	-118.2433	34.0522	2nd story text example…
$ID_{a,t=3}$	Los Angeles, CA	https://www.…	<iframe…	-118.2433	34.0522	3rd story text example…

*This link is called "embedded" for video and audio, as explained in the study and illustrated in Figures 4(a) and 4(b).

5. The resulting HTML map can be shared via email or hosted on a website, such as GitHub or RPubs, for free. See our fully reproducible example here: https://cutt.ly/k7X9tfN.

To provide as much clarity as possible, each of these steps is further detailed in a subsection on the following pages.

Stories or narratives collection

The process starts with the collection of stories or narratives. The data gathering of these stories may follow an ethnographic, qualitative, survey, or mixed methods approach. These stories may also take the form of interview transcripts, essays or reflections, open-ended responses in surveys (see the "Application Example to Survey Responses Involving Campus Safety" subsection below), videographies (i.e., capturing moving images on electronic media and even streaming media)—as in the case of Ajayakumar et al., 2019a who collected videos and then transcribed the voices in those videos for their geonarratives—ethnographic observations (Matthews et al., 2005a), social media posts (e.g., volunteered geographic information, as discussed by Mooney et al. (2013a), and, more generally, any type or form of text or qualitative evidence collected.

In this respect, although GeoStoryTelling is capable of displaying vast amounts of text in the resulting visualizations, we strongly recommend that researchers be selective and purposeful in the selection of the narratives or stories they choose to display in the HTML rendering. From this perspective, we also highly recommend that they practice interactive reviewability (Alexander et al., 2019a; González Canché, 2023a), where they ask their research participants whether the preliminary text selection conducted by the researchers appropriately captures their participants' experiences, or whether there are better narratives or aspects of their participants' episodic memories (Murray et al., 2018a) that more accurately reflect their viewpoints and/or experiences. Relatedly, please note that GeoStoryTelling is a tool that aims to convey ideas and experiences. For that reason, the resulting maps should be treated as guides to interpretation rather than as interpretations in themselves. GeoStoryTelling can render spatially contextualized stories with exceptional clarity, but researchers must still decide what those geostories mean, how place-based details should be weighed, and how spatial structure should be translated into analytic explanation. Accordingly, GeoStoryTelling should be viewed as a contextualizing tool rather than as the end goal of our studies. That is, as Matthews et al. (2005a) stated, this tool should not be the only source of analysis; instead, more qualitative evidence is still needed to reach well-balanced and nuanced analyses, as well as useful conclusions, recommendations, and plans of action.

Database creation

Once we have gathered and selected the texts or narratives that we want to feature in our GeoStoryTelling, we are then ready to start creating our database. This process is as straightforward as adding seven columns to a Microsoft Excel or Google Sheets object and saving it as a comma-separated value (or *.CSV) file or database. Note that this process will be the same if we have large-scale surveys, as we discuss in the subsection called "Application Example to Survey Responses Involving Campus Safety."

As we proceed with our discussion, we will elaborate more on each of these seven columns. Note that we do not have to follow the column names or the order we are showing below and in Table 8.1. However, for the sake of clarity, let us present the seven column names we are displaying in our database example. These names are:

1. Name or ID
2. Location or Address
3. Link to Webpage
4. Video, image, or audio link: Images may be loaded either from our local computer or from the internet. In the case of videos and audios, we need to obtain the embedded versions of the corresponding links, as illustrated below.
5. Longitude
6. Latitude
7. GeoStory

In this list of columns, GeoStory (i.e., the text or narrative we just discussed above) is the last column of that database. The seventh or last position of this column is selected purely for convenience. The text content of this column will typically result in wider columns than most of the other column contents, which may complicate readability if GeoStories are placed earlier in the database. Having said this, as we mentioned earlier in this subsection, the order of the columns is not important for the GeoStoryTelling software to render the HTML interactive outputs.

To illustrate the *step-by-step* process followed to build a database that can be used with GeoStoryTelling, we have added four panels to Table 8.1, each of which can be mapped back to our flow diagram shown in Fig. 8.1.

Panel A in Table 8.1 shows the initial status after the addition of the seven columns we described above has been completed. Panel B represents the addition of the narratives or stories that we want to feature. For illustrative purposes, we are simply populating these cells with generic texts that read "first story text example," "second story text example," finishing with an "Nth story text example." That is, our database may accommodate as many stories as we need or want to feature. The addition of these cases is as easy as copying each of the texts selected from our transcripts and pasting them into the column called "GeoStory."

Also note that in Panel B in Table 8.1, we have added identifiers to each of these stories. To illustrate the flexibility of GeoStoryTelling, we decided to feature the story of the same participant (called "A") at different points in time in the first two rows. This is participant A, whose first selected story took place at time 1 and may be represented as $ID_{A,t=1}$. Moreover, this same participant A also provided us with another story, which, in this case, took place at time 2 ($ID_{A,t=2}$). This representation of time follows *global time stamps*, as described by Artale et al., 2007a and Jensen et al. (1998b)—see also González Canché (2022) and González Canché (2023a). Global time stamps allow us to be flexible in how time is measured. That is, time may be measured in months for participant A in one instance, and it may also be measured in years for that same participant in her second story, based on her storytelling or episodic memories (Murray et al., 2018a).

Moreover, these time elements may not be linear in how they evolved. Instead, the storytelling of our participants may jump back and forth in the space-time continuum, and it is our challenge and responsibility as researchers to capture these jumps—and to consult with our participants to ensure our interpretations are accurate.

Nonlinear temporal dimensions in qualitative, ethnographic, and mixed-methods research enable more flexibility in capturing nuanced understandings of the stories our participants are sharing (Neale, 2015a, 2019). More specifically, and organically speaking, what our participants shared regarding "time X" may have prompted them to share more information regarding "time $X + n$," and this latter story may have prompted them to share events that occurred at "time $X - n$." Here, X is a point in time situated around our participants' experiences, and n represents time units before or after such a point in time took place. Once more, the n time units may be hours, days, years, or decades, for example, before or after a given point in "time X." To minimize confusion and create common understandings, let us assume that a participant shared with us the following narrative:

"I finished college in **1997** back home in Mérida, Spain. Who knew that **10 years later, in 2007**, I would start my graduate coursework in a different country—in Los Angeles, California? In full transparency, those graduate school years were extremely challenging in many aspects of my life. This experience was in stark contrast to my middle school dreams of studying abroad; the real world didn't match my expectations."

Here, this participant initially offered information situated in 1997, but then there was a jump in time ten years into her future with respect to 1997 (time $X + n$), when more information about graduate education was provided. Also, note that in this same example, our research participant followed her train of thought by jumping back to her high school years, which, following this nonlinear time rationale, may be expressed as "time $X - n$."

This brief example serves to showcase two main features. We can accommodate multiple mini-stories of our participants, and the units of time we may depict do not need to follow linearly evolving trends nor even be measured in the same unit of time. Instead, we may capture more nuance in the storytelling of our participants by applying nonlinear, global time stamps. In addition to enabling the inclusion of meaningful identifying information, the content of this column will be displayed as a legend in the HTML rendering and will be automatically mapped with specific colors to add more context and strengthen the navigation process. From this view, all events that share the same ID will have the same color in the GeoStory.

As shown in Table 8.1, the next step in our database-building process consists of geo-referencing those stories or narratives.

Geotagging or situating those stories on earth's surface

Note also that in this brief example, our participant shared with us information that may be situated on Earth's surface. Specifically, she mentioned Mérida, Spain, and Los Angeles, California. Although our participants' stories will likely not be as short as these experiences shared in our example, recall that the selection of texts or narratives is based on the purpose of the study and the decisions of the research team. For convenience, let us assume that $ID_{A,t=1}$ and $ID_{A,t=2}$ in Table 8.1 capture her graduation from college in 1997 and her starting of graduate education 10 years later (and in another country), respectively.

As described earlier, the only required elements that GeoStoryTelling needs to produce HTML outputs are latitude and longitude coordinates, which are the most fundamental pieces of information that enable us to locate units on Earth's surface. Although coordinates may take two main forms, depending on whether we rely on a planar projection of Earth (measured in meters) or a geodesic grid (measured in degrees) (see C. Holdgraf and L. A.Wasser, 2022), for our GeoStoryTelling purposes, let us just note that we may retrieve these coordinates from services like Google Maps (https://www.google.com/maps), Bing Maps (https://www.bing.com/maps) or GPS Visualizer (https://www.gpsvisualizer.com). The latter is an online, free-to-use tool that conveniently allows us to retrieve batches of coordinates at a time (i.e., dozens, hundreds, or thousands of addresses or locations, as we illustrate below). This process of retrieving latitude and longitude coordinates from addresses or locations is called geo-coding or geo-referencing—hence, we will use these terms interchangeably henceforth.

Geo-coding or geo-referencing

To further elaborate on this geo-referencing process, let us discuss Fig. 8.2A. This figure shows the result of retrieving the latitude and longitude coordinates using Bing and Google Maps services. In each case, we may input an address in the search bar or simply right-click on a desired point on the map. In Fig. 8.2, we right-clicked on the same location (i.e., 4341 Baltimore Avenue in Philadelphia, PA, USA). As you can see, in both cases, we have two sets of numbers that, although they show some variations (based on each service's implementation of Earth's projections), in the grand scheme of things, match the representation of the location of interest.

The first number (starting with 39.94) is the latitude, which ranges from – 90 to 90 degrees. A value of 0 represents the Equator, which divides Earth into the northern and southern hemispheres. Hence, positive values indicate that such a point is north of the Equator, and negative values indicate that such a point is south of the Equator (González Canché, 2023d). The second number is the longitude coordinate. Longitudes range from – 180 to 180 degrees. A value of 0 degrees longitude is referred to as the Prime Meridian, located in Greenwich. From the Prime Meridian to the left, up to –180 degrees, we move west—hence all the longitudinal values in the American continent are negative. Similarly, from the Prime Meridian to the right, up to the +180 degrees, we move east (González Canché, 2023d).

For our purposes, we just need to copy these coordinates and paste them into our database under the latitude and longitude columns shown in Table 8.1. Note that this georeferencing process works with specific or more general locations. That is, if we input addresses, the results will be quite precise. However, we can also just input or type city names or even country names, and services like Google or Bing Maps will still render the latitude and longitude coordinates, capturing the centroid (center) of those cities or countries. In going back to our brief narrative, our participant mentioned two places: Mérida, Spain, and Los Angeles, California. This information serves to demonstrate a batch geocoding approach also available free of charge when connected to Bing Maps via GPS Visualizer (https://www.gpsvisualizer.com/geocoder/). To use this service free of charge across the globe, we need to register at Bing (https://www.bingmapsportal.com/). This is a process that takes five minutes and results in a unique access key, like the one we covered in red in our rendering of Fig. 8.2B—note that we should never share our free Bing API key.

Batch geocoding is useful when dealing with multiple locations or large-scale projects. Although, in our brief example, we only have two locations, the same rationale applies to hundreds or thousands of addresses or general inputs like cities and countries. As can be seen in Fig. 8.2B, we may select Bing as the geocoding engine and add our Bing API key. Then, we simply need to copy and paste the addresses or places we need to georeference. This latter process takes two main forms. The first is a raw list with one address per line, and the second is a tabular form where the engine assumes that the first line contains column names and, accordingly, ignores the information contained in this first line for the batch geocoding procedures. For example, let us assume that we want to geocode a version of Table 8.1. If we copy and paste this table or the specific column containing the address or location to be geocoded into GPS Visualizer and select "Type of data: tabular (columns & a header)," the process will render the same result as shown in Fig. 8.2B. However, if we input tabular data and select "Type of data: raw list, 1 address per line," the batch geocoding process will attempt to geocode the line containing column names, which will result in missing cases for this first line—i.e., column names do not have coordinates unless one of the column names is recognized as a place on Earth's surface, like a column called "Rome," for example.

The result of this batch geocoding process will be a comma-separated output that can be copied into Excel. This tabular output will contain the following elements: latitude, longitude, name, desc, color, source, and precision. The column "name" matches our input requests (i.e., Mérida, Spain, and Los Angeles, California, USA, as shown in Fig. 8.2B) to ease the identification of the addresses. The column "desc" shows the actual name of that location as reported, in this case, by Bing. For example, our first input name was "Mérida, Spain," and the "desc" column rendered "Mérida, Extremadura, Spain." That is, Bing Maps also rendered the region in Spain where the city of Mérida is located.

As in the previous example, we need to take the latitude and longitude coordinates and add them to our database. By doing so, we have effectively been able to link our participants' narratives to the spaces they are located on Earth's surface. This result is shown in Panel C in Table 8.1, where we are adding both the location information and the corresponding latitude and longitude coordinates.

This description concludes the minimum elements required to produce interactive maps via GeoStoryTelling. However, although at this point the resulting maps will display the IDs (along with time in this case, as shown in Table 8.1), the locations' names, and the narratives, no other multi-media data have yet been incorporated into our GeoStoryTelling database. This brings us to our last subsection, which, as described in Fig. 8.1, is optional but highly recommended to provide more detail and nuance to our GeoStories.

Additional contextualizing elements: Video, audio, images, and external hyperlinks

Like the process of adding latitude and longitude, the addition of video, audio, or images that can be directly displayed along with our GeoStories does not require specialized software or hardware. Instead, we can retrieve these pieces of information from resources freely available on sites such as SoundCloud (https://soundcloud.com), YouTube (https://www.youtube.com/), or Vimeo (https://vimeo.com/), to mention a few streaming services. In the case of images, we can feature any image hosted online by simply retrieving its direct link from the internet—as we show below. Alternatively, we can also load any image stored on our computer to be displayed as part of our GeoStoryTelling.

Note that, although we programmed GeoStoryTelling's back-end to automatically detect whether the information provided is an image, a video, or audio, the data-gathering process to populate our database follows three main divergent approaches.

1. Audio recordings and videos need to be embedded, as shown in Fig. 8.4A and Fig. 8.4B, respectively.
2. Images may be retrieved from their hosting location on the internet, as we show in Fig. 8.4C.
3. Images can be loaded from our computers locally, as shown in Fig. 8.4D.

Although the gathering or compilation of these elements may be tedious with large-scale projects, note also that these contextualizing elements are optional, for it is only the narratives and the latitude and longitude coordinates that are required for the GeoStoryTelling software to render the HTML outputs.

Once we have collected these audios, videos, or images, we can then populate the column called "Video, Image, or Audio Link" in Table 8.1. As indicated in Table 8.1, the video and audio need embedding, which is a specialized HTML code that allows those objects to become part of our GeoStoryTelling maps. However, following our no-code approach to empower researchers or democratize data science, we can copy the embedding code by selecting "Share" from most public streaming websites with videos or audio, as shown in Fig. 8.4A and Fig. 8.4B, and then selecting the option "Embed." After doing so, as shown in Fig. 8.4A and Fig. 8.4B, we can simply copy and paste those links into our database, as represented in Table 8.1.

As we indicated, we programmed the back-end procedures of GeoStoryTelling to automatically detect whether we are providing videos, audio, or images, so we do not need three columns, but only one, as shown in Table 8.1.

For the sake of completeness, note that we added one extra row to Panel D in Table 8.1. This is to illustrate the three main sources of information that the column "Video, Image, or Audio Link" can handle. The first row in Panel D, by starting with a web address, indicates that the output should be an image. The second shows a local directory located on the disk "C." Note that we do not need to add quotations to any of these formats; we programmed GeoStoryTelling to automatically handle blank spaces. The third row in Panel D starts with < iframe, which suggests

(a) Audio Embedding from SoundCloud https://soundcloud.com

(b) Video Embedding from vime https://vimeo.com

(c) Image Retrieved From the CCATE Website https://www.ccate.org/

(d) Image Retrieved From a Local Directory

FIG. 8.4 Video and Audio Embedding, and Image Linkage. (a) AudioEmbeddingfromSoundCloud https://soundcloud.co (b)VideoEmbeddingfromvime https://vimeo.co (c)ImageRetrievedFromtheCCATEWebsite https://www.ccate.org (d) ImageRetrievedFromaLocalDirectory.

that this is an embedded link to either an audio or a video. With respect to images stored locally, please note that, although our example is demonstrated with a Windows path, the same process to add a local image on a Mac computer is executed, also without having to include quotation marks when pasting this path into our databases.

Finally, the software also allows us to include an external link to a webpage that provides more information, and this link may simply be copied from any web browser address and pasted into our table. Note that in Panel D in Table 8.1, the first cell in the "Link to Webpage" column is empty. This is completely fine, as GeoStoryTelling can handle missing cases if no additional information about a given story or narrative is added—nonetheless, stories with complete information will display this external link. As shown, when this link is present, it is added under the name "Hyperlink for more information." As its name indicates, when users click on this hyperlink, they are taken to a webpage that provides more information about the story. For example, in addition to displaying a short documentary called "Surviving Kensington" on the map via an embedded video, the "Hyperlink for more information" featured in the table takes users to a TIME Magazine story (This is the link to that story: https://time.com/jeffrey-stockbridge-kensington-philadelphia-photographs/). Although this story was not written or produced by the same authors as the video documentary featured in our GeoStoryTelling, it may potentially enrich our understanding of the topic and the stories we are sharing.

This concludes our discussion of the creation of the database as shown on the left-hand side of our flow diagram (Fig. 8.1). Once this database is saved as a "*.csv" file, it can then be loaded into the GeoStoryTelling UI, as we discuss next. Although this description may seem overwhelming, workshop offerings with participants who have no prior experience in GIS or computer programming have consistently indicated that full completion of datasets with 10 cases was achieved in one hour, including the explanation time for each of these elements. Participants in Europe and South America indicated that the creation of *.csv files separates columns with semicolons instead of commas. In these cases, the Microsoft setup must be updated to use commas. Alternatively, Google Sheets may be used to create these files and then save them locally before uploading them to the GeoStoryTelling software.

Geostorytelling user interface

Fig. 8.5 shows the GeoStoryTelling UI. As described earlier, this UI comes preloaded with a toy dataset that can be used for replication.

As illustrated in the flow diagram, we can start by loading a database we have created following our previous discussion. Alternatively, to get familiarized with GeoStoryTelling, we can also choose to load the dataset included with the software. To achieve the latter, we can select the option "Minimal example with six cases" under the section "A. Load your data or the example" and then select the option "Click to load GeoStory data example shown below" under section B in the UI. If, instead, you want to use your own data, you can drag your "*.csv file" to the UI or search for your database directly by clicking on "Click to browse from your CSV," as shown in the GeoStoryTelling UI in Fig. 8.5, also under section B in the UI.

Once the data are loaded, such a database will be automatically displayed in the UI. To execute GeoStoryTelling, you just need to select the columns that correspond to the options in Section C. In the toy database we are providing, the columns of the database match the column names we discussed above on page 54. As indicated in the flow diagram (Fig. 8.1), the only required columns to obtain the interactive maps are the latitude and longitude coordinates. The remaining columns discussed in Table 8.1 are highly recommended but optional.

Note that GeoStoryTelling was programmed to automatically recognize the column names of the databases loaded into the UI. These column names will also match the database displayed within GeoStoryTelling's UI once the databases are loaded. To select these columns in Section C, simply click on each of the options displayed on the left-hand-side of the UI, as shown in Fig. 8.5, and select the specific column name that conceptually matches the option. For example, in the toy dataset, we refer to latitude as "lat." Accordingly, under the option

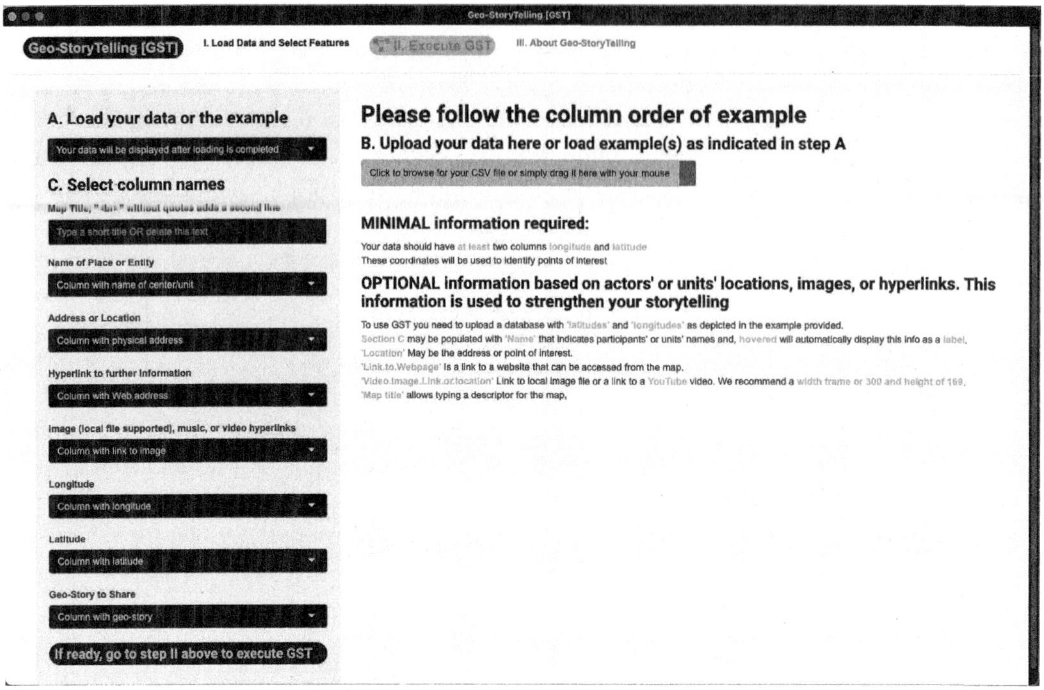

FIG. 8.5 GeoStoryTelling User Interface. Mac users can download GeoStoryTelling herehttps://cutt.ly/E724wAx, Windows users herehttps://cutt.ly/J724f0C.

"Latitude," simply click the dropdown menu and select the column "lat." A similar rationale applies to the rest of the columns in the database.

Finally, note that GeoStoryTelling allows users to provide their own titles for the resulting HTML visualization. The default text for the title (added as a place-holder) reads "Type a short title or delete this text." This implies that if users do not want to add any title to their resulting maps, they should delete this text in the UI—if they do not delete this text, the map will display "GST Title: Type a short title or delete this text." The GST title is included in the back-end of GeoStoryTelling. When users want to modify the title to be displayed in the HTML map, they can type the title they believe best captures the goal or purpose of their GeoStories.

Once all desired options have been populated, we can proceed to generate the HTML interactive maps by clicking on the tab called "II. Execute GST." If the latitude and longitude coordinates are selected, GeoStoryTelling will automatically launch the map in a browser. In addition, GeoStoryTelling is programmed to automatically save the resulting HTML file locally, allowing it to be distributed or hosted on the internet at researchers' discretion. Note that, in addition to a file called "index.html," researchers will also need to distribute the folder called "lib." Finally, every time GeoStoryTelling is executed, this file and folder will automatically be rewritten.

These outputs are also portable across analytic stages. The interactive HTML maps, route traces, screenshots of key moments, exported images, and linked multimedia panels are useful not only for identifying spatial structure, but also for supporting subsequent interpretive work. In the broader ecosystem developed in this book, researchers may bring these outputs into ISARI by uploading screenshots of the maps, copied text from the narrative panels, exported figures, or notes derived from route and distance comparisons. Because ISARI can operate locally and work with both textual and image-based artifacts, these GeoStoryTelling outputs can become the basis for scholarly brainstorming, memoing, comparison of alternative interpretations, and evidence-grounded writing without moving sensitive materials outside the researcher's control.

Distance calculations and mapping editing

The outputs produced by GeoStoryTelling allow users to calculate distances in two main forms. This capability meaningfully extends the methodological range of GeoStoryTelling. By allowing researchers to compare straight-line distances with editable, path-dependent route tracing, the framework can move beyond static geolocation and begin approximating access, mobility constraints, travel burden, and route-conditioned experiences that may shape participants' narratives in important ways. The first and simplest way to calculate distances is assuming that we can "fly from point A to point B" (see González Canché, 2023d). Although easy to achieve, this distance calculation, referred to as "as the crow flies," may be misleading depending on road availability along with many other place-based factors (González Canché, 2023d). The second type of distance included in the visualizations allows users to follow paths and roads and, based on users' knowledge, may be much more precise.

To illustrate these two types of distances, let us discuss Fig. 8.6. This figure zooms into the content presented in Fig. 8.3 and focuses on two points: a home (Manu's, as shown on the map in green) and a place of work (spatial literacy, as shown on the map in magenta). In this map (accessible here: https://cutt.ly/9r2wkXav), if users select "Draw a polyline" (marked with 1 in Fig. 8.6) and click on Manu's location, then click twice on spatial literacy, a line connecting these two points will be drawn, as shown in Fig. 8.6. After this, when users click on "Enable Measurements" (marked with 2 in Fig. 8.6), the "as the Crow Flies" distance will be displayed. In this case, the resulting distance is 3.8 miles. If we repeat these steps once more, another line will be added. However, if after this addition we select "Edit layers" (marked with 3 in Fig. 8.6), we will be able to add nodes to this line, and by doing so, we can follow driving or walking routes. This dotted line is shown in green in Fig. 8.6. After carefully following driving laws, we can see that the dotted line shows a distance of 4.8 miles. This more precise distance calculation results in an increase of one mile compared to the simpler "as the Crow Flies" procedure. Within the broader analytic architecture of this book, such route-based comparisons may also function as prompts for local AI-supported scholarly reasoning in ISARI, where researchers may compare alternative interpretations of movement, access, and spatial burden and begin translating those contrasts into evidence-grounded prose.

Note that the lines are by default magenta in color. This can be adjusted by selecting "Style Editor" (marked with 4 in Fig. 8.6) and then selecting the line or any other object to be edited. Other objects include pins, polygons, and circles, all of which can be added to the maps. Each of these elements may calculate distances following the same rationale described here. That is,

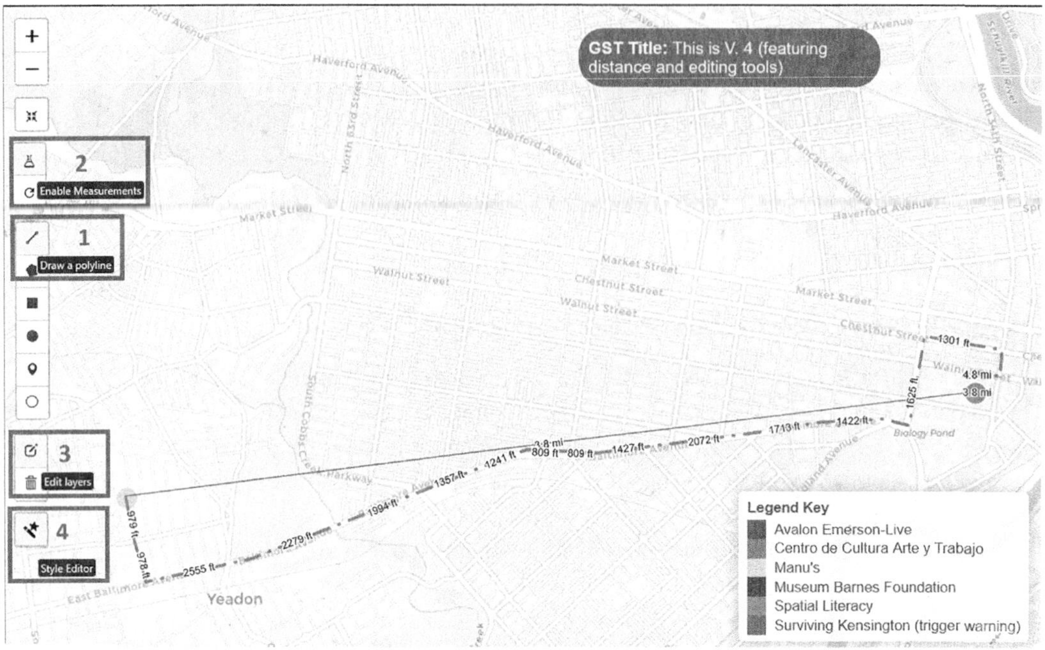

FIG. 8.6 Figure illustrating the addition of distance measures and their different precision levels. In this figure, we illustrate how "as the Crow flies" distance measures, which assume we can travel directly from point A to point B, may result in lower-biased estimates compared to distance measures that follow roads and traffic laws.

after adding an object, we should select "Enable Measurements" to calculate distances. In the case of polygons or circles, their addition may aid our geostories by highlighting areas of interest on the maps.

Toy database included in geostorytelling

For illustrative purposes, the database included with GeoStoryTelling contains six cases. Two of these cases feature embedded videos from different sources. One is retrieved from YouTube and corresponds to a 2016 presentation focused on applications of spatial data modeling to education and social sciences research. In this case, the hyperlink that provides more information about this story simply features a link to the university where that presentation took place. Finally, the text is a short reflection on the meaning of that talk for the presenter.

The second video was retrieved from Vimeo. This video captures the narratives of inhabitants of the Kensington neighborhood in Philadelphia. Considering that the participants featured in this documentary are experiencing addiction to drugs, homelessness, overdoses, as well as injections, we felt compelled to add a trigger warning as part of the name of this story. The link added in this case leads to another report on the same topic conducted in 2012 by TIME magazine, and the text featured in this story is a description of the video documentary as retrieved from Vimeo.

The third story features the live recording of a concert that took place in September 2022. This recording was retrieved from SoundCloud. In this case, the link associated with this story is a website, https://www.1001tracklists.com, containing all tracks from that concert as separate audio files (or videos of the original versions, which can also be played on YouTube, for example), rather than being mixed together by the featured artist. The text depicted in this story simply accounts for the description provided by SoundCloud.

In line with our goal to showcase the flexibility of GeoStoryTelling, we are also including a short reflection on "home" as a during- and postpandemic place where academic work and learning may occur based on technology access and online resources. Returning to our illustrative goal, in this example, we are also demonstrating how images locally stored on our computers may be embedded in our GeoStoryTellings. The hyperlink selected in this case features the participant's job-related website.

The last two stories show two examples of literacy-related places. One features the Barnes Foundation as a traditional and well-established place where culturally-relevant events are realized or exhibited. The image embedded in the map was retrieved from the gallery of images featured by the Barnes Foundation, and the hyperlink takes users to its main website. In this example, the text is an attempt to convey how this museum should not necessarily be considered more important than other places where literacy and cultural activities are happening, such as the last example we are presenting.

The last example features a center where counternarratives of traditional notions of art are taking place. This center, although not exclusively, mostly works with undocumented immigrants from Mexico and South America. As part of their activities, members of this community create different forms of art (music, dance, paintings) and provide spaces for after-school activities and literacy-oriented opportunities for adults. The image depicted in this story shows the role that this center played by serving as a vaccination site, where the (likely undocumented) community members could be vaccinated without fear based on their legal status in the United States. The text featured also highlights that the cultural activities are not limited to immigrants but also reach Americans (mostly academics from neighboring universities) who are investing time and resources in strengthening the scope and presence of this cultural center. Finally, the hyperlink simply leads to the main website of CCATE, which, among other things, features ticket sales for their artistic events.

Please note that this dataset was primarily built to illustrate the types of multi-media elements that can be integrated with GeoStoryTelling. In our discussion section, we will delve more deeply into one of these featured stories to further illustrate how researchers may use GeoStoryTelling, focusing on a more serious health and social issue: the opioid crisis.

Discussion and applications

We were purposefully eclectic in the selection of the stories featured in our toy database, mostly to illustrate, on a generic or "meso-level," the flexibility of featuring narratives that do not necessarily need to be connected. More realistically, however, a study with a clear purpose will be more uniform in the selection of the narratives, which, although individually they will likely present important variations, will also convey information about unifying themes. That is, if the topic is drug addiction in a given neighborhood, as we discuss next, although individual stories may vary, it is quite likely we will find common ground or structure across the multiplicity of voices and experiences being captured.

In the following lines, please allow us to return to the discussion of the selection of narratives or stories featured in our GeoStories. That is, as stated earlier, although GeoStoryTelling is capable of displaying large amounts of text, featuring entire transcripts would be both challenging and likely not very useful. Instead, in the same way qualitative researchers select quotes to illustrate their points in a traditional qualitative paper, researchers relying on GeoStoryTelling may also want to select and geocontextualize a quote or quotes that will become central findings illustrating the points they wish to communicate. In terms of deciding how to proceed with participant selection, although in our example we presented different participants or cases in a single map, researchers may instead decide to feature one participant across their space-time continuum per map.

With these possibilities in mind, the following discussion presents two nonexhaustive analytic approaches that researchers could implement. In the first approach, we illustrate how our analyses may be more precisely contextualized by zooming-in on the specific locations where our participants' stories took place. In the second analytic approach, we illustrate how following the evolution of our participants' stories or experiences over time may further strengthen our understanding and lead to the development of plans of action that may have better prospects of success. Note, however, that GeoStoryTelling is flexible in incorporating a myriad of analytic decisions that will likely evolve organically based on the research team's goals and the topic of study. In the end, then, the strategies featured next are not mutually exclusive; instead, they can be merged based on our research goals and data collection strategies.

Stories with more precise locations

To illustrate this analytic strategy, let us revisit what we consider the most relevant example demonstrating the analytic power of GeoStoryTelling that we included in our toy database. This example corresponds to the presentation of the Kensington neighborhood in Philadelphia. As depicted in the video documentary, there were multiple participants featured. Although all these participants were situated in the Kensington neighborhood, their stories felt somewhat aggregated, as they represented the neighborhood globally. Even though the video includes specific places (e.g., a park, a street corner, a house) that could be "decomposed" to offer more detailed accounts of these participants' experiences, the presentation remained generalized. Accordingly, we could more precisely feature or pinpoint those locations along with the specific geostories being highlighted. In other words, researchers, instead of presenting a video that features multiple stories in an area encompassing multiple blocks, may decide to feature individual stories while situating those stories in the specific places (i.e., addresses) where they were recorded or gathered.

This process of "zooming-in" to provide more detail about location and experiences may increase our understanding of those experiences; and, to the extent that we are able to add more context to those narratives via "hot-links," more nuance in these participants' stories may also be gained. We must, however, note that this is a vulnerable population, and as such, we should also have a clear goal in mind if, or when, deciding to feature these more granular levels of detail. These goals may involve the creation of programs or strategies to address their addiction to drugs, provide them with shelter options, or both. We must also be aware of the added risks we are creating for these participants by not only overtly showing their identities but also revealing their locations. Their addictions may make them targets of many types of abuse, even more than before they agreed to be featured in this type of research or documentary. These risks may become even greater when multimedia elements, hyperlinks, or route traces are added, because the accumulation of contextual detail can make re-identification easier even when names have been removed. For that reason, the public sharing of GeoStoryTelling outputs should always be treated as optional and governed by the same principles of confidentiality, data minimization, participant protection, and contextual sensitivity that shaped the original qualitative or mixed-methods design (see González Canché, 2025). In many cases, local storage, restricted sharing, selective redaction, or the use of screenshots rather than full interactive release may be methodologically and ethically preferable.

Fewer stories followed over time

In addition to decomposing multiple stories, we could also focus on documenting the pathways that brought our participants to the time and place where the interview was taking place. For example, assume we may be interested in identifying current inhabitants of Kensington *who did not grow up in this neighborhood* or even in Philadelphia. Here, the goal may be to understand the processes they experienced to "find Kensington," with an emphasis on highlighting their first experiences in or the points of entry to this neighborhood. That is, we could aim to understand not only what their entry points were but also the processes and reasons that led them to reach those entry points. In this form of data gathering, the contextualizing tools may be their voices, linked to their places, and the text featured in those locations may be our reflections or our analyses of their experiences. Ideally, we may be able to decompose their stories based on global time-stamps (González Canché, 2022, 2023a) where we may reconstruct the temporal evolutions of their stories based on their episodic memories or their recollections of the events that brought them to the time and place where they shared their stories with us.

Far from being a voyeuristic endeavor, the goal of documenting and understanding their trajectories is to determine if there are commonalities or structures that explain entry points into "Kensington." We are using quotation marks when referring to Kensington here because "Kensington-like" places, also known as "open-air narcotics markets," are unfortunately becoming more prevalent across the United States (Percy, 2018a, para. 5). Accordingly, to the extent that we are able to understand entry points, their mechanisms, and the types of structures supporting their prevalence, plans of action may be preemptively designed to offer support to future "Kensington inhabitants or newcomers" before they reach homelessness or more acute drug addiction problems.

This is important because "drug tourists who come [to Kensington-like places] may never leave" (Percy, 2018a, para. 7).

Application example of survey responses involving campus safety

In this section, we provide an illustration of how GeoStoryTelling may be used to gather data via online surveys that are larger in scope. In this example, we request three prompts designed to protect the identity and preserve the anonymity of the survey respondents.

The following lines describe how we recommend phrasing the instructions. Language related to Institutional Review Boards in the United States and Canada, as well as the General Data Protection Regulation in Europe, should be included in actual implementations of this example.

Proposed instructions

With the goal of strengthening our efforts to provide you with the best possible on-campus experience, we are asking for your support in sharing any experiences you may have had while on campus where you felt unsafe, threatened, or unwelcome, while briefly elaborating on the issue or issues that made you feel that way.

All your responses will be completely anonymous, as no names are asked, and your participation is voluntary. However, in order to gain more nuanced understandings, we would appreciate it if you could provide us with your pronouns and ethnicity, if you feel comfortable sharing this information with us. In addition, we are interested in mapping the locations where these incidents took place, which is why we ask you to describe these locations as detailed as possible, as if you were looking for these places using Google Maps. You can be specific; for example, in the case below, the respondent mentioned the second floor of Stiteler Hall at The University of Pennsylvania, which, on Google Maps, pointed to the Graduate School of Education student records office.

The specific fields we are requesting information on are the following. Note that each prompt also contains example responses below:

1. **Pronouns and ethnicity:** They/them, Chinese-American
2. **Place of incident:** Second Floor, Stiteler Hall, The University of Pennsylvania
3. **Incident description:** I felt threatened by the security officer who did not allow me to enter the building because I had forgotten my official ID, even though I had a class to attend. Fortunately, a professor who was passing by corroborated that I am a student at the University of Pennsylvania. I think security officers should be able to locate students' records in the system to corroborate our identities.

In this example, we can see how narratives may be collected, and even if these responses amount to hundreds or thousands of cases, the process of geotagging them is straightforward. That is, once our initial database has been downloaded from services like Qualtrics or SurveyMonkey, for example, we can copy the responses from the "place of incident" field (prompt or question number two in our example) and paste them into GPS Visualizer (https://www.gpsvisualizer.com/geocoder/) as illustrated in Fig. 8.2B. As discussed above, this batch geocoding method allows the effortless retrieval of thousands of latitude and longitude coordinates with free services like Bing Maps—note that we also mentioned that a free API key can be created to access Bing map servers at https://www.bingmapsportal.com/. Please note that this example is for illustrative purposes; its implementation would require Institutional Review Board or General Data Protection Regulation approval. Also note that, although no names are requested, there is the possibility that the narratives may include names of alleged aggressors. Accordingly, the research team must be extremely careful in vetting and curating the data to avoid any mishandling of information or unfairly damaging the reputation of individuals involved in these incidents.

In addition to geomapping the raw responses provided in the original incident descriptions (third question or prompt shown above), in data collections where these responses reach hundreds or thousands of cases, researchers may also apply machine learning text classifications to label these incidents into categories. This process may be useful when trying to label and synthesize or summarize these incidents. This process may be achieved with another no-code software available at Expert Systems with Applications (see González Canché, 2023c) or the International Journal of Qualitative Methods (see González Canché, 2023b). The output of these text classification analyses will have the same information in terms of location that can be geocoded using GPS Visualizer, as depicted above. The main difference is that we may now map these incidents based on a count of similar classes of issues, namely threats or discrimination, rather than listing the original comments. The decision to present original responses or the classes of incidents depends on the goals of the research team, and in all instances, we must prioritize fairness and protect the identity of all parties involved.

Integrating GeoStoryTelling outputs into local scholarly brainstorm

One of the strengths of GeoStoryTelling is that its outputs are portable across analytic stages. The interactive HTML maps, embedded videos and audios, linked external resources, screenshots of specific geostories, distance calculations, and route traces are useful not only for situating narratives in place, but also for supporting subsequent interpretive work. In the broader methodological ecosystem developed throughout this book, these outputs can be brought into ISARI, a local, fully offline, open-source, and multimodal brainstorming environment designed to support scholarly memoing, comparison, synthesis, and evidence-grounded writing.

This possibility is especially important because geocontextualized stories, however compelling, do not determine their own meaning. Researchers still need to ask why certain locations mattered more than others, how route constraints shaped participants' experiences, what spatial patterns recur across stories, whether place-based fears or opportunities cluster around common mechanisms, and how geostories should be translated into qualitative or mixed-methods explanation. Because ISARI can work with textual notes, screenshots, exported figures, scanned pages, and other image-based or document-based artifacts, researchers can upload GeoStoryTelling outputs in multiple forms—for example, screenshots of maps, copied text from story panels, exported distance comparisons, route-trace images, or notes derived from multimedia-linked cases—and use them as the basis for local scholarly reasoning.

This workflow should not be mistaken for automation of interpretation. On the contrary, it extends GeoStoryTelling's methodological purpose. The outputs remain guides, not conclusions. ISARI can help researchers compare geospatial interpretations, generate analytic memos, test explanatory contrasts, and begin drafting evidence-grounded prose, but responsibility for deciding what these geocontextualized narratives mean remains with the research team. In that sense, GeoStoryTelling and ISARI are methodologically aligned: one renders splace-based narrative structure visible while preserving local control; the other helps researchers think with those outputs under conditions of privacy, interpretive accountability, and data sovereignty.

Limitations and future steps

We understand that the opioid crisis and harassment or discrimination issues represent important social and health-related problems that are more dire than many other research topics in the social sciences. In a sense, then, by illustrating how GeoStoryTelling may be used to provide more nuanced understandings based on geocontextualizing these complex cases, we are also demonstrating the salience and flexibility of this methodology for social and health scientists. To illustrate this analytical flexibility, we also provided an example of how researchers may apply GeoStoryTelling to survey responses, either to present "raw stories" or even to first classify those stories and then map the resulting categories.

Despite these illustrations, we are not claiming that our proposed implementations and the operationalization of our methodology are without flaws. In addition to the added risk of disclosing locations—and potentially, but not necessarily, the identities of our participants (we can use pseudonyms or blur their faces, for example)—there is also the issue of selecting the specific narratives we decide to feature in our geostories. As indicated above, whenever possible, we should seek the advice of our participants to validate whether our narrative or story selection actually captures their experiences—via interactive reviewability (Alexander et al., 2019a; González Canché, 2022, 2023a).

Relatedly, whenever possible, we should add, as part of the narratives we are selecting (i.e., the actual text included in the column "Geo-Story" as shown in Table 8.1), place-based indicators that we, or our participants, may have considered worth mentioning to provide more nuance to their resulting geostories. Specifically, we could add, as part of their text narratives, attributes of the locations where those stories took place that may help us better understand their stories. This process ties back to our initial discussion of the relevance of the neighborhoods where we experienced life in influencing our decisions, fears, beliefs, and perceptions of the world (González Canché, 2023d).

Although these place-based indicators could be mapped using GIS methods and data sources, there are two main concerns that deterred us from including these attributes in the current version of GeoStoryTelling.

First main concern

The first concern is that data collection would be much more complicated than the one we illustrated here—which will require some form of GIS training and at least some basic statistical programming—thus defeating the goal of GeoStoryTelling, which aims to avoid both of these barriers, as we further discuss below. See González Canché (2023d) for a set of low-code tools to apply GIS in social science research. That is, in the easiest approach, the data collection will require accessing place-based indicators from the American Community Survey in the United States, for example, or from other agencies in other countries, along with the shapefiles (more on these files below) required to actually map one of these indicators at a time (González Canché, 2023d). That is, visualization of indicators via shapefiles handles one indicator per map, unless several layers of data are added to the resulting HTML visualization. More complicated data collection approaches may consist of gathering our own data with specialized equipment, for example.

Let us briefly discuss shapefiles. First, note that based on OpenStreetMap.org, GeoStoryTelling includes points, lines, and polygons (González Canché, 2023d) to represent buildings (i.e., points or the intersection of a latitude and longitude coordinate), streets or rivers (i.e., lines or the continuous intersection of a collection of points), and blocks, ZIP code tabulated areas, or cities (i.e., polygons or the intersection of lines that end at their starting point). What GeoStoryTelling is not accounting for are place-based attributes (like crime or poverty rates, for example) that may be merged with those shapes to add color, for example, to the resulting points, lines, or polygons.

Even though the process of merging shapefiles with place-based attributes is not complicated per se, this merging process requires feature engineering procedures (i.e., advanced data cleaning and preparation) that may also require at least some form of statistical programming proficiency and/or access to proprietary software like ArcGIS. From this view, then, the addition of place-based features that require their merging with shapefiles, in essence, defeats two of the main goals of GeoStoryTelling. The first is ease of use; that is, we designed GeoStoryTelling to expand access to data science and visualization tools that do not require any computer coding or statistical programming. The second is that we are offering GeoStoryTelling free of charge so that cost is not a barrier to its utilization. In this respect, the work by González Canché (2023d) offers minimal or low-code tools to achieve these integrations and obtain interactive visualizations such as those produced by GeoStoryTelling.

Second main concern

On a more substantive level (i.e., moving beyond technical expertise and monetary costs), depending on the episodic memories of our participants and considering the nonlinear temporal dimensions in ethnographic and qualitative research (Braun and Clarke, 2006), ideas shared by our participants may prompt them to remember other past events that may have shaped their experiences. In these cases, then, the inclusion of place-based attributes across multiple time spans or time periods is not currently feasible to achieve, even with the newest developments in spatial visualization (González Canché, 2023d). Rather, place-based indicators are currently constrained to show a snapshot in time of how certain areas were expected to fare. This expectation is provided by surveys like the American Community Survey in the form of "estimates" of place-based indicators like unemployment or household composition, for example. That is, these estimates are statistical predictions of those place-based attributes, as opposed to the actual measures of such indicators (U.S. Department of Commerce, Economics and Statistics Administration, U.S. Census Bureau, 2020) (hence the relevance of considering margins of error when utilizing these estimates).

How do we address these apparent limitations?

Although we decided not to include traditional place-based estimates in the GeoStoryTelling interactive maps, note that the texts, pictures, videos, sounds, and links to other sites we are adding to each point (i.e., the intersection of latitudes and longitudes) are indeed qualitative attributes that help us understand that point. Within the broader workflow developed in this book, those same multimodal attributes may later be examined in ISARI as prompts for local scholarly reasoning, allowing researchers to compare how qualitative place-based detail complements or complicates more traditional spatial indicators without requiring that all such indicators be embedded in the map itself. From this perspective, our maps do contain attributes, and these attributes are not statistical predictions or estimates but more precise accounts of our participants' experiences.

Nonetheless, we can also more organically address the absence of traditional "quantitative" estimates across time associated with those locations by purposefully being more strategic during the data collection process. That is, if we aim to geocontextualize our participants' experiences in issues related to place-based safety, for example, we should prompt them to include place-based details that may help us better understand their experiences. In returning to one of our previous examples, which featured the experiences of one participant who attended graduate school in Los Angeles, California, as researchers, we could follow up our conversation with them as follows:

[For context, we are adding the last part of this participant's narrative; please see the full quote on page 56.] I was going to start my graduate coursework in a different country—in Los Angeles, California. In full transparency, these graduate school years were extremely challenging in many aspects of my life.

[Here we show how we followed up on this thought with a purposeful intervention to geocontextualize that last statement.] Thank you for sharing. In what ways was your experience challenging? Was it related to just being in a different country, or did it somehow relate to certain aspects of your immediate surroundings, spaces, and places, such as the neighborhood where you used to live or the overall attributes of the city?

[Response] It was a combination of everything, I guess. Coming from a different country and struggling to afford rent in Los Angeles implied that the apartment I was able to afford was both far from campus and in a not-so-great neighborhood. To put it bluntly, I did not feel safe going home at night, and I did not have a car then. This meant that my fear of gangs and violence made me plan my day to either go home before it got dark or try to stay in the library, which, back then, was thankfully open 24 hours. You know, feeling trapped really hurt my mental health. I didn't come to this country to suffer insecurity or poverty; this was supposed to be the best experience of my life.

In this example, we were able to gather two key pieces of information that were space and place-related. In terms of space, as we discussed earlier, its meaning may be more abstract and, in this case, it may imply notions of distance. For this participant's experience, commuting time or distance was quite uncomfortable, not only because of the distance itself but also due to the type of neighborhood where her apartment was located. Specifically, in terms of places, it seems that she not only disliked her neighborhood but also felt quite unsafe, to the extent that she preferred to spend the night at the library rather than risk going home after dark or arriving home after dark. These two pieces of information should be added to the resulting text narratives selected to be featured in the GeoStoryTelling visualizations, as they will help us better understand the source(s) of her challenges during graduate school. Moreover, these details will also address the issue of not including other quantitative place-based estimates for the reasons mentioned above.

This example illustrates what Ajayakumar et al. (2019a) were referring to when describing the relevance of including *context* that goes well beyond merely adding layers with quantitative estimates of sociodemographic and socioeconomic place-based indicators. In their words, the interest is not only in indicating that something occurred but, more importantly, in explaining why it did or how it did (Ajayakumar et al., 2019a).

Once more, we should double-check with our participants to ensure that our interpretations are accurate and, overall, try to give them as much agency as possible in the sharing of their stories. After all, GeoStoryTelling as a methodology is only relevant to the extent that we are accurately capturing the geocontextualized experiences of our participants.

Is geostorytelling just another software tool?

It is worth noting that a reviewer mentioned that most of the features contained in the output rendered by the GeoStoryTelling software can be approximated with Google Earth. In this respect, it is worth noting that, like our software, Google Earth cannot incorporate layers of quantitative data. Unlike our software, Google Earth cannot support audio or video in the resulting visualizations. Like GeoStoryTelling, Google Earth can measure distances using the "as the crow flies" method (that is, drawing a single line assuming we could "fly from point A to point B"), and it can even generate paths to be more precise in distance calculations—see the section "Distance Calculations and Mapping Editing" above. However, even if users prefer to use Google Earth for familiarity, those users may also rely on the methodological framework we are presenting here and create their geostories with Google Earth, even if this platform does not enable users to upload their data, as our GeoStoryTelling software does. In this respect, note that we are not arguing that the software accompanying our framework is better than any other software available on the market. The true value added by our software is that it is *completely code-free, free to use, free to distribute, and runs locally.* This local architecture is methodologically significant because some newer AI-assisted alternatives for generating standalone HTML maps require users to upload their databases to external servers. Even when such tools are visually powerful, they may be less appropriate for qualitative, health, and mixed-methods projects involving sensitive place-based narratives. GeoStoryTelling's commitment to local execution therefore remains one of its most important differentiating strengths. This means we can obtain state-of-the-art interactive visualizations without any coding or programming experience, and we do not need to upload any of our data to any external server. More importantly, we are providing this software as a complement to the methodological framework presented in this paper. That is, GeoStoryTelling *is not just another software* tool, as users may still apply the concepts we are describing in this study and execute them in other software tools that may even require computer and statistical programming proficiency.

Potential learning curve

Despite being user-friendly, researchers might still require time to familiarize themselves with the software's functionalities, especially in integrating various multimedia elements. This is why, in this subsection, we are offering an estimate of the time needed to create a database and become fully familiar with the software's capabilities based on workshop offerings, wherein participants have started with no prior knowledge and have achieved the expected HTML outputs using the GeoStoryTelling software tool.

We have had the opportunity to offer free professional development workshops to undergraduate college students at the University of Guadalajara in Jalisco, Mexico, and to high school students in Shanghai and Guangzhou in China. In each of these workshops, we spent about one hour in total with a group of 100 students (approximately 25 in each workshop offering—with four workshop offerings in total) to create usable databases that rendered the expected interactive HTML outputs. This time included explanations for participants with no prior knowledge of the software's rationale and capabilities, as well as detailed instructions on the processes and steps needed to gather data, retrieve latitude and longitude coordinates, and compile links to connect the GeoStories to other websites. It is important to note that this does not mean complete research projects can be achieved in this amount of time. The exercises involved asking participants to identify places and spaces that were meaningful to them and then create the respective datasets incorporating images, texts, hyperlinks, and video or audio sources. It is worth noting that these participants were assumed to be tech-savvy, given that they grew up with technological tools. Older trainees, who may not have grown up with such tools, may require longer periods to build their datasets. That said, although not in workshop contexts, we have individually trained qualitative researchers in their sixties without any issues. In these few cases (three in total), the resulting training time consisted of two-hour sessions. However, it is important to note that these were individual sessions, as opposed to group training sessions.

Finally, please note that as part of our ongoing collaboration with ethnographers, qualitative, and mixed-methods researchers, we have started collaborating with researchers whose main form of data collection is through stories. Some of these ethnographers work with third and fourth-grade students, and they commented that they have started to follow the guidelines depicted in the paper to geocontextualize the stories generated by their students. They have also mentioned their intentions to teach their students how to build their own datasets to use with the GeoStoryTelling software. In this respect, these researchers are confident that their students have enough technological familiarity to generate and understand the steps required to create these datasets.

Technical support and sustainability

As part of our efforts to democratize data science, we are committed to offering technical support and maintaining the software up to date. In terms of technical support, we will continue to offer yearly professional development workshops at international conferences like the American Sociological Association and the American Educational Research Association. More to the point, in addition to providing our contact information in the UI, we are in the process of recording tutorial and instructional videos focused on troubleshooting common end-user issues that may emerge during database creation and curation, especially when working with large-scale research projects.

With respect to software updating and sustainability, we are committed to maintaining the software updates in both Code Ocean and the Windows and Mac versions. In all instances, we will document all changes made. For example, due to high costs, Stamen (see https://maps.stamen.com) was unable to continue providing its base tiles, which were used by GeoStoryTelling. Although this change impacted the rendering of the HTML visualizations in version 1.0 of GeoStoryTelling, resolving this issue simply required changing the tile provider, which Code Ocean reviewers accepted, and it is now published as version four. See https://doi.org/10.24433/CO.7385885.v4.

Software access and execution

Mac users can download GeoStoryTelling here https://cutt.ly/E724wAx, and Windows users can download it here https://cutt.ly/J724f0C. To execute the application, (1) Extract the zip folder. (2) Open the folder starting with "GeoStoryTelling-." (3) double click on the file called "GeoStoryTelling-Mac" or "GeoStoryTelling-Win." The first time GeoStoryTelling is executed, it may be necessary to select the menu options "View" and "Force Reload" to generate all paths to the newly extracted folder. If, after forcing reload, GeoStoryTelling does not display as shown in Fig. 8.5, close the window, execute GeoStoryTelling again, and, if needed, force reload once more. The interface will then load. This process is only required the first time GeoStoryTelling is launched on a computer. Mac users are also recommended (and in some instances they do need) to place the GeoStoryTelling software (and all software products discussed in this book) in their applications folder for newer versions can only read and write files from this location. The software applications have been comprehensively tested for viruses and are housed in a secure server location. Accordingly, all executable files are virus-free but may require administrative permission to be deployed. Finally, each of these executable files for the Mac and Windows OS contains all the source code required to modify and/or expand the application's capabilities in the file called "app.R." The Code Ocean server also hosts the software's backend code and can be accessed herehttps://doi.org/10.24433/CO.7385885.v4.

Final thoughts

The only requirement for the GeoStoryTelling software to work is that each of the selected texts or narratives needs to be linked to the spaces and places (or splaces, see González Canché (2023d)) we want to incorporate in our analyses. This process requires us to have a clear understanding of where we need or want to situate these narratives. Even in the Kensington neighborhood, where there exists an overt and clearly defined area, the selection of specific splaces to be featured in our GeoStories may include an aggregated depiction of the neighborhood or may instead share more precise locations. In other instances, such as with interview data, the selection of splaces will depend on the episodic memories of our participants. For example, we may be in Spain during some aspects of the story and in Los Angeles during other time periods.

In both cases, we strongly recommend consulting with our participants to not only reach common understandings of where the narrative should be situated but also to validate our interpretations and analytic decisions—i.e., interactive reviewability (González Canché, 2022, 2023a).

As mentioned above, our GeoStories may need to feature the same participant in multiple spaces and across different points in time. If this is the case, our database construction should reflect this space-time continuum, and the ID should reflect this time, which will become part of the resulting interactive map's legend. Recall also that the measure of time is relative to the specific experiences shared by our participants. That is, our reliance on global time stamps provides us with flexibility in the selection of units of time. While for some participants in the same study, *time* may be measured in years or decades, for others, time measurement may be operationalized in days or weeks. Both inputs should be respected and captured as precisely as possible. We hope that we have made it clear that GeoStoryTelling is capable of capturing these global and nonlinear time stamps and that the process to build the resulting databases to be loaded into the UI may be achieved without any need for specialized software. This ease of analysis, however, does not mean that the database construction is not time-consuming, for depending on whether we want to add hot links, images, video, or audio, the compilation of these optional elements may take some time. If, instead, we are satisfied with only featuring narratives that are geocontextualized, as in our survey example, then the dataset construction is less time-demanding, but the resulting GeoStories would be less contextualized, nuanced, or rich. In short, researchers may need to decide what elements they are interested in featuring when building their stories.

We hope that researchers in training and professionals alike will interact with the GeoStoryTelling UI, either by building their own datasets or, at the very least, by engaging with the toy database we are providing. We are convinced that access to these tools may lead to enriching stories being shared and, hopefully, to the discovery of structures and mechanisms that may aid in the development of effective plans of action. Within the broader infrastructure developed in this book, these same geocontextualized and multimedia outputs may also serve as structured prompts for local scholarly reasoning in ISARI, where maps, route traces, screenshots, and linked story elements can be compared, memoed, and translated into evidence-grounded qualitative and mixed-methods explanation. The time has come to remove barriers that continue to prevent ethnographers, qualitative researchers, and mixed-methods researchers from benefiting from the analytic and structure-unveiling capabilities that sophisticated data science and visualization tools bring to social and health sciences research. Finally, note that GeoStoryTelling may also be used for investigative reporting, like those stories featured in the Washington Post or the New York Times, thus going beyond academic research endeavors.

Appendix

Database creation

Once we have gathered and selected the texts or narratives that we want to feature in our GeoStoryTelling, we are then ready to start creating our database. This process is as straightforward as adding seven columns to a Microsoft Excel or Google Sheets object and saving it as a comma-separated value (or *.CSV) file or database. Note that this process will be the same if we have large-scale surveys, as we discuss in the subsection called "Application Example to Survey Responses Involving Campus Safety."

As we proceed with our discussion, we will elaborate more on each of these seven columns. Note that we do not have to follow the column names or the order we are showing below and in Table 1. However, for the sake of clarity, let us present the seven column names we are displaying in our database example. These names are:

1. Name or ID
2. Location or Address
3. Link to Webpage
4. Video, image, or audio link: Once more, images may be loaded from our local computer or from the internet. In the case of videos and audios, we need to obtain the embedded versions of the corresponding links, as illustrated below.
5. Longitude
6. Latitude
7. GeoStory

In this list of columns, GeoStory (i.e., the text or narrative we just discussed above) is the last column of the database. The seventh or last position of this column is selected purely for convenience. The text content of this column will typically result in wider columns than most of the other column contents, which may complicate readability if GeoStories are placed earlier in the database. Having said this, as we mentioned earlier in this subsection, the order of the columns is not important for the GeoStoryTelling software to render the HTML interactive outputs.

To illustrate the *step-by-step* process followed to build a database that can be used with GeoStoryTelling, we have added four panels to Table 1, each of which can be mapped back to our flow diagram shown in Fig. 8.1.

Panel A in Table 1 shows the initial status after the addition of the seven columns we described above has been completed. Panel B represents the addition of the narratives or stories that we want to feature. For illustrative purposes, we are simply populating these cells with generic texts that read "first story text example," "second story text example," finishing with an "Nth story text example." That is, our database may accommodate as many stories as we need or want to feature. The addition of these cases is as easy as copying each of the texts selected from our transcripts and pasting them into the column called "GeoStory."

Also note that in Panel B in Table 1, we have added identifiers to each of these stories. To illustrate the flexibility of GeoStoryTelling, we decided to feature the story of the same participant (called "A") at different points in time in the first two rows. This is participant A, whose first selected story took place at time 1 and may be represented as $ID_{A,t=1}$. Moreover, this same participant A also provided us with another story, which in this case took place at time 2 ($ID_{A,t=2}$). This representation of time follows *global time stamps*, as described by Artale et al., 2007a and Jensen et al. (1998b)—see also González Canché (2022) and González Canché (2023a). Global time stamps allow us to be flexible in how time is measured. That is, time may be measured in months for participant A in one instance, and it may also be measured in years for that same participant in her second story, based on her storytelling or episodic memories (Murray et al., 2018a). Moreover, these time elements may not be linear in how they evolved. Instead, the storytelling of our participants may jump back and forth in the space-time continuum, and it is our challenge and responsibility as researchers to capture these jumps—and to consult with our participants to ensure our interpretations are accurate.

Nonlinear temporal dimensions in qualitative, ethnographic, and mixed-methods research enable more flexibility in capturing nuanced understandings of the stories our participants are sharing (Neale, 2015a, 2019). More specifically, and organically speaking, what our participants shared regarding "time X" may have prompted them to share more information regarding "time $X + n$," and this latter story may have prompted them to share events that occurred at "time $X - n$." Here, X is a point in time situated around our participants' experiences, and n represents time units before or after such a point in time took place. Once more, the n time units may be hours, days, years, or decades, for example, before or after a given point in "time X." To minimize confusion and create common understandings, let us assume that a participant shared with us the following narrative:

"I finished college in **1997** back home in Mérida, Spain. Who knew that **10 years later, in 2007**, I would start my graduate coursework in a different country, in Los Angeles, California? In full transparency, those graduate school years were extremely challenging in many aspects of my life. This experience was in stark contrast to my middle school dreams of studying abroad; the real world didn't match my expectations."

Here, this participant initially provided information situated in 1997, but then there was a jump in time ten years into her future with respect to 1997 (time $X + n$), when more information about graduate education was shared. In addition, note that in this same example, our research participant followed her train of thought by jumping back to her high school years, which, following this nonlinear time rationale, may be expressed as "time $X - n$."

This brief example serves to showcase two main features. We can accommodate multiple mini-stories of our participants, and the units of time we may depict do not need to follow linearly evolving trends nor even be measured in the same unit of time. Instead, we may capture more nuance in the storytelling of our participants by applying nonlinear, global time stamps. In addition, to enable the inclusion of meaningful identifying information, the content of this column will be displayed as a legend in the HTML rendering and will be automatically mapped with specific colors to add more context and strengthen the navigation process. From this view, all events that share the same ID will have the same color in the GeoStory.

As shown in Table 1, the next step in our database-building process consists of geo-referencing those stories or narratives.

GeoTagging or situating those stories on earth's surface

Note also that in this brief example, our participant shared with us information that may be situated on Earth's surface. Specifically, she mentioned Mérida, Spain, and Los Angeles, California. Although our participants' stories will likely not be as short as these experiences shared in our example, recall that the selection of texts or narratives is based on the purpose of the study and the decisions of the research team. For convenience, let us assume that $ID_{A,t=1}$ and $ID_{A,t=2}$ in Table 8.1 capture her graduation from college in 1997 and her starting of graduate education 10 years later (and in another country), respectively.

As described earlier, the only required elements that GeoStoryTelling needs to produce HTML outputs are latitude and longitude coordinates, which are the most fundamental pieces of information that enable us to locate units on Earth's surface. Although coordinates may take two main forms, depending on whether we rely on a planar projection of Earth (measured in meters) or a geodesic grid (measured in degrees) (see C. Holdgraf and L. A.Wasser, 2022), for our GeoStoryTelling purposes, let us just note that we may retrieve these coordinates from services like Google Maps (https://www.google.com/maps), Bing Maps (https://www.bing.com/maps) or GPS Visualizer (https://www.gpsvisualizer.com). The latter is an online, free-to-use tool that conveniently allows us to retrieve batches of coordinates at a time (i.e., dozens, hundreds, or thousands of addresses or locations, as we illustrate below). This process of retrieving latitude and longitude coordinates from addresses or locations is called geo-coding or geo-referencing—hence, we will use these terms interchangeably henceforth.

Geo-coding or geo-referencing

To further elaborate on this geo-referencing process, let us discuss Fig. 8.2A. This figure shows the result of retrieving the latitude and longitude coordinates using Bing and Google Maps services. In each case, we may input an address in the search bar or simply right-click on a desired point on the map. In Fig. 8.2, we right-clicked on the same location (i.e., 4341 Baltimore Avenue in Philadelphia, PA, USA). As you can see, in both cases, we have two sets of numbers that, although they show some variations (based on each service's implementation of Earth's projections), in the grand scheme of things, match the representation of the location of interest.

The first number (starting with 39.94) is the latitude, which ranges from – 90 to 90 degrees. A value of 0 represents the Equator, which divides Earth into the northern and southern hemispheres. Hence, positive values indicate that such a point is north of the Equator, and negative values indicate that such a point is south of the Equator (González Canché, 2023d). The second number is the longitude coordinate. Longitudes range from – 180 to 180 degrees. A value of 0 degrees longitude is referred to as the Prime Meridian, located in Greenwich. From the Prime Meridian to the left, up to – 180 degrees, we move west—hence all the longitudinal values in the American continent are negative. Similarly, from the Prime Meridian to the right, up to the + 180 degrees, we move east (González Canché, 2023d).

For our purposes, we just need to copy these coordinates and paste them into our database under the latitude and longitude columns shown in Table 8.1. Note that this georeferencing process works with specific or more general locations. That is, if we input addresses, the results will be quite precise. However, we can also just input or type city names or even country names, and services like Google or Bing Maps will still render the latitude and longitude coordinates, capturing the centroid (center) of those cities or countries. In going back to our brief narrative, our participant mentioned two places: Mérida, Spain, and Los Angeles, California. This information serves to demonstrate a batch geocoding approach also available free of charge when connected to Bing Maps via GPS Visualizer (https://www.gpsvisualizer.com/geocoder/). To use this service free of charge across the globe, we need to register at Bing (https://www.bingmapsportal.com/). This is a process that takes five minutes and results in a unique access key, like the one we covered in red in our rendering of Fig. 8.2B—note that we should never share our free Bing API key.

Batch geocoding is useful when we are dealing with multiple locations or large-scale projects. That is, although in our brief example we only have two locations, the same rationale applies to hundreds or thousands of addresses or general inputs like cities and countries. As can be seen in Fig. 8.2B, we may select Bing as the geocoding engine and add our Bing API key. Then, we simply need to copy and paste the addresses or places we need to georeference. This latter process takes two main forms. The first is a raw list with one address per line, and the second is a tabular form where the engine assumes that the first line contains column names and, accordingly, ignores the information contained in this first line for the batch geocoding procedures. For example, let us assume that we want to geocode a version of Table 8.1. If we copy and paste this table or the specific column containing the address or location to be geocoded into GPS Visualizer and select "Type of data: tabular (columns & a header)," the process will render the same result as shown in Fig. 8.2B. However, if we input tabular data and select "Type of data: raw list, 1 address per line," the batch geocoding process will try to geocode the line containing column names, which will result in missing cases for this first line—i.e., column names do not have coordinates unless one of the column names is recognized as a place on Earth's surface, like a column called "Rome," for example.

The result of this batch geocoding process will be a comma-separated output that can be copied into Excel. This tabular output will contain the following elements: latitude, longitude, name, description, color, source, and precision. The column "name" matches our input requests (i.e., Mérida, Spain, and Los Angeles, California, USA, as shown in Fig. 8.2B) to ease the identification of the addresses. The column "description" shows the actual name of that location as reported, in this case, by Bing. For example, our first input name was "Mérida, Spain," and the "description" column rendered "Mérida, Extremadura, Spain." That is, Bing Maps also rendered the region in Spain where the city of Mérida is located.

As in the previous example, we need to take the latitude and longitude coordinates and add them to our database. By doing so, we have effectively linked our participants' narratives to the spaces they are located on Earth's surface. This result is shown in Panel C in Table 1, where we add both the location information and the corresponding latitude and longitude coordinates.

This description outlines the minimum elements required to produce interactive maps via GeoStoryTelling. However, although at this point the resulting maps will display the IDs, the locations' names, and the narratives, no other multi-media data have yet been incorporated into our GeoStoryTelling database. This brings us to our last subsection, which, as described in Fig. 8.1, is optional but highly recommended to provide more detail and nuance to our GeoStories.

Additional contextualizing elements: video, audio, images, and external hyperlinks

Like the process of adding latitude and longitude, the addition of video, audio, or images that can be directly displayed along with our GeoStories does not require specialized software or hardware. Instead, we can retrieve these pieces of information from resources freely available on sites such as SoundCloud (https://soundcloud.com), YouTube (https://www.youtube.com/), or Vimeo (https://vimeo.com/), to mention a few streaming services. In the case of images, we can feature any image hosted online by simply retrieving its direct link from the internet—as we show below. Alternatively, we can also load any image stored on our computer to be displayed as part of our GeoStorytelling.

Note that, although we programmed GeoStoryTelling's back-end to automatically detect whether the information provided is an image, a video, or audio, the data-gathering process to populate our database follows three main divergent approaches.

1. Audio recordings and videos need to be embedded, as shown in Fig. 8.4A and Fig. 8.4B, respectively.
2. Images may be retrieved from their hosting location on the internet, as shown in Fig. 8.4C.
3. Images can be loaded from our computers locally, as shown in Fig. 8.4D.

Although the gathering or compilation of these elements may be tedious with large-scale projects, note also that these contextualizing elements are optional, for it is only the narratives and the latitude and longitude coordinates that are required for the GeoStoryTelling software to render the HTML outputs.

Once we have collected these audios, videos, or images, we can then populate the column called "Video, Image, or Audio Link" in Table 8.1. As indicated in Table 8.1, the video and audio need embedding, which is a specialized HTML code that allows those objects to become part of our GeoStoryTelling maps. However, following our no-code approach to empower researchers or democratize data science, we can copy that embedding code by selecting "Share" from most public streaming websites with videos or audio, as we show in Fig. 8.4A and Fig. 8.4B, and then select the option "Embed." After doing so, as shown in Fig. 8.4A and Fig. 8.4B, we can simply copy and paste those links into our database, as represented in Table 8.1.

As we indicated, we programmed the back-end procedures of GeoStoryTelling to automatically detect whether we are providing videos, audio, or images, so we do not need three columns, but only one, as shown in Table 8.1.

For the sake of completeness, note that we added one extra row to Panel D in Table 8.1. This is to illustrate the three main sources of information that the column "Video, Image, or Audio Link" can handle. The first row in Panel D, by starting with a web address, indicates that the output should be an image. The second row shows a local directory located on the disk "C." Note that we do not need to add quotation marks to any of these formats, as we programmed GeoStoryTelling to automatically handle blank spaces. The third row in Panel D starts with < iframe`, which suggests that this is an embedded link to either an audio or a video. With respect to images stored locally, please note that, although our example is demonstrated with a Windows path, the same process to add a local image on a Mac computer is executed, also without having to include quotation marks when pasting this path into our databases.

Finally, the software also allows us to include an external link to a webpage that provides more information, and this link may simply be copied from any web browser address and pasted into our table. Note that in Panel D in Table 8.1, the first cell in the "Link to Webpage" column is empty. This is completely fine, as GeoStoryTelling can handle missing cases if no additional information about a given story or narrative is added—nonetheless, stories with complete information will display this external link. As shown, when this link is present, it is added under the name "Hyperlink for more information." As its name indicates, when users click on this hyperlink, they are taken to a webpage that provides more information about the story. For example, in addition to displaying a short documentary called "Surviving Kensington" on the map via an embedded video, the "Hyperlink for more information" featured in the table takes users to a TIME Magazine story. Although this story was not written or produced by the same authors as the video documentary featured in our GeoStoryTelling, it may potentially enrich our understanding of the topic and the stories we are sharing.

This concludes our discussion of the creation of the database, as shown on the left-hand side of our flow diagram (Fig. 8.1). Once this database is saved as a "*.csv" file, it can then be loaded into the GeoStoryTelling UI, as we discuss next. Although this description may seem overwhelming, workshop offerings with participants with no prior experience in GIS or computer programming have consistently indicated that full completion of datasets with 10 cases was achieved in one hour, including the explanation time for each of these elements. Participants in Europe and South America indicated that the creation of *.csv files separates columns with semicolons instead of commas. In these cases, the Microsoft setup must be updated to use commas. Alternatively, Google Sheets may be used to create these files and then save them locally before uploading them to the GeoStoryTelling software.

References

Agnew, J., 2005. Space: Place Spaces of Geographical Thought: Deconstructing Human Geography's Binaries. SAGE Publications Inc., United States, pp. 81–96. Available from: http://knowledge.sagepub.com/view/spaces-of-geographical-thought/SAGE.xml. doi:10.4135/9781446216293.n5.

Agnew, J., Livingstone, D.N., 2011. The SAGE Handbook of Geographical Knowledge the SAGE Handbook of Geographical Knowledge. SAGE Publications Inc., United States, pp. 1–636. Available from: http://knowledge.sagepub.com/view/hdbk_geoknowledge/SAGE.xml. doi:10.4135/9781446201091.

Ajayakumar, J., Curtis, A., Smith, S., Curtis, J., 2019a. The use of geonarratives to add context to fine scale geospatial research. Int. J. Environ. Res. Public Health 16 (3), 515. doi:10.3390/ijerph16030515.

Alexander, E., Eppler, M.J., Bresciani, S., 2019a. Visual replay methodology: a mixed methods approach for group discussion analysis. J. Mixed Methods Res. 13 (1), 33–51. Available from: http://mmr.sagepub.com/. doi:10.1177/1558689816664479.

Artale, A., Parent, C., Spaccapietra, S., 2007a. Evolving objects in temporal information systems. Ann. Math. Artif. Intell. 50 (1-2), 5–38. doi:10.1007/s10472-007-9068-z.

Braun, V., Clarke, V., 2006. Using thematic analysis in psychology. Qual. Res. Psychol. 1478-0887. 3 (2), 77–101. doi:10.1191/1478088706qp063oa.

Carroll, A., 2020. Story maps. In: Richardson, D., Castree, N., Goodchild, M.F., Kobayashi, A., Liu, W., Martson, R.A. (Eds.), International Encyclopedia of Geography. John Wiley & Sons, Ltd, pp. 1–6. doi:10.1002/9781118786352.wbieg2048.

Chetty, R., Hendren, N., 2018. The impacts of neighborhoods on intergenerational mobility I: childhood exposure effects*. Q. J. Econ. 133 (3), 1107–1162. doi:10.1093/qje/qjy007.

Chetty, R., Hendren, N., Katz, L.F., 2016a. The effects of exposure to better neighborhoods on children: new evidence from the moving to opportunity experiment. Am. Econ. Rev. 106 (4), 855–902. Available from: http://pubs.aeaweb.org/doi/pdfplus/10.1257/aer.20150572. doi:10.1257/aer.20150572.

Chetty, R., Hendren, N., Kline, P., Saez, E., 2014a. Where is the land of opportunity? the geography of intergenerational mobility in the United States. Q. J. Econ. 129 (4), 1553–1623. Available from: Available from: http://www.oxfordjournals.org/our_journals/qje/. doi:10.1093/qje/qju022.

Corbett, J.M., Keller, C.P., 2005. An Analytical Framework to Examine Empowerment Associated with Participatory Geographic Information Systems (PGIS). Cartogr. Int. J. Geogr. Inf. Geovisual. 0317-7173. 40 (4), 91–102. doi:10.3138/j590-6354-p38v-4269.

Cresswell, T., 2008. Place: encountering geography as philosophy. Geography. 0016-7487. 93 (3), 132–139. doi:10.1080/00167487.2008.12094234.

Drozdzewski, D. and Berengueres, J.O., 2024. Developing QualNotes: A collaborative and cross-disciplinary ethnography. Digital Geography and Society 6, 100086. doi:10.1016/j.diggeo.2024.100086.

González Canché, M.S., 2022. Mapping, organizing, and visualizing interdependent events (movie): a rigorous analytic framework and cost-free software application designed to model temporal and dynamic complex realist structures in social research settings. Methodol. Innov. 15 (3), 263–288.

González Canché, M.S., 2023a. Graphical retrieval and analysis of temporal information systems (gratis): an integrative mixed methodology and open-access software to analyze the (non-) linear chronological evolution of information embedded in textual/qualitative data. J. Mixed Methods Res. 15586898231166968.

González Canché, M.S., 2023b. Latent code identification (lacoid): a machine learning-based integrative framework [and open-source software] to classify big textual data, rebuild contextualized/unaltered meanings, and avoid aggregation bias. Int. J. Qual. Methods 22, 16094069221144940.

González Canché, M.S., 2023c. Machine driven classification of open-ended responses (mdcor): an analytic framework and no-code, free software application to classify longitudinal and cross-sectional text responses in survey and social media research. Expert Syst. Appl. 215, 119265.

González Canché, M.S., 2023d. Spatial Socio-econometric Modeling (Ssem): A Low-code Toolkit for Spatial Data Science and Interactive Visualizations Using R, Springer Nature.

González Canché, M.S., 2025. Ethics and trust in social network analysis: how data science and interactive visualizations impact anonymity, confidentiality, consent, and IRB procedures?. In: Handbook of Social Network Analysis and Education. Edward Elgar Publishing, pp. 114–131. doi:10.4337/9781035312191.00013.

Holdgraf, C., Wasser, L.A., 2022. Gis in python: intro to coordinate reference systems in python. Earth Lab. https://www.earthdatascience.org/courses/use-data-open-source-python/intro-vector-datapython/spatial-data-vector-shapefiles/intro-to-coordinate-reference-systems-python/.

Jensen, C.S., Dyreson, C.E., Bohlen, M., Clifford, J., Elmasri, R., Gadia, S.K., Grandi, F., Hayes, P., Jajodia, S., Käfer, W., Kline, N., Lorentzos, N., Mitsopoulos, Y., Montanari, A., Nonen, D., Peressi, E., Pernici, B., Roddick, J.F., Sarda, N.L., Scalas, M.R., Segev, A., Snodgrass, R.T., Soo, M.D., Tansel, A., Tiberio, P., Wiederhold, G., 1998b. The consensus glossary of temporal database concepts - February 1998 version. Lect. Notes Comput. Sci. 16113349. 1399, 367–405. Available from: https://www.springer.com/series/558. doi:10.1007/bfb0053710.

Jung, J.-K., Elwood, S., 2010. Extending the qualitative capabilities of gis: computer-aided qualitative gis. Trans. GIS 14 (1), 63–87. Available from: http://onlinelibrary.wiley.com/journal/10.1111/(ISSN)1467-9671. doi:10.1111/j.1467-9671.2009.01182.x.

Kwan, M.-P., Ding, G., 2008. Geo-narrative: extending geographic information systems for narrative analysis in qualitative and mixed-method research. The Professional Geographer 60 (4), 443–465. doi:10.1080/00330120802211752.

Li, W., Wang, Z., Wang, Y., Weng, D., Xie, L., Chen, S., Zhang, H., Qu, H. (2023a). Geocamera: telling stories in geographic visualizations with camera movements. Proceedings of the 2023 CHI Conference on Human Factors in Computing Systems, 1–15.

Matthews, S.A., Detwiler, J.E., Burton, L.M., 2005a. Geo-ethnography: coupling geographic information analysis techniques with ethnographic methods in urban research. Cartogr. Int. J. Geogr. Inf. Geovisual. 40 (4), 75–90. doi:10.3138/2288-1450-w061-r664.

Mennis, J., Mason, M.J., Cao, Y., 2013a. Qualitative gis and the visualization of narrative activity space data. Int. J. Geogr. Inf. Sci. 27 (2), 267–291. doi:10.1080/13658816.2012.678362.

Mooney, P., Corcoran, P., Ciepluch, B., 2013a. The potential for using volunteered geographic information in pervasive health computing applications. J. Ambient Intell. Humaniz. Comput. 4, 731–745. doi:10.1007/s12652-012-0149-4.

Murray, M., et al, 2018a.Narrative data. Sage Handbook of Qualitative Data Collection, 264–279.

Neale, B., 2015a. Time and the lifecourse: perspectives from qualitative longitudinal research. In: Researching the Lifecourse. Policy Press. pp. 25–42. Available from: http://policypress.co.uk/Researching-the-lifecourse.

Neale, B., 2019. Qualitative longitudinal research. In: Atkinson, S.P., Delamont, A., Cernat, J., Sakshaug, Williams, R. (Eds.), Sage Research Methods Foundations, SAGE Publications Ltd. doi: 10.4135/9781526421036824355

Palis, J., 2022a. Geonarratives and countermapped storytelling. In: The Routledge Handbook of Global Development. Taylor and Francis, Philippines, Routledge. pp. 700–712. Available from: https://www.routledge.com/The-Routledge-Handbook-of-Global-Development/Sims-Banks-Engel-Hodge-Makuwira-Nakamura-Rigg-Salamanca-Yeophantong/p/book/9780367862022. doi:10.4324/9781003017653-66.

Percy, J., 2018a. Trapped by the 'walmart of heroin. N.Y. Times 10. https://www.nytimes.com/2018/10/10/magazine/kensington-heroin-opioid-philadelphia.html.

Richardson, D., Castree, N., Goodchild, M.F., Kobayashi, A., Liu, W., Marston, R.A., 2020. Story Maps International Encyclopedia of Geography: People, the Earth, Environment and Technology. Wiley, Wiley, pp. 1–6. doi:10.1002/9781118786352.wbieg2048.

Tuan, Y.-F., 1977. Space and Place: The Perspective of Experience, University of Minnesota Press.

U.S. Department of Commerce, Economics and Statistics Administration, U.S. Census Bureau, 2020. Understanding and Using American Community Survey Data: What All Data Users Need to Know. U.S. Government Publishing Office, https://www.census.gov/content/dam/Census/library/publications/2020/acs/acs%5C_general%5C_handbook%5C_2020.pdf.

PART V

Interpretation, Synthesis, and Scholarly Brainstorming with Local Generative AI

9

Intelligent Systems for Academic Research Integration (ISARI): A local and fully offline brainstorming partner for ethical scholarly inquiry

Introduction

The methodological frameworks presented throughout this volume were developed in response to a practical and epistemological challenge that qualitative researchers, mixed-methods scholars, and applied social scientists increasingly face. Contemporary research projects often rely on textual corpora that may contain hundreds, thousands, or even hundreds of thousands of pieces of evidence drawn from interviews, ethnographic observations, focus groups, essays, social media posts, open-ended survey responses, policy documents, or historically archived texts. These materials contain patterns, structures, contradictions, and forms of complexity that are often too large to be read exhaustively by one researcher or even by a team without analytic assistance. Yet these same data also contain meanings, nuances, ambiguities, and contextual subtleties that cannot be reduced to statistical summaries alone.

It is within this tension that the frameworks discussed throughout this book were created. LACOID was developed to classify long textual material while reducing latent code aggregation bias and preserving contextualized meaning (González Canché, 2023b). MDCOR was designed to classify short texts, such as open-ended survey responses and social media posts, rapidly and rigorously while preserving participants' original voices (González Canché, 2023c). SENA extended analytic possibilities by identifying the distribution of sentiments and emotions across textual corpora while encouraging attention to within-group and across-group differences rather than aggregate outputs alone (González Canché and Zheng, 2025b). NAQD demonstrated that coded qualitative data can be conceptualized as participant–code networks capable of revealing structural patterns and supporting formal hypothesis testing (González Canché, 2022a). GRATIS introduced the possibility of visualizing the linear and nonlinear chronological evolution of coded evidence (González Canché, 2023a). VERITAS, in turn, made it possible to map and replay synchronous interactions so that communication exchanges, turning points, and shifts in collective knowledge production could be traced over time (Alexander et al., 2019; González Canché, 2026). GeoStoryTelling further demonstrated that qualitative evidence can be interpreted through place as well as through topic, chronology, or relation by

https://doi.org/10.1016/B978-0-443-21961-0.00010-3

geo-contextualizing stories within meaningful spatial settings (González Canché et al., 2025, 2025a, 2022, 2023).

What these frameworks share is not merely their reliance on data science or visualization. Their deeper commonality is epistemological. None of them treats a statistical output, network structure, temporal replay, or visual display as a final conclusion. Instead, all were designed to function as analytic guides that return researchers to the original evidence. Quantitative or visual outputs identify where to look, what to compare, what to revisit, and where qualitative reconstruction of meaning is needed. In that sense, they collectively operationalize a truly equal status design (TESD), in which qualitative evidence and quantitative or computational structure remain interdependent rather than hierarchically ordered (Johnson and Onwuegbuzie, 2004; Fetters and Freshwater, 2015). Across these frameworks, the larger methodological commitment is the exploratory reconstruction of qualitative meaning across scale, relation, time, emotion, visualization, and place.

This chapter is intentionally positioned immediately before the concluding chapter because it addresses a stage that follows all of the frameworks introduced in this book: interpretation, memoing, synthesis, and scholarly writing. Intelligent Systems for Academic Research Integration (ISARI)—enters the ecosystem at precisely this point. It is not another classification framework, nor does it replace the tools presented in the book. Rather, it is best understood as a local, fully offline, open-source brainstorming and thinking partner that helps researchers work more effectively with the outputs already generated by those methods. It assists in comparing topic solutions, describing latent codes, drafting integrative memos, organizing representative quotations, translating visual outputs into prose, and supporting the disciplined movement back and forth between structure and meaning. Because many researchers new to local artificial intelligence (AI) still wonder what it means in practice to use a model as a thinking partner rather than as an answer generator, a later section of this chapter provides concrete example prompts showing how ISARI can be asked to support first-pass pattern description, comparison, memoing, and evidence-grounded synthesis across the analytic outputs generated by the frameworks in this book.

ISARI, however, is not merely a local language-model environment for text-centered interpretation, synthesis, and writing. Natural language processing (NLP) capabilities, although important, are not sufficient. Namely, the frameworks in this volume do not produce textual outputs alone. They also produce *images* in the broad analytic sense of the term: network graphs, temporal plots, diagnostic charts, replay frames, screenshots, spatial displays, map-based storyboards, and other forms of visual evidence that play a central role in interpretation. As a result, the interpretive stage of the workflow increasingly requires a local environment capable of reasoning not only across text but also across image-based analytic outputs. For this reason, ISARI also includes two additional vision-capable brains for image analysis, thereby expanding the local interpretive infrastructure available to researchers without changing the methodological center of the book.

This chapter also addresses a recurring concern observed in methodological training and workshops with scholars across qualitative and mixed-methods communities. Even when researchers begin to embrace machine learning and other data science tools, many remain deeply uncomfortable—ethically, practically, and institutionally—with uploading qualitative data to external servers, as is typically required when using commercial generative AI services.

That discomfort is not incidental; it is reinforced by long-standing concerns in research ethics regarding anonymity, confidentiality, consent, and Institutional Review Board (IRB) procedures whenever data structures or visualizations make participants more visible or more easily re-identifiable (Borgatti and Molina, 2005; Klovdahl, 2005; González Canché, 2025c). In this context, an offline AI brainstorming partner is not simply a technical convenience; it is part of a methodologically coherent and ethically defensible research infrastructure. The same logic applies to visual outputs. If a temporal replay, a network graph, a screenshot from a synchronous interaction, or a place-based display is analytically sensitive, then keeping its interpretation within a local environment is as important as keeping the underlying text local.

Intelligent Systems for Academic Research Integration within the methodological architecture of the book

The title of this volume, *Data Science and Interactive Visualization Tools for the Analysis of Qualitative Evidence*, provides a useful point of entry for locating ISARI conceptually. The phrase *data science and interactive visualization tools* refers to the frameworks that form the methodological backbone of the book. The phrase *for the analysis of qualitative evidence* clarifies that these tools were never developed to displace qualitative inquiry but to strengthen it. Accordingly, the role of ISARI should be framed in a way that respects this architecture.

With ISARI, the center of this volume remains the disciplined use of data science and visualization to enhance, rather than substitute for, qualitative and mixed-methods inquiry. This distinction matters because local language models can be misunderstood if introduced carelessly. They may appear to offer a shortcut around the rigorous processes that LACOID, MDCOR, SENA, NAQD, GRATIS, VERITAS, and GeoStoryTelling were designed to make possible. Such a framing would be inconsistent with the epistemology of the book. A more accurate conceptualization is to position ISARI as an interpretive and writing support tool that becomes most useful after the frameworks in the book have already structured the evidence. ISARI helps give coherence to the book's methodological ecosystem—each framework produces a different form of analytic structure, and ISARI supports the interpretive movement through which those structures become scholarly meaning.

Once texts have been classified, once emotional distributions have been mapped, once participant-code networks have been visualized, once chronology has been reconstructed, once synchronous interactions have been replayed, and once stories have been geo-contextualized, researchers still face a demanding task. They must compare, describe, connect, explain, draft, revise, and validate. It is precisely at this stage that ISARI becomes valuable. In that sense, ISARI broadens the practical reach of the book without changing its methodological center. The frameworks remain the engines that retrieve structure, detect patterns, and produce visual or statistical outputs. ISARI helps researchers work with those outputs more coherently, more reflexively, and more efficiently.

The inclusion of vision-capable brains deepens, rather than disrupts, this logic. It does not mean that ISARI becomes a replacement for visualization or a substitute for the frameworks that generate visual analytic evidence. Instead, it acknowledges a practical truth: researchers often need help *reading* and *writing about* visual outputs after those outputs have been produced. A diagnostic chart from MDCOR, a network map from NAQD, a replay frame from VERITAS, or a temporal plot from GRATIS may be analytically powerful, but it still must be translated into clear prose, connected to qualitative evidence, and situated within a larger argument. The role of the vision-capable brains is to support that downstream interpretive labor while preserving the local, offline, and privacy-preserving design principles that already govern the broader ISARI environment. This positioning also connects ISARI to the book's equal-status logic: outputs matter when they guide researchers back to the evidence.

A truly equal-status design and the role of an offline thinking partner

The idea of a TESD is foundational to the methods discussed in this volume. TESD rests on the argument that rigorous and useful knowledge emerges only when qualitative evidence and quantitative outputs are treated as equally necessary components of the analytic process. This is not a vague call for balance; it is a precise claim that one component cannot carry the full interpretive burden without the other. Quantitative summaries, classifications, and visualizations reveal structure, but they do not speak for themselves. Qualitative evidence provides the meanings that give those structures substance, but when databases become large or complex, unaided human reading may obscure patterns that matter.

The frameworks in this volume instantiate this design differently. LACOID and MDCOR use machine learning and natural language processing to retrieve latent structures while preserving participants' original voices. SENA identifies sentiments and emotions while insisting that researchers attend to individual and group variation rather than only aggregate outputs. NAQD demonstrates that coded data may be treated as relational structures and statistically compared across groups without losing the underlying coded content. GRATIS reminds us that chronology itself is analytic evidence and that qualitative data are always richer when their temporal emergence is made visible. VERITAS shows that synchronous group settings contain dynamic structures that cannot be reduced to static summaries if we wish to understand how ideas, alignments, and conflicts actually evolve. GeoStoryTelling adds the insight that stories unfold in meaningful places and that geo-contextualization can deepen interpretation without requiring GIS specialization (González Canché and Zheng, 2025b).

ISARI does not alter this arrangement. Its legitimate contribution lies in strengthening the movement between output and evidence. When researchers compare alternative code solutions, reconstruct the meaning of codes, explain why one group's emotional structure differs from another's, narrate what changed over time, describe a pivotal moment in a working-group interaction, interpret how place reshapes experience, or translate a complex visual display into prose, they remain engaged in TESD. The offline thinking partner helps them move through that process with greater speed, clarity, and flexibility. It does not replace interpretation; it supports it.

The relevance of vision-capable brains emerges directly from this point. If quantitative, network-based, chronological, and spatial outputs are not conclusions in themselves, then researchers must return to the evidence from those outputs with equal care, whether the output is textual, numerical, or visual. In other words, the need for interpretive support is not limited to words; it extends to graphs, maps, plots, frames, screenshots, and other representations that increasingly mediate the movement between computational structure and qualitative meaning. A local vision-capable model does not change the design from equal-status to model-led; it simply makes the equal-status movement more feasible when the outputs to be interpreted are visual.

Why does local execution matter?

A central feature shared by the frameworks in this book is their commitment to local execution, no-code access, and the democratization of data science. This local orientation matters methodologically, practically, and ethically. Methodologically, local execution allows researchers to work with sensitive qualitative evidence without restructuring their workflows around remote platforms. Practically, it reduces barriers to use in institutions and projects where external data transfer is restricted. Ethically, it helps protect confidentiality and minimizes the unnecessary circulation of participant narratives beyond the immediate research environment.

ISARI should be framed within this same logic. The importance of an offline thinking partner is not merely that it avoids internet dependence; its significance lies in how it extends local control into the reasoning, memoing, and writing stages of the research process. A researcher may classify sensitive interviews with LACOID, analyze open-ended responses with MDCOR, identify group-based emotional structures with SENA, test participant-code similarities with NAQD, reconstruct chronology with GRATIS, replay synchronous processes with VERITAS, and geo-contextualize narrative evidence with GeoStoryTelling—all without sending data to external servers. ISARI preserves that same protective boundary when the time comes to interpret, compare, summarize, draft, revise, and write. The addition of vision-capable brains extends that protective boundary to the visual outputs generated by those frameworks, allowing charts, graphs, maps, replay frames, and screenshots to remain inside the local environment as well.

This point should not be overstated, as if local execution alone resolves every ethical concern. Researchers remain responsible for judgment, abstraction, confidentiality, and fair representation when discussing participant data. Yet, local execution materially reduces one major risk: the externalization of sensitive evidence into systems over which the researcher has little or no control. In a research environment in which trust, privacy, and data stewardship matter profoundly, this is not a secondary technical preference; it is part of the methodological integrity of the workflow. Locality turns ethics into infrastructure: evidence, interpretation, and responsibility remain together.

Technical foundations: Local language models, model size, and functional fine-tuning

ISARI's architecture makes local, evidence-bound interpretation practically attainable. ISARI operates entirely on local large language models that run on the researcher's own machine under macOS and Windows operating systems. Once the compressed folder has been decompressed (i.e., unzipped), these models function without internet connectivity. ISARI is also offered as open source, which means that its code base can be inspected, audited, modified, and extended by the research community. As more powerful models become available, they will be integrated into ISARI's infraestructure, provided these models can still run in computers with modest computing power. New software releases will be posted in ISARI's Code Ocean capsule and users will be able to download the update system from our server (https://cutt.ly/ltJuvTEL).

Importantly, ISARI was designed to remain practically usable on modest local hardware. In its lightest configuration, that is, when running its smallest brain (Phi-4 Mini as described below), ISARI can operate on a computer with 8 GB of RAM and approximately 340 GB of available storage space. This lower-threshold configuration matters because the chapter's commitment to local and privacy-preserving AI support would be considerably weakened if meaningful access to ISARI depended on unusually powerful or specialized machines. Put differently, locality only becomes methodologically democratizing to the extent that it remains technically attainable for researchers working outside highly privileged computing environments.

The numerical labels often attached to these local models—such as 4B, 7B, 8B, or 11B—refer to the number of trainable parameters within each model, expressed in billions. Parameters are the internal weights learned during training that enable a model to represent linguistic patterns, semantic associations, and reasoning tendencies. In general, smaller models require less memory and computational power, respond more quickly, and are better suited for summarization, memo drafting, and concise explanation. Mid-sized models offer a balance between efficiency and structure, making them useful for procedural reasoning, prompt scaffolding, and repeatable workflows. Larger local models are more computationally demanding but often provide stronger long-context reasoning, deeper comparison, and more integrative synthesis across multiple outputs.

Yet model size alone does not determine an analytic role. In ISARI, the practical difference among brains is not simply that some are bigger or smaller; it is that they are *functionally specialized.* Some are better suited for concise conceptual writing, others excel in structured reasoning, code, or template generation, and still others are more effective when the task requires interpretive comparison or deeper synthesis. With the addition of the two vision-capable brains, ISARI can now support forms of local scholarly work that involve images, charts, diagrams, scanned pages, or other visual artifacts rather than text alone.

Language support is also not uniform across brains, and this matters methodologically for researchers who work with multilingual corpora, bilingual coding schemes, Spanish-language survey responses, translated interviews, or geo-contextualized stories that combine English and non-English materials. Gemma 3 (Google DeepMind, 2025), the lightweight multimodal baseline, supports over 140 languages and thus offers unusually broad linguistic flexibility for everyday work. Qwen2.5-VL, the document- and chart-oriented vision brain, inherits the broader multilingual competencies of the Qwen2.5 family and is especially notable for multilingual document parsing, making it attractive when researchers need to work with scanned materials, diagrams, or figures containing labels in languages other than English (Qwen Team, 2025a). Phi-4 Mini supports 24 languages, including English, Spanish, Portuguese, French, German, Arabic, Chinese, Japanese, Korean, Russian, Thai, and several others[a], making it especially useful for concise multilingual reasoning and methods-oriented writing (Microsoft, 2025). Llama 3.2 Vision, by contrast, is more constrained: while its text-only mode officially supports English, German, French, Italian, Portuguese, Hindi, Spanish, and Thai, its image-plus-text mode is officially English-only (Meta Llama, 2024). This does not eliminate its usefulness, but it does mean that its strongest fit inside ISARI is deeper English-language visual reasoning rather than multilingual document analysis. The current Ollama package for DeepSeek R1 (8B) is based on the Qwen3 8B distill (DeepSeek-AI, 2025) and therefore benefits from the broader multilingual reach of the Qwen3 family, although in practical ISARI use it is best understood first as a reasoning model and only secondarily as a multilingual generalist. Qwen 2.5 Coder likewise inherits broad multilingual support from the Qwen family while also covering 92 programming languages, which makes it especially useful for method documentation, scripts, tables, and reproducible prompt structures rather than image-centered work (Qwen Team, 2025b).

For nontechnical researchers, however, the more important point is that meaningful adaptation does not require retraining a model or changing its parameters. In conventional machine-learning language, retraining would be referred to as fine-tuning. That is not what ISARI expects from its users. Instead, ISARI supports what may be called *functional fine-tuning* through prompt design, model selection, and task alignment. In practical terms, researchers adapt ISARI to their needs by specifying: (1) the analytic stage, (2) the type of evidence being supplied, and (3) the epistemic constraints that should govern interpretation. This makes ISARI approachable for scholars who do not have computer science or programming backgrounds. They do not need to modify a model; they need to formulate a rigorous analytic request and select the brain whose combination of computational scale, linguistic coverage, and interpretive role best fits that request. Seen in this light, prompting is not a technical trick. It is a disciplined way of translating scholarly intention into a local, evidence-bound analytic request.

Prompting as analytic memoing

In ISARI, prompts function as structured analytic memos rather than as generic instructions for answer generation. Effective prompts specify: (1) the analytic stage (e.g., postclassification interpretation, comparison of topic solutions, translation of a network output into prose, interpretation of a temporal graph, or description of a chart or screenshot), (2) the type of evidence being provided (e.g., representative quotations, centrality measures, dynamic visual summaries, replay frames, geo-contextualized narratives, or bilingual figures), and (3) the epistemic constraints that must govern the response (e.g., remain close to the supplied outputs, do not

[a]These languages are: **European:** Czech, Danish, Dutch, English, Finnish, French, German, Hungarian, Italian, Norwegian, Polish, Portuguese, Russian, Spanish, Swedish, Ukrainian. **Middle Eastern:** Arabic, Hebrew, Turkish. **Asian:** Chinese, Japanese, Korean, Thai.

infer unsupported meanings, identify where a return to the original text is needed, distinguish what is visually observable from what requires substantive interpretation). In this sense, prompt writing operationalizes long-standing qualitative practices of memoing, constant comparison, and theory building (Miles and Huberman, 1994; Charmaz, 2014; Saldaña, 2021) within a local computational environment rather than introducing an entirely new analytic paradigm.

The practical implication of this framing is that prompts can be written around the outputs of each framework in this book. A prompt may ask ISARI to describe the visible structure of a network, compare two topic solutions, summarize the affective differences across groups, narrate the turning points of a synchronous replay, or draft an initial analytic memo grounded in representative quotations and fieldnotes. In all such cases, the goal is not to outsource interpretation, but to structure the first pass of scholarly reflection in a way that remains transparent, revisitable, and anchored in the evidence supplied by the researcher.

Why is "brainstorming" explicitly built into the interface?

This memoing logic clarifies the language of brainstorming: ISARI helps generate possibilities that remain provisional, evidence-bound, and subject to scholarly judgment. ISARI is explicitly framed as a thinking and brainstorming partner, and this role is visibly reinforced in the interface through the label "Brainstorming... " during execution and output generation. This is more than a usability decision; it signals to users—especially nontechnical researchers—that the system is not intended to deliver authoritative answers or automated conclusions.

Methodologically, brainstorming refers here to a constrained generative process in which the system helps enumerate plausible interpretive paths, identify contrasts, surface uncertainties, and suggest where researchers should return to the underlying evidence. This aligns closely with established qualitative practices such as analytic memoing, iterative categorization, and constant comparison (Braun and Clarke, 2006; Charmaz, 2014; Miles and Huberman, 1994). By foregrounding brainstorming rather than "answering," ISARI helps prevent the epistemic slippage that can occur when generative outputs are mistaken for findings. Its proper role is to assist researchers in thinking more carefully, not to think instead of them.

This is particularly important when visual brains are involved. Visual descriptions can create a false sense of certainty if researchers begin to confuse a model's ability to label visible components with substantive interpretation of those components. The interface language helps counteract this risk. Whether the supplied evidence is textual, graphical, temporal, or spatial, ISARI remains a local brainstorming partner rather than an answer-generating system (Fig. 9.1). In a nutshell, ISARI generates possibilities while the researcher determines meaning.

Alt Caption. ISARI user interface showing the local research workspace, saved chats, model-selection area, file-attachment option, and the visible "Brainstorming..." state, which reinforces the platform's role as a thinking and writing partner rather than an answer-generating system. The figure also illustrates ISARI's practical design orientation toward local, fully offline, and nontechnical scholarly use, including its support for switching among brains with different analytic roles (see Table 9.1).

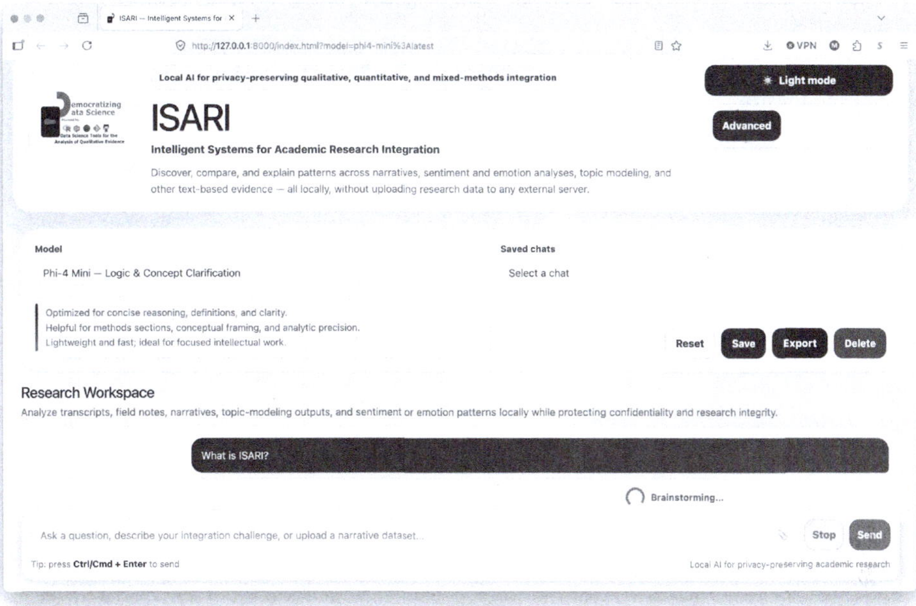

FIG. 9.1 ISARI user interface.

Table 9.1 ISARI brains, analytic roles, computing demand, and language support.

ISARI brain	Approximate model size	Computing demand	Primary analytic role	Best-fit scholarly uses	Language support/ notes
Gemma 3	4B	Low	Lightweight multimodal baseline	Summaries, memo drafting, prose translation, quick visual grounding, and lower-resource everyday tasks	Broad multilingual reach (140 + languages); useful for English, Spanish, and multilingual everyday prompting
Qwen 2.5 Coder	7B	Moderate	Structured reasoning and method organization	Tables, scripts, prompt templates, reproducible workflows, and methodological appendices	Broad multilingual text support, inherited from Qwen2.5 (29 + languages), plus 92 programming languages.
Phi-4 Mini	~4B	Low	Conceptual clarification and concise reasoning	Methods text, bridge paragraphs, definitional work, and short analytic explanations	24 supported languages, including English, Spanish, Portuguese, French, German, Arabic, Chinese, Japanese, Korean, Russian, Thai, and others

Continued

Table 9.1 ISARI brains, analytic roles, computing demand, and language support—cont'd

ISARI brain	Approximate model size	Computing demand	Primary analytic role	Best-fit scholarly uses	Language support/ notes
DeepSeek R1	8B	Moderate–High	Deep reasoning and synthesis	Comparative interpretation, theory building, integrative discussion, and cross-framework synthesis	Current local 8B package is Qwen3-based; multilingual behavior extends beyond English, but its strongest fit in ISARI is reasoning-first interpretive prose
Vision brain (document-oriented)	~7B	Moderate	Document, chart, and layout interpretation	Scanned pages, charts, tables, diagrams, bilingual figures, and screenshots of analytic outputs	Broad multilingual document parsing is especially attractive when labels or page structures are not English-only
Vision brain (high-detail)	~11B	High	Higher-detail general visual reasoning	Complex screenshots, dense figures, nondocument images, and visually rich outputs on stronger hardware	Text-only multilingual support includes English, Spanish, French, German, Italian, Portuguese, Hindi, and Thai; image + text mode is best treated as English-first

The current ISARI brains and their analytical roles

ISARI currently supports multiple local models, referred to in the interface as "brains," each aligned with distinct analytic needs. As summarized in Table 9.1, Gemma 3 (4B) is especially useful for summarizing representative texts, drafting code descriptions, translating technical outputs into accessible academic prose, and providing quick visual support when researchers are working with screenshots, basic charts, or simple image-based evidence. Qwen 2.5 Coder (7B) is useful for structured reasoning, procedural documentation, prompt reproducibility, and method-oriented organization. Phi-4 Mini is efficient on lower-power machines and is well-suited for concise conceptual explanation, multilingual clarification, and short-form methodological writing. DeepSeek R1 (8B) is best suited for extended reasoning, deeper comparison of analytic alternatives, and integrative synthesis across multiple outputs.

ISARI also includes two additional vision models for image analysis. The first is a document- and chart-oriented visual brain designed for scanned academic pages, tables, diagrams, layout-sensitive figures, screenshots of software interfaces, and Optical Character Recognition (OCR) heavy materials. The second is a higher-detail visual brain designed for stronger machines and more demanding image-centered work, particularly when researchers need deeper

English-language visual reasoning across complex screenshots, dense figures, or visually rich outputs that exceed the lighter multimodal capacity of Gemma 3. These additions do not alter ISARI's role as a downstream interpretive partner; they expand the range of *outputs* with which that partner can work locally.

The practical value of this architecture is that it allows researchers to match a model to a task rather than expecting one local model to perform equally well across every form of scholarly work. This is particularly important in qualitative and mixed-methods research, where the same researcher may require concise explanations, structured organization, interpretive comparisons, multilingual support, chart reading, screenshot descriptions, and long-form writing assistance at different moments of the project. Thus, the multiple-brain design gives ISARI methodological flexibility: different forms of evidence require different forms of interpretive support.

The next section translates this integrative logic into practice through example prompts. These prompts show how researchers can use ISARI to move from structured outputs toward disciplined, evidence-grounded brainstorming.

Integrating ISARI with the frameworks in this book

This section shows how ISARI works across the methodological ecosystem of the book. Each framework structures a different dimension of qualitative evidence; ISARI helps researchers turn those structures into grounded interpretation, memoing, and prose. The practical prompt examples shown in the subsection called "Purposeful Brainstorming with ISARI: Example Prompts Across the Methodological Ecosystem," illustrate ISARI's integrative role more concretely.

ISARI and LACOID

LACOID was designed to address a consequential problem in the analysis of long qualitative texts: the tendency of conventional machine-learning classifications to aggregate entire documents into single topics, even when those documents contain multiple meanings, stances, or shifts in emphasis. By decomposing documents into sentences or paragraphs before classification and preserving the original location and wording of those text segments, LACOID reduces latent code aggregation bias and more closely mirrors the line-by-line logic of qualitative coding (González Canché, 2023b).

ISARI becomes especially useful after LACOID has already produced its outputs. It can help researchers compare alternative code solutions, articulate what is gained or lost when moving from a more aggregated to a more granular classification, summarize each latent code after the research team has examined representative texts, and organize quotations into draft analytic tables or memos. The sequence matters. ISARI should not be asked to discover the latent structure independently of LACOID. The classification work remains prior methodological labor. ISARI strengthens the interpretive stage that follows.

The addition of vision-capable brains matters here because LACOID's interpretive stage increasingly involves visual outputs as well as textual summaries. Researchers may generate distribution charts, screenshots of model outputs, or diagnostic figures while comparing alternative code solutions. They may also work with scanned archival documents or photographed field materials that become part of the evidentiary conversation after classification. The document-oriented vision brain is especially useful in these contexts because it can help describe charts, compare layout-sensitive materials, or summarize the visible structure of scanned pages without requiring researchers to leave the local environment. In multilingual projects—for example, those involving English- and Spanish-language archival materials or interview excerpts embedded in figures—its broader document-centered multilingual support is especially useful.

ISARI and MDCOR

MDCOR addresses a different but equally important challenge: how to classify short texts, such as open-ended survey responses or social media posts, rapidly and rigorously while preserving original voices and avoiding the burdens of manual coding at scale (González Canché, 2023c). Once MDCOR has identified a defensible code solution, ISARI can assist in summarizing patterns across codes, drafting comparative tables, generating early narrative descriptions for reports, and highlighting where group-based differences may require deeper attention. In applied policy and survey contexts, this role is especially valuable because researchers often need to move quickly from code structures to usable interpretation.

The addition of vision-capable brains extends this usefulness to visual diagnostic outputs associated with short-text classification. Researchers frequently summarize open-ended response classifications through bar charts, code-distribution plots, cross-tabulated figures, or screenshots from dashboards and no-code interfaces. These visuals may become especially important in reporting contexts, where policymakers, practitioners, or institutional audiences first encounter findings through charts before they read narrative explanations. The document-oriented vision brain can help translate such visuals into disciplined prose, identify the visible structure of a comparative figure, or support memoing when labels, legends, or axis annotations are partially bilingual. The higher-detail visual brain becomes useful when a classification figure is visually dense, multipanel, or embedded within a more complex interface screenshot.

ISARI and SENA

SENA contributes an analytical perspective focused on sentiment and emotion rather than latent topical classification alone. By combining lexicon-based approaches, group comparisons, and network-based representations of emotional structure, SENA expands the use of sentiment and emotion analysis beyond simplistic aggregate polarity scores (González Canché, 2025a). Once SENA has produced its visualizations and summaries, researchers still need to translate those outputs into prose that is precise, cautious, and useful.

ISARI can help describe the overall emotional structure of a corpus, compare emotional distributions across groups, explain why aggregate and disaggregated views may differ, summarize emotional entropy, and connect emotional findings to the thematic structures previously identified through LACOID or MDCOR. This is especially useful when SENA reveals something an aggregate analysis might otherwise conceal—for example, when a corpus appears predominantly positive overall, while one subgroup expresses more fear, sadness, or inconsistency than another.

The vision-capable brains strengthen this integration because SENA often produces outputs that are not simply numeric tables but visual displays of emotional structure: network graphs, cluster diagrams, heatmaps, multigroup plots, or comparison charts. These visuals can be analytically rich but verbally difficult to render. A local vision-capable brain can assist researchers in describing the visible contrasts among emotional categories, the relative prominence of nodes or clusters, or the way a multigroup comparison is visually organized, all while preserving the principle that substantive interpretation must still be grounded in the underlying texts. This is especially important in multilingual sentiment work, where legends, labels, or excerpts embedded in a figure may involve English and Spanish or other language combinations.

ISARI and NAQD

NAQD expands the methodological horizon of the book by demonstrating that coded qualitative data can be conceptualized as participant-code networks (González Canché, 2022a). Through two-mode network modeling, centrality measures, community detection, and QAP-based group comparison, NAQD makes it possible to identify central issues, peripheral concerns, clustered perspectives, and statistically meaningful differences across groups.

After NAQD has produced its structural outputs, ISARI can serve as an interpretive bridge between the sociogram and the chapter draft. It can help researchers explain what centrality means in substantive terms, identify the implications of community detection, compare aggregated and disaggregated network structures, and draft prose that moves carefully from statistical patterns to qualitative interpretation. This is an especially productive integration because NAQD already foregrounds the idea that structural outputs should guide qualitative reanalysis rather than replace it.

The addition of vision-capable brains deepens this role. NAQD outputs are often visually decisive: the configuration of a sociogram, the visible clustering of participants and codes, the placement of central versus peripheral nodes, and the contrast between group-specific network structures may all matter interpretively. Researchers frequently need to move from what the graph *looks like* to what that structure *means* substantively. The higher-detail vision brain is especially useful when the network visual is dense, multilayered, or visually complex. The document-oriented vision brain becomes useful when the network is embedded in a report, when legends or annotations are extensive, or when the output appears as a chart-heavy static figure. In bilingual or multilingual qualitative projects, the broader language support of Gemma 3 and the document-oriented vision brain also become relevant when labels, categories, or figure annotations do not appear in English alone.

ISARI and GRATIS

GRATIS addresses a major blind spot in much qualitative and mixed-methods work: the tendency to analyze textual evidence as though it were static rather than temporally evolving. By operationalizing temporal information systems through dynamic visualizations, text retrieval tools, and global timestamps, GRATIS enables researchers to replay the chronological or storytelling-based evolution of coded information (González Canché, 2023a).

Once GRATIS has generated time-aggregated and time-evolving outputs, ISARI can help researchers write about temporal structure in ways that are often difficult to achieve from visuals alone. It can assist in comparing the aggregated network with the dynamic one, describing the analytic gain obtained by restoring chronology, drafting explanations of early, middle, and late phases, and supporting the interpretation of linear versus nonlinear temporal coding decisions. GRATIS makes time visible. ISARI helps make that visibility legible in prose.

This integration becomes even stronger with vision-capable brains because temporal outputs are often first encountered visually rather than numerically. Researchers may work with line graphs, event-phase diagrams, frame-by-frame replays, or screenshots of dynamic visualizations frozen at analytically meaningful moments. The document-oriented vision brain can help describe the visible structure of temporal plots, especially when legends, labels, or summary boxes need to be translated into prose. The higher-detail visual brain becomes especially useful when the temporal output is visually dense, when the key interpretive difference lies in the spatial arrangement of events or phases, or when screenshots from evolving displays must be compared across time. In projects involving bilingual or multilingual time-stamped material, the broader language support of Gemma 3, Qwen2.5-VL, and Phi-4 Mini becomes particularly useful for describing labels or narrative fragments embedded in a temporal output.

ISARI and VERITAS

VERITAS extends the temporal and structural logic of the book into synchronous group settings. Grounded in visual replay methodology and dynamic temporal thinking, it was developed to map the emergence, stabilization, and decay of communication exchanges and coded information in real-time or time-lapse form (Alexander et al., 2019). Once VERITAS has generated actor–code or actor–actor replays, ISARI can help researchers narrate the development of the discussion, identify turning points, summarize the role of influential actors, describe how a given code or claim became central, and distinguish among early contributions, pivotal shifts, and closing phases of group interaction.

VERITAS is perhaps the clearest case in which the addition of vision-capable brains materially enlarges ISARI's downstream usefulness. Replay frames, screenshots of evolving interaction maps, and visual traces of synchronous participation are often central to the interpretive argument itself. Researchers need to describe *what changed visually* across phases, how interactional density shifts, which actors become central, and where turning points become visible. The higher-detail vision brain is particularly useful for reading these replay images and helping researchers verbalize the structure they contain. The document-oriented vision brain is useful when replays are embedded in annotated reports or when explanatory overlays, legends, or figure labels need to be parsed. Because Llama 3.2 Vision is strongest in English-language image-plus-text work, it is particularly useful when the replay outputs are English-dominant; when the replay or annotations are multilingual, Qwen2.5-VL offers a stronger local fit.

ISARI and geostorytelling

GeoStoryTelling extends the methodological ecosystem of the book by demonstrating that sophisticated, place-based, and multimedia forms of qualitative interpretation can also be achieved through local, no-code, and cost-free software (González Canché and Zheng, 2025b). Its contribution is particularly important because it reminds researchers that experiences are not merely thematic or temporal; they are also spatially situated. Stories unfold in neighborhoods, institutions, streets, homes, campuses, clinics, and communities that shape meaning.

When GeoStoryTelling produces geo-contextualized narrative outputs, ISARI can function as a second-stage interpretive partner that assists in translating spatial visualizations into analytic prose, comparing narratives across places or time periods, and memoing how spatial context reshapes thematic or emotional interpretation. In this respect, GeoStoryTelling strengthens the case for ISARI because both tools share a broader design philosophy: advanced analytic affordances without requiring researchers to upload their data to external servers.

The relevance of vision-capable brains is especially clear here. Place-based outputs are often inherently visual: maps, route traces, multimedia storyboards, annotated place markers, or layered spatial displays in which text and image coexist. The document-oriented vision brain can help interpret legends, place labels, and map-embedded explanatory text, while the higher-detail visual brain can assist in describing more complex spatial layouts or visually rich storyboards. When place names, narrative fragments, or annotations are bilingual—as is often the case in educational, migration, or community-based research involving English and Spanish—broader multilingual support becomes more than a convenience. It becomes part of determining whether a local interpretive environment can work coherently with the evidence the framework has produced (Fig. 9.2).

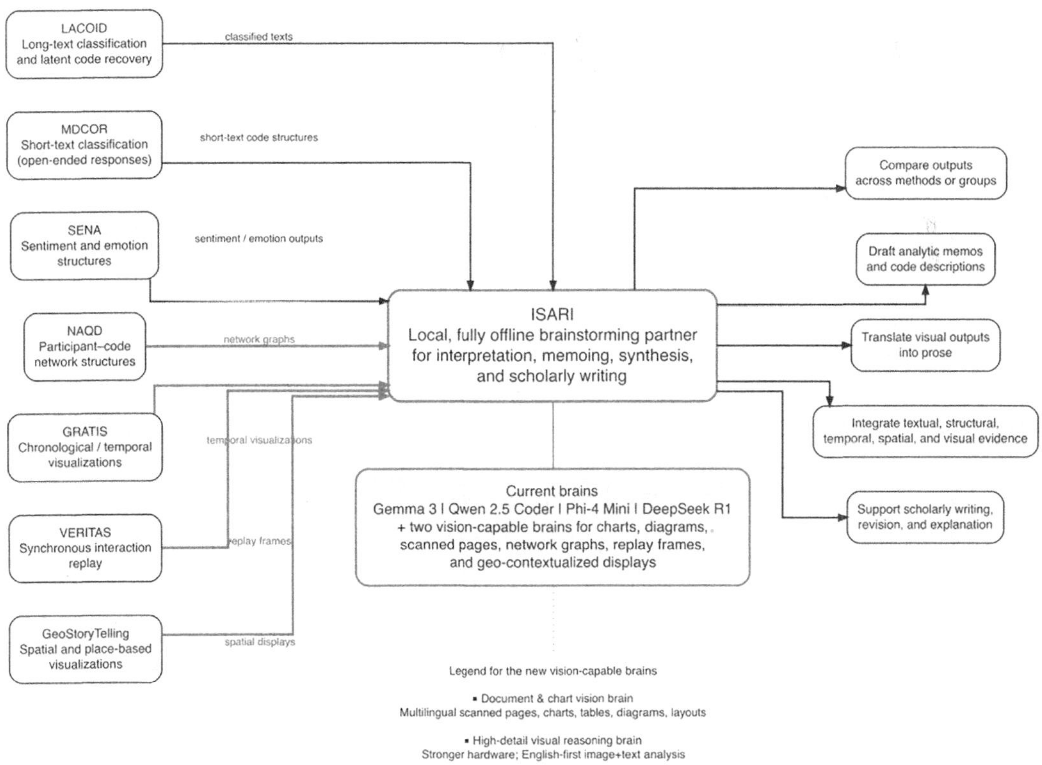

FIG. 9.2 Conceptual workflow positioning ISARI in the analytic process.

Alt Caption. Conceptual workflow illustrating that ISARI follows LACOID, MDCOR, SENA, NAQD, GRATIS, VERITAS, and GeoStoryTelling in the analytic sequence. The methodological frameworks in this book retrieve structure, emotion, chronology, relational patterns, and spatial context from qualitative evidence. ISARI operates downstream of those frameworks as a second-stage interpretive and writing partner, supporting comparison, memoing, synthesis, and scholarly prose. Magenta, or "Mexican pink," highlights the outputs that especially benefit from the addition of the two vision-capable brains, including network graphs, temporal visualizations, replay frames, and geo-contextualized displays.

To make this integration more operational and pedagogically useful, the next section provides concrete example prompts showing how ISARI can be asked to support first-pass pattern description, comparison, memoing, and evidence-grounded synthesis using the outputs generated by this framework. These examples are intended to help researchers move from abstract ideas about AI-assisted scholarly support to more disciplined and practically usable forms of local brainstorming that remain anchored in the evidence already structured by the methodological tools presented in this book.

Purposeful brainstorming with ISARI: Example prompts across the methodological ecosystem

Having described how ISARI can be integrated with each of the major frameworks in this volume, a practical question naturally follows: what exactly should researchers ask ISARI to do when those outputs are in hand? This question matters because, if ISARI is framed only in broad terms as a local thinking or brainstorming partner, some readers may still be uncertain about how to operationalize that role in relation to the actual outputs generated by LACOID, MDCOR, SENA, NAQD, GRATIS, VERITAS, or GeoStoryTelling. The purpose of the present subsection is therefore to make that role more concrete by providing example prompts that show how ISARI can be used to support first-pass pattern description, comparison, memoing, and evidence-grounded synthesis while remaining fully accountable to the outputs and original materials supplied by the researcher.

One of the most practical questions raised by researchers encountering ISARI for the first time is deceptively simple: what exactly should we ask it to do? This question matters because, if ISARI is framed only in abstract terms as a "thinking" or "brainstorming" partner, some users may still not know how to operationalize that role in relation to the actual outputs generated by the frameworks in this book. For that reason, this subsection provides a set of concrete example prompts that show how ISARI can be asked to support first-pass pattern description, memoing, comparison, and evidence-grounded synthesis after another framework has already structured the evidence.

The prompts below are not scripts to be copied mechanically, but are exemplars of how researchers can formulate requests that remain faithful to the epistemology of this volume. In each case, the role of ISARI is downstream and bounded. The model is not being asked to generate findings independently of the prior methodological work. Rather, it is being asked to help the researcher describe patterns already made visible by LACOID, MDCOR, SENA,

NAQD, GRATIS, VERITAS, or GeoStoryTelling, and to do so in a way that remains close to the supplied evidence. In every case, the strongest prompts tend to do four things clearly: (1) they name the framework already used, (2) they specify what outputs are being supplied, (3) they request a bounded interpretive task, and (4) they ask for evidentiary grounding, uncertainty, and caution where appropriate. To illustrate this logic, the following examples move framework by framework.

Example prompts for LACOID

LACOID is typically used after longer texts such as interviews, fieldnotes, observations, or essays have been decomposed and classified into latent codes. At that stage, ISARI can be asked to support the transition from machine-learned structure to first-pass description of code meaning.

Prompt example 1: First-pass code description

"I relied on LACOID to classify a long-form textual database into three latent codes. I am uploading the top 20 most representative original texts for each code, together with my field-notes and coding notes. Can you provide an initial description of the main patterns you see within each code? Please be as clear as possible, keep your interpretation close to the texts I am supplying, and include representative quotations that support your descriptions. Also indicate where my fieldnotes seem to enrich or complicate the interpretation of the code."

Prompt example 2: Comparing competing code solutions

"I ran LACOID twice and obtained two plausible solutions: one with three codes and another with five codes. I am uploading the most representative texts for both solutions, along with my notes about the project. Can you help me compare what appears to be gained and lost when moving from the more aggregated to the more disaggregated solution? Please focus on interpretive clarity rather than on declaring one solution 'correct,' and point out where I should return to the original texts before deciding."

Example prompts for MDCOR

MDCOR is often applied to short open-ended responses or social media posts, where researchers need a rigorous and efficient way of organizing large volumes of brief texts.

Prompt example 1: Thematic summary of classified short texts

"I relied on MDCOR to classify a database of short open-ended survey responses into four classes. I am uploading the top 20 most representative responses for each class and a short memo describing the context of data collection. Can you provide an initial thematic description of each class, identify any overlaps or contrasts across classes, and quote some responses that best illustrate your reading? Please keep your response grounded in the supplied texts and indicate where the short length of the responses makes interpretation less certain."

Prompt example 2: Comparing groups across classified outputs

"I used MDCOR to classify responses and then compared results across two participant groups. I am uploading the most representative responses by class and group, along with my own notes.

Can you help me draft an initial analytic memo describing how the classes appear similar or different across groups? Please avoid overinterpreting frequency differences and instead focus on the qualitative content of the representative responses."

Example prompts for SENA

SENA produces sentiment and emotion distributions, emotion networks, and group-based comparisons that often require careful translation into prose.

Prompt example 1: Describing emotional structure

"I used SENA to analyze the sentiment and emotion distribution of a set of texts and I am uploading screenshots of the resulting plots, the summary tables, and several representative original texts. Can you provide an initial description of the main emotional patterns you see, including which emotions appear more central, whether there seems to be emotional coherence or contradiction, and how those patterns relate to the representative texts? Please include quotations that support your observations and distinguish clearly between what is visible in the plots and what still requires return to the original corpus."

Prompt example 2: Comparing emotional differences across groups

"I used SENA to compare two groups and obtained group-specific emotion distributions and a QAP comparison summary. I am uploading the figures, tables, and a few representative texts from each group. Can you help me draft a memo describing where the emotional differences between groups seem most substantively meaningful? Please stay close to the supplied evidence and note where the visual differences are clear, where they are ambiguous, and how the representative texts support or complicate the comparison."

Example prompts for NAQD

NAQD outputs often include network sociograms, centrality measures, and group-comparison outputs that are visually rich but verbally difficult to render.

Prompt example 1: First-pass sociogram interpretation

"I used NAQD to analyze a participant-by-code network and I am uploading screenshots of the time-aggregated sociogram, the centrality output, and some representative quotations connected to the most central codes. Can you provide an initial description of what patterns are visible in the network? Please comment on central actors, central codes, possible communities or clusters, and any apparent structural contrasts. Also include quotations where useful and make clear what is directly visible in the network versus what still requires substantive interpretation from me."

Prompt example 2: Comparing two relational structures

"I used NAQD to compare participant-code structures across two subgroups. I am uploading both network screenshots, group-specific centrality information, and the quotes associated with the most central codes. Can you help me draft a comparative memo describing how the structure differs across groups? Please focus on interpretive possibilities rather than final conclusions and indicate where I should go back to the coded excerpts."

Example prompts for GRATIS

GRATIS helps reconstruct the chronological or nonlinear evolution of coded evidence, which often produces dynamic outputs that need to be translated into analytic narrative.

Prompt example 1: Narrating temporal evolution
"I used GRATIS to visualize the evolution of coded information across time and I am uploading screenshots of several time points, the aggregated output, and selected original excerpts corresponding to those moments. Can you help me describe the main temporal phases that seem visible in these outputs? Please identify what appears to emerge early, what stabilizes, what shifts, and what decays, and connect those observations to the excerpts where possible."

Prompt example 2: Comparing chronology and story
"I used GRATIS to reconstruct the evolution of a narrative using global time stamps. I am uploading the dynamic visualization, selected screenshots, and the original qualitative excerpts associated with several turning points. Can you help me draft a memo that compares the chronological structure visible in the output with the storyline that seems to emerge from the excerpts? Please note where the temporal visualization clarifies the narrative and where more interpretive work is still needed."

Example prompts for VERITAS

VERITAS outputs often involve replayable synchronous interactions, quote-linked actor–code or actor–actor maps, and turning points in the flow of communication.

Prompt example 1: Identifying turning points in synchronous interaction
"I used VERITAS to replay a synchronous focus-group interaction, and I am uploading screenshots of several key moments, the aggregate output, and quotations associated with the main turning points. Can you help me draft an initial description of how the interaction evolved, including which actors became more central, where possible inflection points occurred, and how the quotes help explain those shifts? Please stay close to the visible structure of the replay and the supplied quotations."

Prompt example 2: Moderator influence and participant agency
"I used VERITAS to analyze a moderated group discussion and I am uploading replay frames, actor-level summaries, and selected quotations from moderators and participants. Can you help me compare whether the interaction appears to have been participant-driven or moderator-driven across different phases? Please identify what is visible in the replay outputs, what is supported by the quotations, and where I should remain cautious before drawing stronger conclusions."

Example prompts for GeoStoryTelling

GeoStoryTelling outputs often include maps, route traces, spatial clusters, and multimedia-rich storyboards that benefit from descriptive and comparative memoing.

Prompt example 1: Describing geocontextualized stories
"I used GeoStoryTelling to create an interactive map of participant narratives and I am uploading screenshots of the map, the story panels, and several notes I took during data collection. Can you provide an initial description of the main spatial patterns you see? Please identify whether certain stories appear concentrated in similar locations, whether route or movement patterns seem relevant, and how place appears to shape the narratives I am supplying. Please also quote from the story texts where/when useful."

Prompt example 2: Comparing space, narrative, and burden
"I used GeoStoryTelling to visualize routes, locations, and participant narratives connected to experiences of access and constraint. I am uploading map screenshots, distance calculations, route traces, and my fieldnotes. Can you help me draft an initial analytic memo on how place, movement, and narrative seem connected in these outputs? Please distinguish clearly between what is visually observable from the map and what interpretive claims still need to be checked against the original stories."

Example prompts for cross-framework integration

Because ISARI sits downstream of the full ecosystem, some of its most productive uses arise when outputs from more than one method are considered together.

Prompt example 1: Integrating classification and emotion
"I used MDCOR to classify short texts into four classes and SENA to analyze the emotional distribution of each class. I am uploading the top 20 most representative texts, sentiment/emotion screenshots, and my analytic notes. Can you help me draft a memo describing how the thematic and emotional structures appear to reinforce or complicate one another? Please remain close to the supplied evidence and indicate where the emotional outputs seem to add something the classification alone did not reveal."

Prompt example 2: Integrating structure, time, and place
"I used NAQD to identify central actors and codes, GRATIS to examine their evolution over time, and GeoStoryTelling to contextualize some of these narratives spatially. I am uploading screenshots and selected quotes. Can you help me organize an initial integrative memo that explains how structural importance, temporal development, and spatial context may be connected in this project? Please identify where the outputs seem mutually reinforcing and where I should be cautious before making stronger integrative claims."

Taken together, these examples illustrate that prompting in ISARI is most useful when it remains method-aware, evidence-grounded, and task-bounded. The goal is not to ask the model to "interpret the data" in a general sense. The goal is to ask it to help the researcher think more clearly with the outputs already produced, while preserving the principle that meaning must remain accountable to the original evidence and to the researcher's own methodological judgment.

In sum, we can synthesize these prompting principles as sharing five features:

1. They name the prior framework(s) used.
2. They specify exactly what outputs are being supplied (mono or multi-frameworks).

3. They define a bounded interpretive task (describe, compare, memo, summarize, translate into prose).
4. They request evidentiary grounding, such as quotations or direct reference to visible structure.
5. They invite caution by asking the model to distinguish what is directly supported from what still requires researcher verification.

Offline AI, trust, and the realities of Institutional Review Board review

The significance of an offline AI brainstorming partner becomes even clearer when viewed through the lens of data stewardship ethics. Research on network analysis and interactive visualization has shown that, even when researchers believe data have been de-identified, anonymity and confidentiality may still be jeopardized by the structure of the data, the attributes attached to participants, or the visual displays used to represent relationships (Borgatti and Molina, 2005; Klovdahl, 2005; González Canché, 2025c). Similar concerns apply to qualitative and mixed-methods projects, which increasingly use data science tools to retrieve patterns, model relationships, or generate interactive outputs.

As these methods become more common, IRB offices are likely to scrutinize not only data collection but also data processing and interpretation. The point is not merely whether names have been removed; the point is whether identifiable or sensitive qualitative material is being transferred to external servers, stored by third parties, or processed in systems outside the researcher's control. In many institutional settings, especially when working with interviews, focus groups, observational records, or vulnerable populations, the answer will often be that such uploading should be restricted or blocked.

This is why an offline AI tool like ISARI matters. It allows researchers to retain interpretive assistance while respecting the same protective boundary that local classification, local visualization, and local network analysis already preserve. The trust that participants place in researchers does not end when the interview or focus group is complete. That trust extends to how their data are stored, transformed, interpreted, and written about. ISARI supports that continuity. The same is true for visual outputs. A network graph, map, replay frame, or dashboard screenshot may be less obviously sensitive than a verbatim transcript, but it can still encode identifiable relations, sequences, settings, or attributes (González Canché, 2025c). Keeping the interpretation of those visuals local is therefore not incidental; it is ethically continuous with the local handling of the underlying evidence.

ISARI as part of an ethical research infrastructure

Ethics and trust are not ancillary concerns in contemporary qualitative research; they are infrastructural. Our research cannot happen without participants' contributions, and the expansion of analytic power through data science intensifies rather than diminishes our obligation to protect them. Interactive reviewability, de-identification procedures, and careful consent practices remain essential, but so too does the responsible design of the software environment in which analysis takes place (Alexander et al., 2019; González Canché, 2025c).

ISARI contributes to ethical research practices by keeping interpretive work local, supporting transparency through saved and exportable analytic sessions, and reinforcing the distinction between analytic assistance and analytic authority. In this sense, it is not simply another software tool; it is part of a methodologically coherent infrastructure that aligns with participant trust, IRB expectations, and the broader commitments of qualitative and mixed-methods inquiry. The addition of vision-capable brains strengthens this contribution by ensuring that local interpretive support does not collapse when the output to be interpreted is graphical, chart-based, map-based, or image-based rather than textual alone.

A cross-framework workflow for researchers

Taken together, the frameworks in this volume suggest a broader workflow in which ISARI functions as a local companion to the full methodological arc of the book. A project may begin with long interviews or observations classified through LACOID, or with open-ended survey responses analyzed through MDCOR. It may then move to SENA to understand how sentiments and emotions vary across those same texts or groups. If coded data are of interest as participant–code structures, NAQD may be used to identify central issues, communities, and group-based similarities. If chronology matters, GRATIS may reconstruct the temporal evolution of coded content. If the setting involves synchronous group interaction, VERITAS may replay the process by which exchanges and alignments evolved. If place matters, GeoStoryTelling may geo-contextualize the stories being studied. At each of these later stages, ISARI can assist in comparing, memoing, organizing, drafting, and revising.

This workflow highlights an important conceptual point: ISARI does not compete with the methods of the book; it scales their practical use. It makes it easier for researchers to compare outputs across frameworks, explain them to others, produce clearer teaching materials, write interpretive sections more efficiently, and remain within a local environment while doing so. The gain is not merely speed; it is also one of coherence. ISARI can help researchers build continuity across analytic phases that might otherwise remain disconnected in the final written product.

The addition of vision-capable brains expands this continuity. A researcher can move from textual classification to chart-based comparison, from network structure to screenshot-based memoing, from temporal replay to prose explanation, and from geo-contextualized visual display to analytic synthesis without leaving the same local interpretive environment. The result is not universal multimodal mastery, and it should not be described that way. Rather, it is a more coherent local environment for working with the actual diversity of outputs that qualitative data science and interactive visualization now produce.

ISARI's execution steps

ISARI is designed to function as a local scholarly brainstorming partner rather than as an answer-generating authority. As such, ISARI is not a conversational shortcut but a local scholarly workflow. Executing ISARI in a methodologically defensible way involves more than opening a model and issuing a prompt. It begins with selecting the appropriate brain for the evidentiary task, supplying materials in a form that matches the analytic stage, and treating the resulting output as a bounded prompt for thought rather than as an authoritative conclusion. In this sense, ISARI functions less like a general-purpose answer system and more like a local infrastructure for recursive scholarly reasoning.

Fig. 9.3 represents ISARI's workflow. The workflow begins when the researcher launches ISARI locally and confirms that the desired engines are available in the offline environment. This opening step matters conceptually as much as it does technically. The local launch

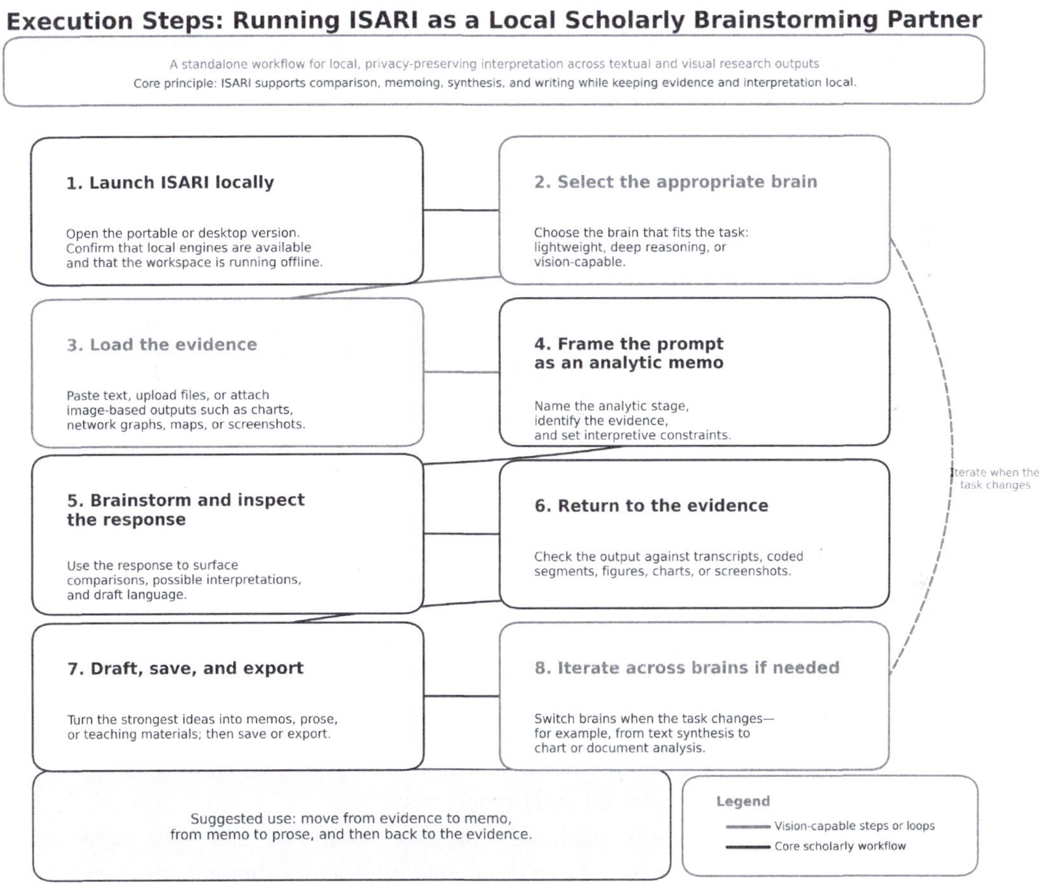

FIG. 9.3 Execution Steps for ISARI as a Local Scholarly Brainstorming Partner. The figure conceptualizes ISARI as a repeatable local workflow for evidence-grounded scholarly reasoning rather than a one-step answer system. The sequence moves from local launch and brain selection to evidence loading, prompt framing, constrained brainstorming, evidentiary checking, drafting, and iterative reuse across brains. Blue marks the core scholarly workflow, whereas Mexican pink highlights the multimodal or vision-capable stages and the iterative loop that becomes especially relevant when the analytic object shifts from text-centered reasoning to chart-, screenshot-, or document-centered interpretation. In this formulation, ISARI is not positioned as a substitute for scholarly judgment but as a privacy-preserving infrastructure for moving from evidence to memo, from memo to prose, and back again to the evidence.

establishes the protective boundary within which interpretation will occur. It signals that the ensuing interaction will remain within a privacy-preserving environment rather than an externally managed cloud system. For a scholarly workflow that may involve institutionally sensitive, relationally identifiable, or otherwise regulated materials, this is not a peripheral convenience; it is part of the methodological logic of the system itself.

Once the workspace is active, the researcher selects the brain that best fits the task. This is the point at which ISARI's internal ecology becomes methodologically relevant. A lightweight multimodal brain may be sufficient when the immediate need is concise memoing, quick summarization, or everyday interpretive support on modest hardware. A deeper reasoning brain becomes more appropriate when the task requires extended comparison, alternative explanations, or later-stage synthesis. When the object of analysis is visual—such as a chart, scanned page, diagram, screenshot, network graph, or other image-based output—a vision-capable brain becomes the stronger choice. Brain selection is therefore not an implementation detail but part of the analytic method: it aligns evidentiary form, interpretive demand, and computational capacity in a way that makes the workflow both more rigorous and more transparent.

The next step is to load the evidence. In some instances, this evidence will be textual: interview excerpts, fieldnotes, open-ended responses, coding summaries, or draft memos. In others, it will be image-based or mixed: charts, screenshots, replay frames, tables, maps, or figure exports from other analytic tools. What matters here is that ISARI does not receive data in the abstract; it receives evidence in relation to a clearly framed interpretive task. This emphasis is central to the scholarly design of the system. It keeps the workflow anchored in specific evidentiary objects rather than in generic conversation. The researcher is not asking the system to generate free-floating answers; the researcher is asking the system to assist in reasoning with supplied materials under stated constraints.

Prompting then functions as analytic memoing. Rather than writing a generic query, the researcher frames the prompt by naming the analytic stage, specifying the evidence under consideration, and clarifying the epistemic boundaries that should govern interpretation. Effective prompts ask ISARI to stay close to the supplied materials, identify contrasts or possible interpretive paths, and distinguish what is directly supported from what still requires a return to the evidence. In this respect, prompting in ISARI operationalizes long-familiar scholarly practices—memoing, comparison, and provisional theorizing—within a local computational environment. The system becomes useful not because it erases methodological work, but because it supports that work in a more structured and iterative way.

The next phase of the workflow is constrained brainstorming. The researcher inspects the response not as a result to be accepted but as a disciplined prompt for thought. The output may help surface possible comparisons, sharpen language, suggest a clearer distinction, or make a visual output more verbally legible. Yet, this usefulness remains conditional on the following step: returning to the evidence. The researcher compares the output against the transcript, coded segment, chart, screenshot, figure, or replay frame and evaluates whether the suggested interpretation is adequately grounded. This return movement is methodologically decisive. It prevents the epistemic slippage through which generated prose is mistaken for a finding and reasserts the principle that the authority of the workflow remains with the researcher, not the model.

Once the strongest ideas have been identified and checked against the evidence, the workflow moves into drafting, saving, and export. At this stage, ISARI supports the conversion of provisional insights into analytic memos, methods notes, draft prose, or teaching materials. Sessions can be saved or exported so that the interpretive trail remains visible and reusable. This is important not only for convenience but also for scholarly accountability. A research-facing AI infrastructure becomes more defensible when the steps through which interpretation emerged can, *or even must,* be revisited, revised, and documented.

Fig. 9.3 below presents ISARI as an eight-step workflow that moves from local launch to iterative scholarly use. The structure alternates across two columns to make the sequence easy to follow while preserving enough room for each step to remain readable in print. Blue marks the core scholarly workflow, whereas magenta or Mexican pink is reserved for the multimodal or vision-capable stages and the iterative loop that becomes relevant when the task shifts from text-centered reasoning to chart-, screenshot-, or document-centered analysis.

Alt caption. A two-column workflow diagram titled *Execution Steps: Running ISARI as a Local Scholarly Brainstorming Partner*. A top rounded banner contains two lines: one states that the workflow is a standalone, local, privacy-preserving process for textual and visual research outputs; the second states the core principle that ISARI supports comparison, memoing, synthesis, and writing while keeping evidence and interpretation local. Below the banner, eight rounded boxes are arranged in two columns and four rows. The left column contains Steps 1, 3, 5, and 7, while the right column contains Steps 2, 4, 6, and 8. The workflow begins with launching ISARI locally, selecting the appropriate brain, loading the evidence, framing the prompt as an analytic memo, brainstorming and inspecting the response, returning to the evidence, drafting/saving/exporting, and iterating across brains if needed. Penn blue is used for the core scholarly workflow, while Mexican pink is used for the vision-capable steps and the curved dashed iteration loop that returns from Step 8 toward earlier stages when the task changes. At the bottom, a blue note box states that the suggested use is to move from evidence to memo, from memo to prose, and then back to the evidence. A legend in the bottom-right corner explains the Penn blue and Mexican pink line meanings.

The visual design reinforces the methodological argument that ISARI is not an answer engine, but a repeatable local workflow that moves from evidence to memo, from memo to prose, and back again to the evidence. In practice, a project may begin with quick summarization, move into deeper reasoning for synthesis, and later switch to a vision-capable brain when the key analytic object becomes a figure or screenshot rather than a passage of text. Executing ISARI, then, is not a linear act of querying a model; it is a recursive and local method of scholarly work in which brain selection, evidence type, prompt framing, interpretive verification, and iterative writing remain conceptually and procedurally tied together.

Transition to the concluding chapter

This chapter foregrounds interpretation as an essential and ethically consequential stage of qualitative and mixed-methods inquiry. The concluding chapter builds on this foundation by synthesizing the methodological ecosystem presented throughout the volume and reflecting on how local, transparent, and rigorously integrated analytic infrastructures—such as those exemplified by ISARI—reshape the future practice of qualitative data science.

Conclusion

The methods collected in *Data Science and Interactive Visualization Tools for the Analysis of Qualitative Evidence* were created to address a shared analytic problem: how to work rigorously and efficiently with textual data whose scale and complexity exceed what unaided manual procedures can reasonably sustain, without sacrificing context, nuance, participant voice, or interpretive transparency. LACOID, MDCOR, SENA, NAQD, GRATIS, VERITAS, and GeoStoryTelling each address one part of that challenge. Their shared contribution is a broader one: they provide a methodological ecosystem for classification, emotional analysis, structural mapping, temporal reconstruction, process replay, and spatial contextualization.

ISARI belongs within this ecosystem as a local and fully offline brainstorming partner. Its proper role is not to replace the frameworks of the book, nor to offer a shortcut around the disciplined movement between quantitative output and qualitative evidence. Its value comes from helping researchers work more effectively with the outputs those frameworks generate. It assists in reading maps, comparing alternatives, drafting interpretations, translating technical outputs into prose, and strengthening continuity across analytic phases—all while keeping sensitive work inside the local research environment.

The incorporation of two vision-capable brains should be understood within the same logic. These models do not transform ISARI into an autonomous visual analyst; rather, they make it possible for researchers to remain within the same local interpretive infrastructure when the outputs to be explained are charts, diagrams, scanned pages, replay frames, screenshots, temporal visualizations, network graphics, or geo-contextualized displays. The ISARI architecture helps researchers *work across visual and textual outputs with greater coherence* while preserving the ethical and methodological commitments that qualitative and mixed-methods scholars most need to uphold.

The example prompts included in this chapter help operationalize that role by showing how ISARI may be asked to support first-pass pattern description, comparative memoing, visual translation, and evidence-grounded synthesis across the outputs generated by the frameworks in this book. ISARI's strongest contribution is practical and methodological at once: ISARI helps researchers think more carefully, write more clearly, and remain within a privacy-conscious, ethically defensible, and methodologically coherent environment. In that sense, a local and fully offline brainstorming partner—especially one capable of assisting with both textual and visual outputs—may become one of the most practical ways to expand the use of **local large language models in academic research** without undermining the commitments that qualitative and mixed-methods scholars most need to preserve. ISARI points toward that future: more deliberate, portable, and integrative forms of scholarly reasoning.

References

Alexander, E., Eppler, M.J., Bresciani, S., 2019. Visual replay methodology: a mixed methods approach for group discussion analysis. J. Mixed Methods Res. 13 (1), 33–51. doi:https://doi.org/10.1177/1558689816664479.

Borgatti, S.P., Molina, J.L., 2005. Toward ethical guidelines for network research in organizations. Soc. Netw. 27 (2), 107–117.

Braun, V., Clarke, V., 2006. Using thematic analysis in psychology. Qual. Res. Psychol. 3 (2), 77–101. doi:https://doi.org/10.1191/1478088706qp063oa.

Charmaz, K. (2014). Constructing Grounded Theory, 2nd ed. Sage.

DeepSeek-AI, 2025. DeepSeek-R1-0528-Qwen3-8B. Model card, Hugging Face.

Fetters, M.D., Freshwater, D., 2015. The 1 + 1 = 3 integration challenge. J. Mixed Methods Res. 9 (2), 115–117.

González Canché M.S., (2022). Post-purchase Federal Financial Aid: How (in)Effective is the IRS's Student Loan Interest Deduction (SLID) in Reaching Lower-Income Taxpayers and Students? Res High Educ, 63(6), 933–986. doi:10.1007/s11162-021-09672-6.

González Canché M.S., (2026). Network Analysis of Qualitative Data (NAQD): An Analytic Framework and Software for Qualitative Data Science Visualization and Hypothesis Testing. International Journal of Qualitative Methods, 25. doi:10.1177/16094069261439942.

González Canché, M.S., 2022a. Network analysis of qualitative data: an integrative software application to visualize and assess similarities in participants' qualitative contributions. J. Mixed Methods Res. 16 (3), 373–377. doi:https://doi.org/10.1177/15586898211051584.

González Canché, M.S., 2023a. Graphical Retrieval and Analysis of Temporal Information Systems (GRATIS): an integrative mixed methodology and open-access software to analyze the (non-)linear chronological evolution of information embedded in textual/qualitative data. J. Mixed Methods Res. 18 (1), 71–103. doi:https://doi.org/10.1177/15586898231166968.

González Canché, M.S., 2023b. Latent Code Identification (LACOID): a machine learning-based integrative framework [and open-source software] to classify big textual data, rebuild contextualized/unaltered meanings, and avoid aggregation bias. Int. J. Qual. Methods 22. https://doi.org/10.1177/16094069221144940

González Canché, M.S., 2023c. Machine driven classification of open-ended responses (MDCOR): an analytic framework and no-code, free software application to classify longitudinal and cross-sectional text responses in survey and social media research. Expert Syst. Appl. 215, 119265. doi:https://doi.org/10.1016/j.eswa.2022.119265.

González Canché, M.S., (2023). Spatial socio-econometric modeling (SSEM): A low-code toolkit for spatial data science and interactive visualizations using R. Springer Nature. doi:10.1007/978-3-031-24857-3.

González Canché, M.S., 2025a. GeoStoryTelling: an integrative multi-media methodology and no-code, cost-free software to geo-contextualize ethnographic, health, and survey evidence. Int. J. Qual. Methods 24, 1–23. doi:https://doi.org/10.1177/16094069251325629.

González Canché, M.S., 2025c. Ethics and trust in social network analysis: how data science and interactive visualizations impact anonymity, confidentiality, consent, and IRB procedures? In: Handbook of Social Network Analysis and Education Nina Kolleck. Edward Elgar. pp. 114–131.

González Canché, M.S., Zheng, K., 2025b. Sentiment and Emotion Network Analysis—SENA: an analytic framework and no-code software to integrate natural language processing, linguistic analysis, and data science with qualitative and mixed-methods research. International Journal of Qualitative Methods 24 .

González Canché, M.S, Zheng, K., Song, Y. and Liang Y., (2025). Standardized Testing for Diverse Talent Identification: A Framework to Address Geographical Bias in Standardized Testing and Increase Diversity in College Admissions in the Post-Affirmative Action/Race-Neutral Admissions Era. Research in Higher Education, 66(2). doi:10.1007/s11162-024-09824-4.

Google DeepMind, G. (2025). *Gemma 3* [Model documentation].

Johnson, R.B., Onwuegbuzie, A.J., 2004. Mixed methods research: a research paradigm whose time has come. Educ. Res. 33 (7), 14–26.

Klovdahl, A.S., 2005. Social network research and human subjects protection: toward more effective infectious disease control. Soc. Netw. 27 (2), 119–137.

Meta Llama (2024). *Llama 3.2 vision model card.*

Microsoft. (2025). *Phi-4-mini-instruct model card.*

Miles, M.B., Huberman, A.M. (1994). Qualitative Data Analysis, 2nd. Sage.

Qwen Team. (2025a). *Qwen2.5-VL technical report.*

Qwen Team. (2025b). *Qwen3 documentation.*

Saldaña, J. (2021). The Coding Manual for Qualitative Researchers, 4th. Sage.

10

Closing thoughts, truly equal status design, and the next phase of democratizing data science

Introduction: Truly equal status design (TESD) as the epistemological spine of the book

Guided by the Truly Equal Status Design (TESD), understood throughout this book as a genuinely equal-status mixed methods and multimodal design, our analytic journeys have highlighted and prioritized a balanced interweaving of qualitative inputs and quantitative outputs. This balance aligns with the mixed-methods integration challenge and remains vital for researchers seeking to construct deep understandings of the structure and meaning embedded in qualitative and multimodal sources of evidence (Fetters and Freshwater, 2015; Fetters et al., 2013). This qual/quant/qual/quant interdependence—or interweaving—helps researchers maintain clarity and transparency in the rationale guiding the interpretation of data and findings. In other words, following TESD, quantitative outputs are analytically incomplete unless researchers return to the original qualitative inputs from which those outputs emerged. It is only when we integrate and intermingle quantitative outputs with unaltered textual, interactional, emotional, temporal, and geocontextualized evidence that deeply powerful and nuanced understandings may be gained and conveyed.

This interweaving is important not only for knowledge building and discovery, but also because the use of quantitative outputs serves as a map for researchers, participants, and readers alike to observe the rationale and evidence through which conclusions, recommendations, and plans of action are reached. In TESD, interactive reviewability—where participants can see whether they agree with our representations of their narratives, trajectories, emotional distributions, interactional patterns, or geocontextualized stories—becomes possible, as does the possibility that readers may directly inspect the interpretive path that produced the final argument (Alexander et al., 2019; González Canché, 2023 , 2024, 2025). From this view, the "analytic black box" of qualitative inquiry need no longer remain opaque. The time has come not only to empower qualitative and mixed methods researchers to take advantage of data science without needing to learn computer and statistical programming, but also to empower participants and readers to validate, question, refine, or even contest the rationale and conclusion-making process. In short, quantitative outputs are to be used as maps that help researchers, participants,

Data Science, Interactive Visualizations, and Generative AI Tools for the Analysis of Qualitative, Mixed-Methods, and Multimodal Evidence. https://doi.org/10.1016/B978-0-443-21961-0.00007-3

and readers gain better, clearer, more nuanced, and more transparent understandings of participants' experiences as expressed in their unmodified narratives, coded contributions, dynamic interactions, and contextualized traces.

It is therefore important to emphasize that TESD is not simply a design preference but a methodological philosophy that equally values qualitative and quantitative methods, using each to enhance the other. Given the vast availability of unstructured qualitative data in contemporary society, TESD employs quantitative tools to clarify and extend insights from participants' stories, rather than to replace them. TESD keeps large-scale analysis transparent, grounded in lived experience, and accountable to the original voices that gave rise to the evidence in the first place (Hesse-Biber, 2010; Tanweer et al., 2021). From this perspective, although we believe that mathematics, network theory, natural language processing, temporal modeling, and geospatial visualization can help reveal structure embedded in qualitative or textual data, the strength of our understanding will only be as nuanced and solid as the extent to which it remains informed by our participants' unaltered voices, stories, trajectories, exchanges, emotions, and places.

One of the most important lessons that emerged from the methodological chapters is that these frameworks do not merely produce outputs; they produce different kinds of analytic maps. This closing realization returns us directly to the book's opening proposition: the value of computational, statistical, and visual outputs lies not in replacing interpretation, but in making the movement from evidence to meaning more ethically-conscious, purposeful, visible, revisitable, and accountable In practice, some of those maps identify themes and latent codes. Others reveal the distribution of sentiments and emotions. Others make visible the relational structure of coded contributions, the temporal evolution of asynchronous data, the replayable dynamics of synchronous interactions, or the geocontextualized trajectories of stories in place and time. The significance of this realization is that, once we accept that the outputs of these methodologies are maps rather than endpoints, we can also understand why a local interpretive environment becomes so consequential. Such an environment allows researchers to bring these maps together, compare them, memo their implications, and translate them into coherent argumentation without sacrificing the researcher's role as the ultimate meaning-maker. It is precisely in this context that ISARI becomes central to the architecture of the book.

The book as a layered local analytic ecosystem

After studying the methodological chapters, it becomes clear that the approaches presented in this book can stand alone; however, they can also be integrated into single studies, grant proposals, dissertations, evaluation projects, and policy analyses. To understand how such integration may occur, it is useful to move away from thinking about the methods as a simple list of seven tools and to instead conceive of them as a **layered local analytic ecosystem**. In this ecosystem, each methodology contributes a different form of analytic clarification, and these clarifications may be stacked, combined, or revisited in iterative ways depending on the needs of the project. From this perspective, the book does not simply offer a toolkit. It offers a methodological infrastructure for truly equal-status qualitative and mixed-methods inquiry.

Foundational coding and classification layer

At the foundational layer are the methods that help researchers identify, classify, and organize the meaning-bearing units of large textual corpora. **LACOID** was designed for long-form texts—interviews, fieldnotes, essays, policy documents, and other complex narrative materials. By decomposing documents into smaller units and classifying those units into latent codes while preserving original wording, cleaned text, and original document location, LACOID minimizes aggregation bias and makes large-scale machine classification accountable to the actual language of participants (González Canché, 2023). MDCOR, in turn, extends this classificatory power to short-form texts such as open-ended survey responses and social media posts while preserving respondents' original voices and rendering downloadable outputs that can be integrated into more traditional quantitative analyses (González Canché, 2023). SENA adds yet another foundational layer by revealing the emotional and affective texture of texts. Instead of simply telling us what themes emerged, it tells us how those themes feel, whether they are emotionally coherent, contradictory, divergent across groups, or marked by especially high entropy that warrants closer interpretive attention (Canché González and Zheng, 2025). Taken together, LACOID, MDCOR, and SENA form the book's foundational classificatory and affective layer: the part of the ecosystem that identifies what is being said and, in the case of SENA, with what emotional tone or complexity.

Structural and relational layer

Once codes, classes, sentiments, or emotions have been identified, researchers often need to understand the **relational structure** emerging from those outputs. This is the domain of **NAQD**, which uses network analysis to reveal which actors, groups, codes, or categories occupy central positions, which clusters or communities emerge, which ties are strongest, and whether the distribution of relations differs significantly across groups. NAQD adds a structural and relational analytic layer to the ecosystem by showing that the significance of a code or actor is not reducible to frequency alone; significance is also a function of how contributions are connected, shared, clustered, or bridged within the larger web of qualitative evidence. From this perspective, NAQD transforms coded qualitative data into relational maps that can reveal influence, consensus, marginality, subgroup formation, and statistical differences across the structure of the network.

Temporal and process layer

The next layer in the ecosystem is temporal and processual. **GRATIS** enables researchers to analyze the chronological or nonlinear evolution of coded information in asynchronous settings. It shows how codes emerge, stabilize, decay, recur, or shift meaning over time and does so through replayable dynamic HTML outputs that preserve context and support interactive reviewability (González Canché, 2024). VERITAS, by contrast, focuses on synchronous social settings where interaction occurs in real time and participants may directly influence one another. It traces communication exchanges, the co-generation of knowledge, and the evolution of coded contributions through replayable temporal sociograms and interaction maps. Together, GRATIS and VERITAS form the temporal-process layer of the ecosystem. Where

NAQD shows us structure, these methodologies show us how structure becomes—how it unfolds, changes, and sometimes breaks. They therefore make visible the processes and turning points that static analysis alone would likely miss.

Spatial and contextual layer

Finally, **GeoStoryTelling** adds a spatial and contextual layer to the book's analytic ecosystem. Stories are never merely abstract sequences of words. They happen in places, in neighborhoods, in institutions, in geographies of opportunity, threat, mobility, and inequality. GeoStoryTelling allows researchers to situate narratives in the splace-time continuum and to enrich those narratives through multimedia elements such as text, images, video, audio, and external hyperlinks. In doing so, it allows researchers to geocontextualize qualitative evidence and to show how the places where life was experienced may have shaped the stories that participants tell. GeoStoryTelling therefore extends TESD into the spatial domain by reminding us that the meaning of a narrative is often inseparable from where the relevant events unfolded (González Canché, 2025).

Although not discussed in the content of the book, note that outputs produced by other methods, may also be integrated into ISARI. For example, Canché González et al. (2025) presented a map inspired by GeoStoryTelling but bult with only quantitative data to identify high achievers in standardized math scores and also plot where they resided. Even if these outputs are 'quantitative in nature' they can also be integrated with narratives and stories coming from our participants' experiences. Procedurally, this brainstorming or integrative process may start with either taking a screenshot of the resulting map (see interactive version here https://cutt.ly/hwvraSQD) or even load the entire interactive plot and uploaded to ISARI's local server to continue with the analyses. The point here is to simply highlight that in the same way each output discussed as part of this can be integrated into the analyses, external outputs may also be included. See also González Canché, (2023) for a comprehensive sets of analyses, all of which can be analyzed with the TESD tenets discussed in this book.

Interpretive orchestration layer

Once these layers are visible—classification, affect, structure, process, and spatial context—the need for an interpretive orchestration layer becomes unavoidable. The outputs of the methodologies in this book are rich, but richness alone does not guarantee coherence. Researchers still need to compare patterns across outputs, ask whether alternative explanations fit better, memo how apparently distinct findings may connect, and draft interpretations that remain accountable to participants' original voices while being transparent and analytically robust. This is precisely the role now played by **ISARI**. Rather than functioning as "tool number eight," ISARI serves as the **interpretive orchestration layer** of the ecosystem: the local, offline, multimodal scholarly reasoning environment in which maps generated by the other methodologies may be brought together and worked through. From this perspective, the book is best understood not as a collection of isolated methods, but as a coordinated analytic ecology in which each methodology generates maps and ISARI helps the researcher think across them.

The approaches presented across the book can be seen as interconnected elements within a broader analytical system. Each contributes a different way of engaging textual, relational, spatial, and multimodal evidence while remaining open to adaptation across contexts. Together, they suggest how analysis can move across layers of structure, language, and integration while continuing to support interpretation as an ongoing and evolving process, including forms of AI-assisted synthesis and reflection without displacing human interpretive authority. In this way, the methods retain their individual strengths while also participating in a more coordinated and portable system of inquiry.

ISARI and the next phase of truly equal status design

The integration of ISARI into the methodological ecosystem described above is not an incidental technological convenience; it is the natural next phase of Truly Equal Status Design. TESD has always insisted that quantitative outputs are necessary yet incomplete, that structure must remain answerable to meaning, and that original qualitative and multimodal evidence remains the ultimate ground of interpretation (Fetters et al., 2013; Fetters and Freshwater, 2015). What ISARI adds to this framework is not an automatic answer engine, but a local scholarly reasoning space in which researchers may work with the maps produced by the other methods in a deliberate, memo-driven, and privacy-preserving way. In this sense, ISARI does not replace TESD. It operationalizes one of its most important implications: if outputs are maps rather than meanings, then researchers need a serious interpretive environment in which to think with those maps.

It is therefore important to be explicit: **ISARI is neither just another methodological chapter alongside the others nor imply "tool number eight."** Unlike LACOID, MDCOR, SENA, NAQD, GRATIS, VERITAS, or GeoStoryTelling, ISARI does not primarily generate its own domain-specific map. Rather, it helps researchers compare and work across maps generated elsewhere. It can ingest screenshots of interactive HTMLs, figures, route maps, network diagrams, sentiment distributions, replay frames, quote-linked interaction panels, and downloaded tables from the other methods. It can then be used to compare possible explanations, generate analytic memos, identify unresolved contrasts, and draft evidence-grounded prose that remains accountable to the original data. In this sense, ISARI is best understood as the local interpretive companion to the seven primary analytic frameworks presented across the book: the environment where researchers remain fully responsible for meaning, but where that meaning-building process becomes more organized, traceable, and productive.

The fact that ISARI operates **locally and offline** is especially important. Across the methodological chapters, one of the book's strongest and most consistent commitments has been the insistence that sophisticated analytics should not require researchers to upload sensitive data to cloud-based infrastructures. LACOID, MDCOR, SENA, GRATIS, VERITAS, and GeoStoryTelling all emphasize local execution as both a technical and an ethical virtue. ISARI continues that same ethic (see González Canché, 2025). It enables downstream interpretation to remain under the researcher's control rather than forcing the researcher to move outputs into server-based systems in order to compare, synthesize, or draft with them. In this respect, ISARI is not merely consistent with the methods in the book; it is the logical extension of their privacy-conscious design philosophy.

A further reason why ISARI matters in the context of TESD is that it strengthens **interpretive accountability**. When researchers can bring the outputs of multiple methodologies into a single local reasoning environment, the path from evidence to conclusion becomes more visible and more documentable. Screenshots of a temporal replay can be compared with a file-to-code strength table. A geostory map can be read alongside a sentiment distribution. A network sociogram can be revisited next to its most representative original quotes. A researcher's memos can then track why one interpretation was favored over another. In this way, ISARI extends the logic of interactive reviewability and the opening of the analytic black box into the final stages of synthesis and writing. It does not make interpretation automatic; rather, it makes interpretation more **explicit, revisitable, and ethically grounded**. In a nutshell, qualitative evidence can be organized into coded or relational databases (either manually or through tools such as LACOID and MDCOR), then examined through relational and network-based analyses using NAQD, GRATIS, or VERITAS, and ultimately integrated within ISARI's local environment to support an exploratory, interpretive process of understanding participants lived experiences. When the incorporation of space makes sense, we can also visually analyze where those experiences took place, as well as how or if their affective or emotional states diverge or remain similarly distributed across groups.

Given that interpretation is not automatic, the next phase of **Democratizing Data Science** should not be understood merely as the future development of more single-purpose tools. The next phase for ethically- and trustworthy-fueled data science qualitative and mixed-methods scholarly analyses is better understood as the emergence of **local analytic ecologies**: infrastructures in which classification, affect, structure, time, geography, and interpretation can all be pursued under researchers' full local control. In such ecologies, the goal is not to displace human expertise but to amplify it. The methods presented in this book bring structure to complexity. ISARI makes it easier for researchers to think across that structure without forfeiting their interpretive responsibility. That is why ISARI belongs in the conclusion not as an appendix to the project, but as one of the clearest signals of where TESD and DDS are headed next.

From analytic maps to integrated research workflows

To further illustrate how the layered ecosystem of this book can function in practice, it is useful to consider not only how the methods can be used individually, but how they may be linked in **integrated workflows**. Such workflows are not meant to prescribe a single correct sequence of analyses. Rather, they are intended to demonstrate that different analytic maps can be generated in sequence or in parallel, and that ISARI can serve as the local reasoning environment through which those maps are compared and woven together.

Consider first a **policy or social media discourse workflow**, such as the release of a plan to reduce (or cancel) student debt burden by then President Joe Biden on August 24, 2022. Before Elon Musk acquired Twitter, there was a free service called AcademicTwitter that allowed researchers to access up to 10 million Tweets per month while specifying social movements (via hashtags) or users, both with specific time frames. Using this tool, we were able to capture all Twitter posts taking place within two months before the release of this plan (June 24 to August 23) and two months after the release (August 24 to October 23) of such a plan (see Vo et al., 2025). With this setup in mind, a study could begin by using **MDCOR** to identify the themes emerging in social media posts before and after the release of a major policy intervention. Those classified outputs could then be carried into **SENA** to examine whether the emotional composition of those themes shifted across the two time frames—perhaps revealing greater anger, hope, resentment, relief, or contradiction after the policy was announced. The same classified data could also be analyzed with **NAQD** to identify whether the most influential actors or communities within the network of discourse remained stable or shifted after the event, and whether those actors clustered around different themes in the pre- and post-policy periods. If location is meaningful or available, **GeoStoryTelling** could then be used to situate the most central actors or narratives geographically, thereby revealing whether the discourse is more concentrated in certain regions, institutional settings, or political geographies. At that point, the study would no longer consist of one result but of several analytic maps: thematic, emotional, structural, and spatial. **ISARI** would then become the place where these outputs are brought together locally—through screenshots, HTML exports, summary tables, and memos—to compare candidate interpretations and draft the integrated argument.

A second workflow may be built around **institutional or working-group decision processes**, such as the NSF-sponsored focus group analyzed in the VERITAS chapter. Here the core question is not only what participants said, but how knowledge emerged, who drove the conversation, when the most consequential turning points occurred, and how moderator influence intersected with participant agency. **VERITAS** allows those interactions to be replayed and studied through time, making visible the emergence of hot zones, inflection points, brokers of communication, and quote-linked trajectories of knowledge production. Those same coded contributions could also be fed into **NAQD** if the researcher wanted a more aggregate or structural account of which codes, actors, or communities became central in the larger network of meaning. Once again, **ISARI** would serve as the space where replay frames, interaction maps, most representative quotations, and researcher memos could be compared to determine whether a turning point should be interpreted as leadership, persuasion, resistance, agenda control, marginalization, or collective convergence. In this case, the integration is not simply between two tools; it is between dynamic replay, relational structure, and local interpretive reasoning.

A third workflow may focus on **longitudinal narrative evolution**. Imagine a corpus of interview transcripts, open-ended survey responses, or periodic reflections collected across multiple time points. **LACOID** or **MDCOR** could first identify the latent code structure embedded in these texts. **GRATIS** could then reveal how those codes emerged, receded, clustered, or reappeared over time, while **SENA** might show whether the emotional coherence or contradiction of those themes shifted alongside their temporal trajectories. If certain codes became increasingly central while also becoming more fear-laden or more contradictory, the interpretive significance of those changes would likely warrant closer attention. The role of **ISARI** in such a workflow would be to bring together the temporally disaggregated maps, the affective summaries, the representative passages, and the researcher's own notes so that change and continuity can be compared more carefully and translated into evidence-grounded explanation. This is where TESD becomes most visibly alive: the temporal and affective outputs do not explain themselves; they require human comparison, contextualization, and judgment.

These examples demonstrate that the methodologies presented in the book are not only compatible; they are **interoperable at the level of output and interpretation**. A researcher may begin with one method and stop there. But a researcher may also proceed iteratively—moving from classification to emotion, from emotion to structure, from structure to time, from time to place, and from all of these maps into ISARI for local synthesis and writing. The point is not to use all methods all the time. The point is that the book now supports more than isolated analyses; it supports the development of integrated research workflows in which different kinds of maps can be created and then meaningfully brought into conversation.

From Aanalytic maps to actionable research

One of the most important implications of the ecosystem described above is that it makes possible not only richer analysis, but more **actionable research**. In many qualitative and mixed-methods traditions, researchers already know that a compelling quotation, a carefully built theme, or a vivid narrative can illuminate complexity. What the methodologies in this book add is the possibility of moving from those illuminating fragments toward a more systematic understanding of structure, contradiction, temporality, affect, and context—without sacrificing qualitative depth. In other words, these methodologies allow researchers to move from **thick description** toward **evidence-grounded intervention planning**, while still honoring the centrality of lived experience.

This matters because plans of action are strongest when they are informed by both **pattern visibility** and **participant-grounded meaning**. A campus safety study, for example, may use **MDCOR** or **LACOID** to classify narratives of threatening or exclusionary incidents, **SENA** to understand whether those incidents are marked more by fear, anger, humiliation, or emotional contradiction, and **GeoStoryTelling** to map where those events occur and whether they cluster around specific sites. At that point, the study has already moved beyond anecdote: it has identified recurring patterns, emotional texture, and spatial concentration. But the question of what should be done next—what intervention might matter, what pattern is most urgent, whether certain areas require environmental changes or training or accountability procedures—still requires interpretation. It is precisely there that **ISARI** can help. By bringing the classified cases, emotional summaries, geospatial maps, and selected original narratives into one local reasoning environment, the research team can more carefully compare possible explanations and draft plans of action that remain accountable to both structure and story.

A similar logic applies in institutional planning. A working group or policy meeting analyzed with **VERITAS** may reveal that a particular moderator dominated the agenda early on, that one participant consistently became the broker through whom others' perspectives moved, or that a crucial conversation about gender, race, or institutional culture briefly emerged and then disappeared. Those patterns matter not simply because they are analytically interesting, but because they may explain why certain policies never develop, why some issues remain peripheral, or why some interventions continue to reproduce the exclusions they seek to solve. Once such replayable evidence has been generated, researchers and practitioners can use **ISARI** to compare alternative readings of those inflection points and to draft more grounded recommendations: perhaps changes in facilitation practice, perhaps more equitable turn-taking structures, perhaps more explicit mechanisms for keeping critical but easily marginalized topics in the center of the conversation. In this way, the ecosystem of methods in the book can help transform qualitative and mixed-methods analysis into a more visible and actionable form of institutional learning.

The same is true in health and community research. **GeoStoryTelling** can help reveal how neighborhood conditions, routes, local fears, or place-based constraints shape participants' narratives. **GRATIS** can show how those narratives evolve over time, especially when nonlinear or episodic trajectories matter. **SENA** can indicate whether a trajectory becomes more emotionally coherent or more conflicted. **NAQD** can show whether certain experiences are more structurally central or peripheral across participants. The resulting maps, when brought together in **ISARI**, can support a much richer conversation about intervention design: not simply what happened, but why, when, where, with what emotional burden, and through what recurring mechanisms. This is precisely the sort of action-oriented knowledge that TESD makes possible when it is treated as a living epistemology rather than as an abstract design label.

From this perspective, the value of the book is not merely that it democratizes advanced analytics. Its value also lies in making those analytics usable for **decision-making, intervention planning, evaluation, and accountable action**. The maps generated by the methodologies in this book are not meant to remain beautiful but inert. They are meant to help researchers, practitioners, and communities think more clearly, decide more responsibly, and act more effectively. TESD ensures that such action remains accountable to original voices. DDS ensures that the analytic tools needed to support such action are no longer restricted to those with money, programming expertise, or cloud-dependent infrastructures. The intention of the book is to help researchers complete that DDS movement by ensuring that the path from analytic map to action-oriented interpretation can remain local, transparent, and humanly governed.

Closing thoughts and the future of democratizing data science

The main goal of this book has always been to provide researchers who, although they may not have had the opportunity to be trained in computer and statistical programming, nonetheless possess content expertise and methodological seriousness that can benefit enormously from data science tools within a truly equal-status epistemological framework. In this respect, the book remains deeply committed to **Democratizing Data Science (DDS)** as a movement. Yet, after the methodological and conceptual expansions carried out across the chapters, it is perhaps more accurate to say that DDS is now becoming more than a movement. It is becoming a **research-facing infrastructure** for qualitative and mixed-methods integrative inquiry.

This emphasis is particularly relevant because the social sciences continue to suffer from segmentary epistemological postures in which qualitative and quantitative researchers too often speak past one another. The purpose of grounding this book in TESD has been to move beyond those extreme and, frankly, increasingly archaic postures. When qualitative and mixed-methods researchers' content expertise is paired with rigorous and innovative data science tools, it becomes possible to build deeper, more transparent, more replicable, and more actionable understandings. The methods presented in this book were never intended to quantify away the qualitative. Rather, they were intended to make the qualitative more visible, more structured, more inspectable, and more powerfully interpretable.

As discussed throughout the chapters, textual data remain enormously prevalent and will not stop expanding. Textual data has long been described as constituting the overwhelming majority of newly generated data, and the challenge of making sense of such data has only become more urgent as digital communication has accelerated (Andre, 2022; Davis, 2019; Kumar & Bhatia, 2013; Raconteur, 2020). But textual abundance is only one aspect of the contemporary analytic challenge. Contemporary social research also confronts images, videos, spatial traces, interaction logs, geolocated narratives, and multimodal evidence that are increasingly difficult to analyze with traditional qualitative tools alone. The first phase of DDS therefore focused on creating **no-code, local, cost-free methods** that could help researchers classify, map, replay, and contextualize such data. That phase is represented by the methodologies in this book. The next phase, however, is now becoming visible: the development of **local analytic ecologies** in which different forms of outputs can be brought into conversation, interpreted across tools, and translated into action without forfeiting confidentiality, researcher control, or interpretive nuance. In this regard, **ISARI** is one of the clearest signs of where DDS is headed.

A key aspect of our approach remains **data protection, confidentiality, and anonymity**. We continue to believe that the most ethically responsible tools for many qualitative and mixed-methods projects are those that work locally and do not require data to be uploaded to online servers or commercial clouds. This is a common feature of all the major methodological tools discussed in this book, and it is a defining feature of ISARI as well. From this perspective, the challenge moving forward is not merely how to develop more powerful tools, but how to do so while preserving the local, privacy-conscious, and human-in-the-loop ethos that has guided DDS from the start. The future of democratizing data science therefore lies not only in no-code development, but also in **privacy-preserving interoperability, local interpretive support, and researcher-governed AI augmentation**.

This also means that DDS is not limited to the analysis of purely qualitative data. As the current chapter already points out, earlier work on **Spatial Socio-econometric Modeling (SSEM)** demonstrated that the goal of opening access to data science tools can extend into other advanced analytic domains as well (González Canché, 2023). The broader lesson is that democratization is not method-specific; it is an orientation toward infrastructure, pedagogy, and access. The methods in this book happen to focus primarily on qualitative, mixed-methods, temporal, relational, affective, and geocontextualized data. But the underlying movement can and should continue to expand toward other domains of analysis where methodological barriers remain artificially high.

Accordingly, the future of DDS should be imagined along at least three interconnected directions. First, we need **more local, cost-free, no-code analytic tools** that respond to the actual needs of qualitative, mixed-methods, health, educational, ethnographic, and policy researchers. Moreover, these tools require not to have the strongest or more powerful brains, for these are again only affordable for a selected few, which clearly defeats our DDS purpose and goal. Second, we need **better interoperability among those tools**, so that outputs can move more easily from classification to structure to time to place to interpretation. Third, we need to continue building **local interpretive environments**—of which ISARI is an early but consequential example—that allow researchers to reason across those outputs without outsourcing either their data or their interpretive agency. This is what a mature democratizing data science infrastructure would look like: not isolated tools, but a locally governed ecosystem for seeing, comparing, and understanding complex evidence.

If there is one final message to leave with the reader, it is this: the time has come not only to remove programming and financial barriers to advanced analytics, but to remove the false choice between computational sophistication and interpretive depth. TESD shows that quantitative outputs and qualitative meaning can and must remain interdependent. DDS shows that access to advanced methods need not depend on money or programming expertise. The methodological chapters of this book show that thematic, emotional, structural, temporal, synchronous, and geospatial maps can all be generated locally and transparently. ISARI now shows that those maps can also be brought into a local interpretive orchestration environment where they can be compared, memoed, and translated into ethically-fueled and evidence-grounded, actionable scholarship without displacing human interpretive authority. If Chapter 1 opened the possibility of a new scholarly infrastructure, this concluding chapter closes by inviting readers not only to use that infrastructure, but to extend it, challenge it, and help shape its next phase. This **democratizing data science movement** will only become stronger with the ideas that come from experts like you. *If you are interested in becoming part of this democratizing data science movement, do not hesitate to share your ideas with us.*

References

Alexander, E., Eppler, M.J., Bresciani, S., 2019. Visual Replay Methodology: A Mixed Methods Approach for Group Discussion Analysis. Journal of Mixed Methods Research, 13 (1), 33–51. doi:10.1177/1558689816664479.

Andre, L., 2022. Important statistics about how much data is created every day. Available from: https://financesonline.com/how-much-data-is-created-every-day/.

Canché González, M.S., Zheng, K., 2025. Sentiment and Emotion Network Analysis—SENA: An Analytic Framework and No-Code Software to Integrate Natural Language Processing, Linguistic Analysis, and Data Science with Qualitative and Mixed-Methods Research. International Journal of Qualitative Methods, 24. doi:10.1177/16094069251346929.

Canché González, M.S., Zheng, K., Song, Y., Liang, Y., 2025. Standardized Testing for Diverse Talent Identification: A Framework to Address Geographical Bias in Standardized Testing and Increase Diversity in College Admissions in the Post-Affirmative Action/Race-Neutral Admissions Era. Research in Higher Education 66 (2). doi:10.1007/s11162-024-09824-4.

Davis, D., 2019. AI unleashes the power of unstructured data. Cio. WWW-dokumentti. Available from: https://www.cio.com/article/3406806/ai-unleashes-the-power-of-unstructured-data.html.

Fetters, M.D., Curry, L.A., Creswell, J.W., 2013. Achieving Integration in Mixed Methods Designs—Principles and Practices. Health Services Research, 48 (6pt2), 2134–2156. doi:10.1111/1475-6773.12117.

Fetters, M.D., Freshwater, D., 2015. The 1 + 1 = 3 Integration Challenge. U.S. J. Mixed Methods Res. 15586901. 9 (2), 115–117. Available from: http://mmr.sagepub.com/. doi:10.1177/1558689815581222.

González Canché, M.S., 2023. Latent Code Identification (LACOID): A Machine Learning-Based Integrative Framework [and Open-Source Software] to Classify Big Textual Data, Rebuild Contextualized/Unaltered Meanings, and Avoid Aggregation Bias. International Journal of Qualitative Methods, 22. doi:10.1177/16094069221144940.

González Canché, M.S., 2023. Machine driven classification of open-ended responses (MDCOR): An analytic framework and no-code, free software application to classify longitudinal and cross-sectional text responses in survey and social media research. Expert Systems with Applications, 215, 119265. doi:10.1016/j.eswa.2022.119265.

González Canché, M.S., 2023. Spatial Socio-econometric Modeling (SSEM): A Low-Code Toolkit for Spatial Data Science and Interactive Visualizations Using R Springer Texts in Social Sciences. Springer International Publishing, Springer International Publishing. 10.1007/978-3-031-24857-3.

González Canché, M.S., 2024. Graphical Retrieval and Analysis of Temporal Information Systems (GRATIS): An Integrative Mixed Methodology and Open-Access Software to Analyze the (Non-)Linear Chronological Evolution of Information Embedded in Textual/Qualitative Data. Journal of Mixed Methods Research 18 (1), 71–103. doi:10.1177/15586898231166968.

González Canché, M.S., 2025. GeoStoryTelling: An Integrative Multi-Media Methodology and No-Code, Cost-Free Software to GeoContextualize Ethnographic, Health, and Survey Evidence. International Journal of Qualitative Methods, 24. doi:10.1177/16094069251325629.

Hesse-Biber, S., 2010. Qualitative Approaches to Mixed Methods Practice. Qualitative Inquiry, 16 (6), 455–468. doi:10.1177/1077800410364611.

Kumar, L., Bhatia, P.K., 2013. Text mining: concepts, process and applications. J. Glob. Res. Comput. Sci. 4 (3), 36–39.

Raconteur. (2020). A day in data. Available from: https://www.raconteur.net/infographics/a-day-in-data.

Tanweer, A., Gade, E.K., Kraft, P.M., Dreier, S., 2021. Why the data revolution needs qualitative thinking. Harvard Data Science Review, 3 (3). Available from: https://hdsr.mitpress.mit.edu/pub/u9s6f22y/release/4.

Vo, D., Zheng, K., Zhang, C., González Canché, M.S., Rios-Aguilar, C., 2025. Network analysis of big data for education policy: challenges, limitations, and opportunities. In: Handbook of Social Network Analysis and Education (pp. 62–84). Edward Elgar Publishing. doi:10.4337/9781035312191.00010.

Index

Note: Page numbers followed by *f* indicate figures, *t* indicate tables.

9780443219610